GAME THEORY AND DECISION THEORY IN AGENT-BASED SYSTEMS

MULTIAGENT SYSTEMS, ARTIFICIAL SOCIETIES, AND SIMULATED ORGANIZATIONS

International Book Series

Books in the Series:

CONFLICTING AGENTS: ***Conflict Management in Multi-Agent Systems***, edited by Catherine Tessier, Laurent Chaudron and Heinz-Jürgen Müller, ISBN: 0-7923-7210-7

SOCIAL ORDER IN MULTIAGENT SYSTEMS, edited by Rosaria Conte and Chrysanthos Dellarocas, ISBN: 0-7923-7450-9

SOCIALLY INTELLIGENT AGENTS: ***Creating Relationships with Computers and Robots***, edited by Kerstin Dautenhahn, Alan H. Bond, Lola Cañamero and Bruce Edmonds, ISBN: 1-4020-7057-8

CONCEPTUAL MODELLING OF MULTI-AGENT SYSTEMS: ***The CoMoMAS Engineering Environment***, by Norbert Glaser, ISBN: 1-4020-7061-6

GAME THEORY AND DECISION THEORY IN AGENT-BASED SYSTEMS

Edited by

Simon Parsons
Brooklyn College, City University of New York

Piotr Gmytrasiewicz
University of Illinois at Chicago

Michael Wooldridge
University of Liverpool

KLUWER ACADEMIC PUBLISHERS
Boston / Dordrecht / London

Distributors for North, Central and South America:
Kluwer Academic Publishers
101 Philip Drive; Assinippi Park
Norwell, Massachusetts 02061 USA
Telephone (781) 871-6600
Fax (781) 681-9045
E-Mail: kluwer@wkap.com

Distributors for all other countries:
Kluwer Academic Publishers Group
Post Office Box 322
3300 AH Dordrecht, THE NETHERLANDS
Telephone 31 786 576 000
Fax 31 786 576 474
E-Mail: services@wkap.nl

Electronic Services <http://www.wkap.nl>

Library of Congress Cataloging-in-Publication Data

Game theory and decision theory in agent-based systems / edited by Simon Parsons, Piotr Gmytrasiewicz, Michael Wooldridge
p. cm. -- (Multiagent systems, artificial societies, and simulated organizations)
Includes bibliographical references and index.
ISBN: 1-4020-7115-9
1. Game theory. 2. Statistical decision. 3. Intelligent agents (Computer software) I. Parsons, Simon. II. Gmytrasiewicz, Piotr. J. III. Wooldridge, Michael. IV. Series.

QA269 .G353 2002
519.3--dc21 2002067473

Printed on acid-free paper. Printed in the United States of America.

Contents

List of Figures ix

List of Tables xiii

Preface xv

Contributing Authors xix

Introduction xxv
Simon Parsons, Piotr Gmytrasiewicz, and Michael Wooldridge

1
Game Theory and Decision Theory 1
Simon Parsons and *Michael Wooldridge*
1 Introduction 1
2 Decision Theory 2
3 Game Theory 10
4 The Prisoner's Dilemma 20
5 Summary 26

2
Selecting Partners 29
Bikramjit Banerje and *Sandip Sen*
1 Introduction 29
2 A Sample Problem 30
3 Coalition Formation 31
4 Payoff-structures of Partnerships 32
5 Selecting the Potentially most Beneficial Partnership for Limited Interactions 32
6 An Example Evaluation 34
7 Comparative Evaluation with a Decision Mechanism for Infinite Interactions 36
8 Evaluation Scenario 37
9 Experimental Results 37
10 Conclusions 39

3
Evolution of Agents with Moral Sentiments in an IPD Exercise 43

Ana L. C. Bazzan, *Rafael H. Bordini* and *John A. Campbell*
1 Introduction 44
2 Moral Sentiments in the Iterated Prisoner's Dilemma 45
3 The Unwitting Benefits of Altruism 48
4 An Evolutionary Setting for the Simulations 51
5 Playing with Moral Sentiments Secures the Survival of the Group 52
6 A Case for Benevolence in Agents 56
7 Conclusions 60

4
Dynamic Desires 65
Mehdi Dastani, *Zhisheng Huang* and *Leendert van der Torre*
1 Introduction 66
2 Utilities, Desires and Goals 67
3 An Example from Negotiation 72
4 Dynamic Logic 74
5 Future Research 79
6 Concluding Remarks 79

5
Emotions and Personality 81
Piotr J. Gmytrasiewicz and *Christine L. Lisetti*
1 Introduction 82
2 Decision-Theoretic Preliminaries 84
3 Emotional States and Personality 87
4 Conclusions and Future Work 91
5 Acknowledgements 93

6
Decision-Theoretic Approach to Game Theory 97
Piotr J. Gmytrasiewicz and *Sanguk Noh*
1 Background on Decision-Theoretic Approach to Game Theory 98
2 Knowledge Base Design 101
3 RMM's Solution Method 106
4 Rational Communication 109
5 Experiments in Air Defense Domain 110
6 Conclusions 114

7
Shopbot Economics 119
Jeffrey O. Kephart and *Amy R. Greenwald*
1 Introduction 120
2 Model 121
3 Sellers 123
4 Buyers 128
5 Shopbot 138
6 Related Work 150
7 Conclusions and Future Work 152
Appendix: No Pure Strategy Nash Equilibrium 153

8
Finding the Best Way to Join in 159

László Á. Kóczy
1 Introduction 159
2 Game Theoretic Foundations 160
3 The Accession Game 163
4 Solving More General Games 168
5 Conclusions 174

9
Shopbots and Pricebots in Electronic Service Markets 177
Panos M. Markopoulos and *Lyle H. Ungar*
1 Introduction and Definitions 177
2 Our Model of an Electronic Service Market 180
3 Comparisons 186
4 Related Literature 189
5 Conclusions 191
Appendix: The Derivation of Equilibrium for the XX Market Configuration 192

10
Polynomial Time Mechanisms 197
Thomas C. O'Connell and *Richard E. Stearns*
1 Introduction 198
2 Mechanism Design 200
3 Polynomial Time Mechanisms 203
4 Dominant Strategy Implementation 206
5 Nash Implementation 208
6 Upper Bounds on Approximability 212
7 Conclusions 213

11
Multi-Agent Q-learning and Regression Trees 217
Manu Sridharan and *Gerald Tesauro*
1 Introduction 218
2 Model Agent Economies 219
3 Single and Multi-Agent Q-learning 222
4 Q-learning with Regression Trees 225
5 Results 227
6 Conclusions 231

12
Satisficing Equilibria 235
Wynn C. Stirling, Michael A. Goodrich and *Dennis J. Packard*
1 Introduction 236
2 A Comparative Paradigm 237
3 Praxeic Utility 243
4 Multi-Agent Systems 247
5 Satisficing Games 252
6 Summary and Discussion 258

13
Investigating Commitment Flexibility in Multi-agent Contracts 267
Vanessa Teague and *Liz Sonenberg*
1 Introduction 267

2 The Scenario 270
3 Some Analysis of Special Cases 272
4 Empirical Work 280
5 Conclusions and Further Work 288

14
Pricing in Agent Economies using Multi-agent Q-learning 293
Gerald Tesauro and *Jeffrey O. Kephart*
1 Introduction 294
2 Model Agent Economies 297
3 Single-agent Q-learning 301
4 Multi-agent Q-learning 305
5 Conclusions 310

15
Using Hypergames to Increase Planned Payoff and Reduce Risk 315
Russell Vane and *Paul Lehner*
1 Introduction 315
2 Discussion 317
3 Results 327
4 Conclusions 333

16
Bilateral Negotiation with Incomplete and Uncertain Information 337
Julita Vassileva and *Chhaya Mudgal*
1 Introduction 338
2 The Application Area 338
3 The Negotiation Mechanism 340
4 Evaluation 343
5 Related Work on Negotiation 350
6 Conclusions 351

17
Robust Combinatorial Auction Protocol against False-name Bids 355
Makoto Yokoo, *Yuko Sakurai* and *Shigeo Matsubara*
1 Introduction 356
2 Generalized Vickrey Auction Protocol (GVA) 358
3 Robust Protocol against False-name Bids 361
4 Proof of Incentive Compatibility 366
5 Evaluation 368
6 Discussion 369
7 Conclusions 371

Index 375

List of Figures

1.1	An example Bayesian network.	4
1.2	The relationship between money and utility.	7
1.3	An example influence diagram.	9
1.4	Determining a mixed strategy for Agent i.	18
1.5	Determining a mixed strategy for Agent j.	19
2.1	Simulation of benefit (1).	39
2.2	Simulation of benefit (2).	40
3.1	Effects of altruism and of the use of the MS strategy.	50
3.2	Scores (number of points) of groups in 10,000 steps.	53
3.3	Simulations with 3 groups, 4 agents each.	54
3.4	Simulations with 3 groups, 20 agents each.	55
4.1	Utilities, desires and goals.	67
4.2	The smoking example.	68
4.3	Enriching BDI with utilitarian desires.	72
4.4	The washing machine example.	73
4.5	A model $\mathcal{S}$ for the washing machine.	77
5.1	An example taxonomy of emotional states.	87
5.2	Simple personality model of a tit-for-two-tats agent.	89
6.1	A simulation of the air defense domain.	102
6.2	A high-level view of a KB design for the air defense domain scenario.	103
6.3	Influence diagram spanning the KB of Battery1.	103
6.4	Three alternative models of Battery2's state of knowledge Battery1 maintains.	105
6.5	Hierarchy of payoff matrices corresponding to nested models of agents.	106
6.6	Average total expected damage (over 100 runs).	111
7.1	The shopbot model.	122
7.2	Probability density functions for three buyer strategy vectors.	127

7.3 Probability distributions over prices paid by buyers and expected prices. 130
7.4 Benefits and costs of search for linear search costs. 132
7.5 Evolution of buyer strategy. 135
7.6 Evolution of buyer strategy and cross-section of basin of attraction. 137
7.7 Optimal linear shopbot price schedule and equilibrium values of buyer strategy vector. 143
7.8 Average buyer surplus, total seller profits, and total search cost. 144
7.9 Example of how a shopbot can adjust its prices dynamically. 146
7.10 Example of how a shopbot can adjust its prices dynamically using an adaptive procing algorithm. 148
9.1 Average prevailing price in a market with shopbots and pricebots as the number of sellers increases. 185
9.2 Price distribution in a market with 2 servers with shopbots and pricebots. 186
9.3 Servers and clients utility. 187
9.4 System load for different market configurations. 190
11.1 Sample profit landscape for seller 1 in Price-Quality model, as a function of seller 1 price p_1 and seller 2 price p_2. 221
11.2 Results of simultaneous Q-learning with lookup tables in the Shopbot model. 224
11.3 Results of single-agent Q-learning with regression trees in the Shopbot model. 228
11.4 Results of varying Minobjs, the minimum no. of cases per leaf node. Single-agent Q-learning with regression trees in the Shopbot model. 229
11.5 Plot of the Q-function for single-agent Q-learning vs. a myopic opponent in the Shopbot model. 230
12.1 Decision regions for (a) bilateral decisions and (b) unilateral decisions. 258
13.1 Average group income vs. "brownie point" weight. 281
13.2 Average group income vs. decommitment penalty. 282
13.3 Running average group income over time for fixed reliability weight. 283
13.4 Average group income vs. reliability weight. 284

13.5 Average group income vs. decommitment penalty, for Figure 13.1 scenario. 285
13.6 Average group income vs. "brownie point" weight, for Figure 13.2 scenario. 286
13.7 Average group income vs. decommitment penalty, for different "brownie point" weights. 287
13.8 Average group income vs. number of runs, for agents using only decommitment penalties and agents using only reliability. 288
14.1 Sample profit landscape for seller 1 in Price-Quality model, as a function of seller 1 price p_1 and seller 2 price p_2. 300
14.2 Results of single-agent Q-learning in the Shopbot model. 304
14.3 Results of simultaneous Q-learning in the Price-Quality model. 306
14.4 Results of simultaneous Q-learning in the Shopbot model. 308
14.5 Results of multi-agent Q-learning in the Information-Filtering model. 309
15.1 An $m \times n$ normal game. 318
15.2 Generalized hypergame without selection strategies. 319
15.3 A hypergame with hyperstrategies. 323
15.4 Outguessing and strategy selection. 325
15.5 Plots of various $EU(H, C_\Sigma)$ to $EU(H, G)$. 326
15.6 Relationships of $EU(H, *)$ to $EU(R_0, *)$. 326
15.7 This figure represents uncertainty in the range of 0–1 as g. Both a partially effective hyperstrategy, H, and the NEMS are plotted. 327
15.8 3×3 subgame in a 4×4 full game. 328
15.9 The attack right row. 330
15.10 An excursion: 3×3 in a 6×6. 332
15.11 Utility with outguessing: 3×3 in a 6×6. 333
16.1 An influence diagram for the decision model. 342
16.2 An influence diagram for the opponent model. 344
17.1 Examples of leveled division sets. 363
17.2 Comparison of social surplus ($p = 0.5$). 370
17.3 Comparison of social surplus ($p = 0.7$). 370

List of Tables

1.1	A payoff matrix with dominant strategies.	13
1.2	A payoff matrix for a game with no pure equilibrium.	17
1.3	A payoff matrix for the Prisoner's Dilemma.	21
3.1	A payoff matrix for the Prisoner's Dilemma.	46
3.2	The tactics used by altruistic and egoistic agents.	47
6.1	The performances of RMM, human, and mixed teams.	113
7.1	Search strategies that co-exist with *Search-1*.	138
7.2	Optimal linear shopbot price schedule.	142
7.3	Shopbot's profit, total seller profits, and average buyer surplus.	143
9.1	The different market configurations that we explore.	179
12.1	Payoff array in ordinal form for traditional Prisoner's Dilemma game.	238
12.2	Selectability and rejectability functions.	254
12.3	Conditional selectability for the satisficing Prisoner's Dilemma.	257
12.4	Frameworks for decision making.	259
13.1	Agents' payoffs for the two-player game.	279
16.1	Rejections in the various configurations.	346
16.2	Comparison of the percentage of better deals achieved by using a different strategy.	347

Preface

This book is a collection of papers drawn from the series of workshops on Game Theoretic and Decision Theoretic (GTDT) Agents which we have been running yearly since 1999. The decision to run a workshop in this area was taken because it seemed to us that models from Game Theory and Decision Theory[1] were becoming quite widely used by researchers interested in agent-based computing, but that the work on the application of these theories had no one outlet. It therefore seemed to be a good idea to provide an event where the people applying these models could get together to discuss their work and learn from each other.

Now, when we were thinking about running a workshop, it seemed to us that there were two distinct constituencies which might be interested in the event. The first was the agents community, which, in our estimation, was largely concerned with the application of game theory to analyse interactions between agents. The second was the decision making/reasoning under uncertainty community, which was beginning to get interested in applying techniques like Bayesian networks to handle the inherent uncertainty that individual agents face in dealing with the world. This seemed to create a further aim for the workshop—to bring these two separate groups together. As a result we held the first GTDT workshop at the European Conference on Symbolic and Quantitative Approaches to Reasoning and Uncertainty, which took place in July 1999 in London.

Running this workshop demonstrated two things. It showed that there was indeed scope for such a workshop both in terms of people wanting to submit papers and come and present them, and also in terms of providing a forum for some communication between different communities. However, it also showed that we had somewhat underestimated the range of topics under the umbrella of "Game theory and decision theory". For instance it seems that the group of people who consider themselves to be working on Game theory and Decision theory also includes people from the knowledge representation community who are interested in qualitative decision making[2]. Though unexpected, we are more than happy to find that this is the case.

The success of GTDT '99 led us to run the workshop once again, this time in conjunction with the International Conference on Multi-Agent Systems in Boston in July 2000, and the most recent iteration was in the form of a AAAI Spring Symposium at Stanford in March 2001.

Neither of the first two workshops had a formal proceedings, though at both we distributed the papers to all participants and they are available on-line at `http://www.csc.liv.ac.uk/~sp/events/events.html`. Following GTDT '99 we began to organise a special issue of *Autonomous Agents and Multi-Agent Systems* which will feature revisions and extensions of selected papers from that first workshop[3]. Around the time of GTDT '00, while we were wondering whether we should do the same, we were invited to put this book together tusing the papers from GTDT '00 as a basis, something were we happy to undertake. Thus many of the papers contained herein first appeared at GTDT '00 though they have subsequently been revised and extended. In addition, since in our opinion some of the papers from the special issue were particularly nice examples of work in this area, we chose to include these as well—the papers which were originally presented at GTDT '99 are those by Kephart and Greenwald, Stirling *et al.*, Tesauro and Kephart, and Vane and Lehner. The result, we believe, is a collection which reflects, as well as a sample ever can, both the depth and breadth of work in applying techniques from game theory and decision theory to design agent-based systems.

As a result of this merging of work from two versions of the workshop, there is a curious relationship between some of the papers, in particular those on computational economies. At GTDT '99, Amy Greewald presented some of her work with Jeff Kephart on the use of shopbots and their effects on electronic market places (which appears as Chapter 7). This was complemented at the workshop by a talk by Gerald Tesauro on work, also with Jeff Kephart, on the use of machine learning techniques in pricebots within exactly the same kind of markets (this work is described in Chapter 14). Subsequently, at GTDT '00, Panos Markopoulos presented a paper (Chapter 9) which takes another look at the behaviour of markets with both shopbots and pricebots, and the workshop also featured a paper (Chapter 11) which extends Tesauro and Kephart's work to cover other types of learning.

This book has only been possible thanks to the success of the GTDT workshops, and, of course, the workshops have only succeeded because of the authors who have submitted papers and turned up to present them, and we are grateful to them for doing so. We have also relied a great deal upon the program committee members, and we are very much in their debt also. They are, in alphabetical order, Klaus Fischer, Amy Greenwald, Jeff Kephart, Sarit Kraus, Jeff Rosenchein, Tuomas Sandholm, Wynn Stirling, Gerald Tesauro, Leon van der Torre, Russ Vane, Nir Vulkan, and Frederik Ygge. We are also grateful to Peter McBurney for his help in proof-reading parts of this manuscript.

SIMON PARSONS, PIOTR GMYTRASIEWICZ, AND MICHAEL WOOLDRIDGE

Notes

1. Which, of course, are, to a large extent, two sides of the same coin.
2. And this range has continued to expand through the more recent workshops.
3. At the time of writing, it has still to appear.

Contributing Authors

Bikramjit Banerjee is a Knowledge-Base researcher and developer at RightNow Technologies. His main research interests are in multi-agent learning, decision theory, natural language processing, information retrieval, and multi-agent systems in general.

Ana L. C. Bazzan is a Professor in the Informatics Institute at the Federal University of Rio Grande do Sul (UFRGS) in Porto Alegre, Brazil. Her research interests are related to artificial intelligence and multi-agent systems. Current research projects include agent-based simulation (e.g., traffic systems, artificial life), agent coordination, and bioinformatics.

Rafael H. Bordini is an Invited Professor in the Informatics Institute at the Federal University of Rio Grande do Sul (UFRGS), and has a Ph.D. in Computer Science from University College London. His research interests are on the foundations of distributed artificial intelligence, agent oriented programming languages, and the application of multi-agent systems for social simulation.

John A. Campbell is a Professor of Computer Science at University College London. Within multi-agent systems, his main interests are in quantitative modelling of collective behaviour, and in how to achieve effective performance when computing time and communications bandwidth are limited. His other research areas in artificial intelligence include time-bounded reasoning, representation and reasoning for mathematical knowledge, assessment of similarity, and the automated processing of multilingual material.

Mehdi Dastani is a researcher at the Institute of Information and Computing Sciences, Utrecht University. Currently, he works on agent theories and applications. His research focus is the topic of specification and implementation languages for cognitive agents and he has contributed to the development of a cognitive agent architecture called BOID.

Piotr Gmytrasiewicz is an Associate Professor in the Computer Science Department at the University of Illinois at Chicago. His research interests include rational artificial agents, distributed artificial intelligence, multi-agent coordination and intelligent communication, and decision and game theories.

Michael A. Goodrich is an Assistant Professor in the Department of Computer Science at Brigham Young University. His research interests include human-robot interaction and multi-agent decision theory.

Amy Greenwald is an Assistant Professor of Computer Science at Brown University in Providence, Rhode Island. Her primary area of interest is multiagent learning on the Internet, which she approaches using game-theoretic models of computational interactions. She was recently awarded an NSF Career Grant entitled "Computational Social Choice Theory." Previously, she was employed by IBM T. J. Watson Reesearch Center, where she was involved in the Information Economies Project. Dr. Greenwald completed her Ph.D. at the Courant Institute of Mathematical Sciences of New York University, where she was funded in part by a dissertation fellowship awarded by the American Association of University Women.

Zhisheng Huang is a researcher at the Vrije Universiteit Amsterdam, interested in logics of practical rationality and their application in agent technology. He has contributed to the development of the BOID architecture. Dr. Huang is now working on the project of development and implementation of web agents for virtual environments.

Jeffrey Kephart manages the Agents and Emergent Phenomena group at the IBM Thomas J. Watson Research Center in Yorktown Heights, New York. At present, his main research interest is the dynamical behavior of large collections of interacting software agents, particularly in the realm of electronic commerce.

László Á. Kóczy is a Ph.D student at the Catholic University of Leuven, in the Centre of Economic Studies in the Faculty of Economics and Applied Economics. His research focuses on the domination-based solution concepts of cooperative coalition formation games and their relations.

Paul Lehner is the Chief Engineer for the Center for Enterprise Modernization (CEM) of the MITRE Corporation. CEM is a Federally Funded Research and Development Center for the US Department of Treasury. He received his Ph.D. in Mathematical Psychology in 1981 from The University of Michigan.

Christine Lisetti is an Assistant Professor in Computer Science at the University of Central Florida. From 1996 to 1998, she was a post-doctoral fellow at Stanford University in the Department of Psychology and the Department of Computer Science. She has focussed her research interests on computational models of affective processes to: (1) enhance human-computer interaction by enriching it with affective contextual knowledge, and (2) develop more intelligent artificial agents by endowing them with functions of the emotional system.

Panos M. Markopoulos is a Ph.D. candidate at the Computer and Information Science department of the University of Pennsylvania. He is interested in information markets, E-commerce automation, software agents and combinatorial auctions.

Shigeo Matsubara is a research scientist in NTT Communication Science Laboratories, NTT Corporation. His main research interests are in the area of rational decision making in autonomous agents and game theory.

Chhaya Mudgal received her MS in Physics and MS in Computer Science. She is currently working as a lead software developer at the Institute of Coastal and Marine Science at Rutgers University, New Jersey. Her interests involve studying collaborative decision making and reasoning among semi-autonomous underwater vehicles.

Sanguk Noh is an Assistant Professor in the Department of Computer Science at the University of Missouri-Rolla. He was a post-doctoral fellow in the Center for Human-Computer Communication at the Oregon Graduate Institute, where he was developing robust agents in condition monitoring settings, and received his Ph.D. degree in Computer Science from the University of Texas at Arlington. He has led the intelligent agent group and served as a research investigator in the Intelligent Systems Center at the University of Missouri-Rolla. He is currently collaborating in research on the flexible decision-making framework enabling agents to be practical in time-critical situations. His research interests lie in the areas of decision theory, multi-agent systems, knowledge management, and real-time systems.

Thomas O'Connell is an Assistant Professor of Computer Science at Skidmore College. His main research interests are in bounded rationality, multiagent systems and games.

Dennis J. Packard is a Professor in the Philosophy Department at Brigham Young University. His research includes social and individual decision theory, epistemology, and American pragmatism.

Simon Parsons is an Associate Professor in the Department of Computer and Information Science at Brooklyn College of the City University of New York. His main research interests are in the area of rational decision making in autonomous agents. He is also a Visiting Professor in the Department of Computer Science at the University of Liverpool where he works in the Agent Applications, Research and Technology (Agent ART) group.

Yuko Sakurai is a research scientist of NTT Communication Science Laboratories, NTT Corporation. Her main research interests are in the area of mechanism design in multi-agent systems, especially auctions.

Sandip Sen is an Associate Professor in the Mathematical and Computer Sciences Department at the University of Tulsa. His research interests include multi-agent systems, intelligent agents, genetic algorithms, and machine learning. He is the recipient of an NSF CAREER award.

Liz Sonenberg is a Professor of Information Systems and Co-Director of the Intelligent Agent Laboratory at the University of Melbourne, Australia. Her main research interest is in agent collaboration and teamwork, with complementary interests in other aspects of synthetic and human reasoning.

Manu Sridharan is a graduate student in Computer Science at MIT. His current research interests include software engineering and programming languages and tools.

Richard E. Stearns is a Distinguished Professor Emeritus of Computer Science at the University at Albany, State University of New York. In 1993, he was awarded the Turing Award jointly with Juris Hartmanis in recognition of their work in establishing the field of computational complexity theory. His research interests include computational complexity, automata theory, analysis of algorithms, and game theory.

Wynn C. Stirling is a Professor of Electrical Engineering in the Department of Electrical and Computer Engineering at Brigham Young University. His research interests include estimation and control theory with emphasis on the distributed control of autonomous multiple-agent systems.

Vanessa Teague is a Ph.D. student in the Computer Science Department at Stanford University. She is a member of the security group and her current research interests are at the intersection of cryptography, economics and networking. The work reported here was completed when Vanessa was a student at the University of Melbourne, Australia.

Gerald Tesauro is a Research Staff Member at the IBM T.J. Watson Research Center. His current research interests are in applications of multiagent learning, reinforcement learning, and nonlinear function approximation, particularly in the development of intelligent software agents for electronic commerce.

Leendert van der Torre is a researcher at the Vrije Universiteit Amsterdam, interested in logics of practical reasoning and their application in agent technology. He has developed logics of obligations, desires and goals—such as the input/output logics—and has contributed to the development of the BOID architecture. He is now looking into simulation of normative systems and specification and verification of cognitive agent systems.

Lyle H. Ungar is a Professor of Computer and Information Science at the University of Pennsylvania. His research areas include data mining, reinforcement learning, combinatorial auction design, and e-commerce automation.

Julita Vassileva is an Associate Professor at the Computer Science Department at the University of Saskatchewan, Canada. Her main research interests are in user modelling, user-adaptive systems and in multi-agent based software applications.

Michael Wooldridge is a Professor of Computer Science in the Department of Computer Science at the University of Liverpool, and head the Agent Applications, Research, and Technology group (Agent ART) group. His main research interests are in the use of formal methods of one kind or another for specifying and reasoning about multi-agent systems.

Makoto Yakoo is a distinguished technical member of NTT Communication Science Laboratories, NTT Corporation. His current research interests include multi-agent systems, especially, mechanism design and constraint satisfaction among multiple agents.

Introduction

Simon Parsons, Piotr Gmytrasiewicz, and Michael Wooldridge

Agent-based systems are, by their very nature, situated in some environment. Their function is to interact with that environment, perceive the state of the environment and make decisions about how to respond to it. When we are designing agents we usually want them to make good decisions—typically we want them to make the best decision that they can, given what is known. For this reason, decision making, in some form or other, is at the very heart of building agent-based systems.

Now, in simple environments, such as the kind of blocks-world scenario studied in early work on artificial intelligence, making the right decision is relatively easy. An agent's knowledge of the world is complete and correct, it has a set of actions which have a single possible outcome, it has a single goal which it is trying to achieve, and there are no other agents disrupting things. As a result, all the blocks-world agent has to do is to figure out a string of actions which will take it from the current, known, position to the goal. Then it is guaranteed that simply executing those actions in sequence will lead to the goal being achieved.

Unfortunately most environments are a good deal more complex than the blocks world. Typically an agent's knowledge, filtered as it is through some set of sensors, is rather incomplete and uncertain. The actions the agent has at its disposal are non-deterministic, having a range of possible outcomes. The agent will have a number of (possibly conflicting) goals which it is trying, more or less urgently, to achieve. In addition, there may be other agents operating in the environment with which the first agent might have to interact. It is these aspects that can be handled by the application of techniques from game theory and decision theory. The ubiquity of these aspects is the reason why game theory and decision theory have an important role to play in agent-based computing.

Decision theory

Broadly speaking, decision theory provides a means of handling the uncertainty of an agent's knowledge and the non-determinism of its actions[1]. Uncertain information is handled by constructing a probability distribution over the set of possibilities, and there are mechanisms for updating this information when new information is perceived. Thus, for example, when building an agent which needs to reason about the weather, rather than, as in the blocks world, knowing that either "it will rain tomorrow in Liverpool", or "it will not rain tomorrow in Liverpool", it is possible to represent information such as there is a probability of 0.4 that "it will rain tomorrow in Liverpool" and 0.6 that "it will not rain tomorrow in Liverpool". This information can then be updated when the weather forecast predicts that there are rain clouds heading in over the Atlantic.

Non-determinism is handled by building a probability distribution across the set of possible outcomes of an action, and doing this across all actions makes it possible to make a choice of actions based upon the set of all possible outcomes and their likelihood of occurring. So, if planning an outing tomorrow, where the choice of actions are to "go to the zoo" or to "go to the beach", and going to the zoo has outcomes like "visit the monkey house" and "picnic in the park" (which happens depending on whether or not it rains), while going to the beach has outcomes like "visit the arcade" and "walk on the sand" (again depending on whether or not it rains), this non-determinism about whether it rains or does not rain can be handled by calculating the overall probability of the four outcomes.

However this is only half of the picture. Different outcomes have different values to the agent making the decision, and this can affect the choice of action quite strongly. Thus in the example above, we might find that the agent in question places a high value on walking on the sand, but a very low value on going to the arcade, and places a relatively high value on both going to the monkey house and picnicing in the park. As a result, that agent will tend to choose going to the beach if it is likely not to rain but the zoo if it is likely to rain. However a second agent might place a high value on both the arcade and the walk, and so want to go to the beach whatever the weather. Decision theory factors in this aspect by assigning each possible outcome a utility (a utility is a numerical representation of an agent's preferences with nice mathematical properties), and then using the probability information to compute the expected utility of every action—essentially the average utility of the outcomes of that action weighted by likelihood. The action with the highest expected utility is then chosen.

One way of looking at decision theory which fits in with this description, is that decision theory models a game between the agent in question and nature. The agent is trying to get to some position that is good for it and has some set of moves it is making with that end in mind while nature makes moves at random. In the example, the agent is trying to have a good day out and nature complicates this decision by making a random move to make it rain (or not). Now, consider what would happen if the game wasn't against nature, but against another agent. In this case the first agent wouldn't be playing against a set of random moves but against a set of carefully thought out moves, indeed against a set of moves which were thought out by an agent which was trying to anticipate it.

Game theory

As an example, consider a variation on the above theme. Two agents, A and B, who are prevented from communicating, are trying to plan the same day out. Both can either go to the beach or to the zoo. Since they are friends, they are glad if they are both in the same place, since then they'll most likely run into each other, and, since the zoo is closer than the beach, for both of them it is much better to go to the zoo than the beach. Thus the four possible outcomes, in order of utility for A, is "A and B go to the zoo", "A goes to the zoo and B goes to the beach", "A and B go to the beach", and "A goes to the beach and B goes to the zoo"—those for B are symmetrical.

Given this information, consider how A might reason. Whichever action it choses a_b (going to the beach) or a_z (going to the zoo), the utility of that action depends upon what B chooses, and it doesn't know what B will do. Similarly B's choice between b_z and b_b depends on what A does. This makes it seem as though neither agent can make a decision until it knows what the other knows. However, whatever choice B makes, A is better off by choosing the zoo, so A can just choose "zoo" and let B choose the option that is best for it. B can reason in a similar way and also choose "zoo". Furthermore, either A or B can reason that the other will choose "zoo" because that is best for it, and then make its decision on that basis.

Game theory provides a theoretical framework for analysing problems like this, or, alternatively, for working out what kind of reward structure one should put in place for encouraging agents to behave in a particular way—think of the example not in terms of A and B figuring out what to do, but of the designer of the system in which A and B will operate working out how to get A and B to the same place. In case this example seems so simple that it isn't clear why game theory is complex enough to be interesting, so consider a variation on the above example.

Consider that rather than A and B, who like to be together, we are interested in C and D who annoy each other if they get too close. If they both go to the beach then, well the beach is big and though they might bump into each other that won't be too annoying. If they both go to the zoo then they will be in a much more confined space and will annoy each other more. Since the zoo is close and the beach is a long way from where both of them live, the best outcome for both of them is if they go to the zoo and the other goes to the beach and the worst outcome is if they go to the beach and the other goes to the zoo.

The problem here is that C can't think in the same way as A and get to the best solution for both, because the zoo is only the best option provided D doesn't go there—if both C and D go to the zoo it is worse than if they both go to the beach. But if they both go to the beach, the outcome is worse for one of them than if that one had gone to the zoo. However, if one goes to the beach and the other goes to the zoo, the one who is at the beach is worse off than if they had gone to the zoo. The result of all this is that each one leans towards going to the zoo (better than going to the beach and finding the other went to the zoo) even though this is worse than both going to the beach.

These examples illustrate the difference between the Pareto optimal solution (the best for both agents together[2]) and the Nash equilibrium solution (the point that the agents get to if they just consider their own best interests[3]). For A and B, the Pareto optimal solution is the same as the Nash equilibrium solution—both agents go to the zoo. In the case of C and D, the two solutions differ. The Pareto optimal solution is for them both to go to the beach, while the Nash equilibrium solution is for both to go to the zoo.

In other variations on this same problem, it can be impossible for the two agents to even come to a direct choice between the two options available to them, because as they think about what the other agent will do in response to whatever their best option is, their best option continues to change. To illustrate this, consider agents E and F. E and F hate each other and this leads E to want to avoid F, but leads F to want to disrupt E's day by being in the same place. Thus for E the joint actions $\langle e_b, f_b \rangle$ and $\langle e_z, f_z \rangle$ have a much lower utility than $\langle e_b, f_z \rangle$ and $\langle e_z, f_b \rangle$ but for F it is the other way around. As before, both prefer the zoo because it is closer.

So, consider E's thinking. Left alone, E would choose the zoo, but F knows this, and this makes F want to go to the zoo as well. Knowing this E should choose to go to the beach. However, E knows that F can follow this chain of reasoning as well, and so will choose to go to the beach, so E should go to the zoo. But F can figure this out as well, and so E should go to the beach .... The situation is the same from F's side, and there is no stable solution if each agent is looking to just pick a single option. However, as described in Chapter 1, there are solutions in which each agent picks a "mixed strategy"—basically, it assigns each single option a probability and then picks an option randomly on the basis of this probability.

The aim of this book

From the above discussion, it is hopefully clear that game theory and decision theory can be used to handle important problems in the design of agent-based systems, and this is certainly the belief of the editors. Broadly speaking, the aim of this book is to back up that position by providing a number of illustrations of the way in which this may be done.

Thus, for example, Chapters 7 and 9 consider the use of game theoretic techniques to identify the behaviour of interacting agents in big electronic markets, while Chapter 3 looks at the kinds of behaviour that can be adopted when agents in the classic Prisoner's Dilemma problem[4] interact with each other many times over (which encourages them to trust each other to work towards the Pareto optimal solution). Similarly, in Chapter 16, decision theory is used to help agents decide what bids to make in an negotiation, building a model of the various uncertain factors which influence the decision.

However, there is more to the application of game theory and decision theory to agent-based systems than just using decision theory to handle uncertain information and non-deterministic actions, and using game theory to analyse interactions between agents. Illustrating work on some of these less obvious issues is a second aim of this book.

Here, for example, classical decision theory (which includes the bits of decision theory described above) makes some fairly strong assumptions about the way in which utilities are constructed[5] and these can be challenged. One alternative approach is described in Chapter 12. This construction of utilities also depends heavily on an agent's set of preferences, and thus these preferences have a role in determining what action a decision theoretic or game theoretic agent chooses to make. In this sense the preferences define the priority that an agent places on the various goals it may have (going to the beach, or going to the zoo in our running example), and these preferences are assumed to remain invariant over time. However, this is not always a valid assumption, and Chapter 4 considers some of the issues that arise when preferences (encoded as an agent's desires) are allowed to change.

Thus, what we aim to achieve here is to present a broad view of the many ways that game theory and decision theory can be applied in agent-based systems, from the standard application of the core elements of the theory to the more cutting edge developments, through the example of some sample applications.

The structure of this book

Because the book takes such a broad view, and because it is difficult at this stage of development of the field to come up with a definitive structure into which all the work fits, we have imposed very little structure on the format of the book. Instead we let the papers speak for themselves—with the exception of the first chapter, which provides a brief introduction to the main ideas from game theory and decision theory—the chapters are just given in alphabetical order of the first author's name. To help the reader navigate we give a brief synopsis of each chapter here.

Chapter 1 is an introduction to the basic concepts of decision theory, giving more technical detail than the discussion above, and hopefully covering enough ground to make the remaining chapters accessible even to readers who have not previously encountered these ideas.

Chapter 2 considers how agents make decisions about collaboration, in particular the formation of coalitions between agents. When agents make one-off decisions about how to work together (or not work together) classical techniques from game theory can be used to decide the best way to act (as with our beach and zoo going agents above). However, when these interactions will be carried out several times, the agents might benefit by making more long term decisions which take into account future possible collaborations, and techniques from decision theory can be used to do this. These matters are discussed in this chapter.

Chapter 3 deals with a similar kind of problem, though one that is set in the context of the Prisoner's Dilemma, and considers both how strategies other than the standard game theoretic solution can benefit agents over the course of a number of interactions, and how such solutions can be determined using evolutionary techniques.

Chapter 4 is concerned with an agent's desires, and, in particular, how to deal with desires that change as opposed to remaining constant as they are assumed to in much of the agents literature.

Chapter 5 looks at the modelling of emotion in agents—a topic that is becoming increasingly important, not least as agent-based interfaces become more sophisticated—examining how this can be done in a way that is rigorous from a decision theoretic point of view.

Chapter 6 looks at how decision theory, in the form of influence diagrams, can be used to model an agent's decisions when taking into account the possible decisions of other agents (thus allowing the kind of analysis we find in game theory). The particular advantage of the approach taken in this chapter is that it allows one agent's model to include another agent's model of what the first agent will do (which itself includes a model of the second agent). Such recursive models can be extremely useful in building complex agent-based systems.

Chapter 7 considers electronic markets, such as those found on the internet, in which a set of shoppers make purchases from a set of sellers. In particular, the chapter studies the way that these markets perform when certain of the shoppers make use of intermediaries—shopbots—which perform price comparisons between sellers thus allowing the shoppers to obtain goods at the lowest price offered by the sellers surveyed by the shopbots.

Chapter 8 studies the formation of coalitions between agents. Whereas most work in this area concentrates on the question of how coalitions are formed from a number of individual agents, this chapter considers the case in which there are some individual agents and a set of existing coalitions. In particular, the chapter examines the case in which a number of individual agents are seeking to join the dominant coalition.

Chapter 9 studies electronic markets similar to those considered in Chapter 7, but also considers the use of a second type of price-surveying agent. These agents—pricebots—are used by sellers to make price comparisons between sellers offering competing products, and allow the seller to automatically update their prices in order to offer better deals to potential shoppers.

Chapter 10 is concerned with the computational complexity of the design of mechanisms for making decisions in a multi-agent setting. The chapter concentrates on situations in which the agents are making decisions about optimisation problems—problems which are, in general, computationally intractable—and shows that allowing the agents to make decisions which approximate the optimal solution makes the computation tractable. However, the chapter shows that, in general, the approximation mechanisms for multiagent optimisation problems are likely to be less successful than approximation algorithms for the same optimisation problems in a non-multiagent setting.

Chapter 11 investigates electronic markets in which sellers use pricebots to fix their prices. One of the problems with such markets is that, when the pricebots seek to maximise sales by undercutting competitors, "price wars" can result in which sellers repeatedly undercut each other until further discounting is counter-productive. This behaviour can be mitigated if the pricebots set prices by considering the long-term impact of a price choice, and this chapter investigates some ways in which this long-term impact can be computed.

Chapter 12 investigates no less than an alternative to classical decision theory. Without changing the basic idea of maximising expected utility, the work presented in this chapter fundamentally alters the nature of the theory by proposing an alternative means of determining utility. This is done by factoring in aspects such as the utility of an outcome to the group of agents including the decision maker rather than just considering the decision maker itself.

Chapter 13 studies two key aspects of automated contracting. The first is that of commitment to contracts, typically handled by allowing agents to decommit from a contract on the payment of a suitable penalty. The second is that of reputation, models for which can take account of both an agent's own reputation and that of other agents in an interaction. The chapter considers how these aspects may be brought together in a single model.

Chapter 14 investigates electronic markets with selling prices set by pricebots. The scenarios are similar to those investigated in Chapter 11 but different methods are used for predicting the long-term reward for setting a particular price.

Chapter 15 is concerned with hypergames, a method which explicitly combines game theoretic and decision theoretic information. Whereas in game theory it is assumed that both agents are playing the same game, and choose their strategies on that basis, in hypergames each agent assumes that their opponent is playing one of a range of games, and wants to choose the best strategy from this set of games. Broadly speaking, hypergames are a means of using decision theory to pick the best strategy from the best strategies of the individual games.

Chapter 16 Looks at an application of decision theory to represent an agents' beliefs in a negotiation scenario. While the use of decision theory is similar to that reported in Chapter 6, the work described here differs in two ways. The decision theoretic model is somewhat simpler than that from Chapter 6, but the application can be considered somewhat more complex.

Chapter 17 considers a very specific problem from the game-theory literature. Some forms of auction are vulnerable to false-name bids, that is where bidders make many bids under different names in order to secure the goods being auctioned. This chapter examines the problem when the auction is combinatorial (allows participants to bid on combinations of goods), and proposes an auction protocol which defeats false-name bidding.

Notes

1. See Chapter 1 for a more precise and detailed discussion.
2. In case this sounds as if we somehow have to sum the values of the outcomes to different agents in order to establish which is the Pareto optimal solution, let us briefly and informally give the definition of Pareto optimality. A solution to a two player game is a Pareto optimal solution if there is no outcome of the game which makes one player better off without making the other player worse off.
3. A formal definition of Nash equilibrium is given in Chapter 1.
4. Which is essentially the problem that C and D are facing above.
5. Again these are described in more detail in Chapter 1.

Chapter 1

AN INTRODUCTION TO GAME THEORY AND DECISION THEORY

Simon Parsons
Department of Computer Science, University of Liverpool, Liverpool L69 7ZF, United Kingdom.
s.d.parsons@csc.liv.ac.uk

Michael Wooldridge
Department of Computer Science, University of Liverpool, Liverpool L69 7ZF, United Kingdom.
m.j.wooldridge@csc.liv.ac.uk

Abstract In the last few years, there has been increasing interest from the agent community in the use of techniques from decision theory and game theory. Our aim in this article is to briefly summarise the key concepts of decision theory and game theory and explain why they are useful when building agent-based systems.

Keywords: Game theory, decision theory, multi-agent systems

1. INTRODUCTION

Decision theory (Raiffa, 1968) is a means of analysing which of a series of options should be taken when it is uncertain exactly what the result of taking the option will be. Decision theory concentrates on identifying the "best" decision option, where the notion of "best" is allowed to have a number of different meanings, of which the most common is that which maximises the expected benefit to the decision maker. Since self-interested entities are assumed to be acting best when maximising expected benefits, decision theory is often claimed to be able to make the most *rational* choice. Overall, decision theory provides a powerful tool with which to analyse scenarios in which a decision must be made.

Now, it is widely believed that the crucial issue in designing autonomous agents is how to provide those agents with the ability to select the best action from a range of possible actions. Frequently the agents in question are operating in an unpredicatble, and hence uncertain, environment, and therefore decision theory seems a natural tool to use to analyse their behaviour.

Game theory (Binmore, 1992) is a close relative of decision theory, which studies interactions between self-interested entities. In particular, it studies the problems of how interaction *strategies* can be designed that will maximise the welfare of an entity in an encounter, and how *protocols* or *mechanisms* can be designed that have certain desirable properties. In the same way that decision theory can be claimed to provide a means of making rational decisions under uncertainty, so game theory can be claimed to provide a rational means of analysing interactions. Notice that decision theory can be considered to be the study of *games against nature*, where nature is an opponent that does not seek to gain the best payout, but rather acts randomly.

In multi-agent systems, the issue of designing interaction strategies and mechanisms is very important, and so it comes as no surprise to learn that game theory has been widely applied. Many of these applications have been to analyse negotiation and co-ordination mechanisms.

2. DECISION THEORY

Classical decision theory, so called to distinguish it from a number of non-classical theories which have grown up in the last few years, is a set of mathematical techniques for making decisions about what action to take when the outcomes of the various actions are not known. Although the area grew up long before the concept of an intelligent agent was conceived, such agents are canonical examples of the decision makers which can usefully employ classical decision theory.

2.1. PROBABILITY THEORY

An agent operating in a complex environment is inherently uncertain about that environment; it simply does not have enough information about the enviroment to know either the precise current state of its environments, nor how that environment will evolve. Thus, for every variable X_i which captures some aspect of the current state of the environment, all the agent typically knows is that each possible value x_{i_j} of each X_i has some probability $\Pr(x_{i_j})$ of being the current value of X_i.

Writing $\mathbf{x}$ for the set of all x_{i_j}, we have:

$$\Pr : x \in \mathbf{x} \mapsto [0, 1]$$

and

$$\sum_j \Pr(x_{i_j}) = 1$$

In other words, the probability $\Pr(x_{i_j})$ is a number between 0 and 1 and the sum of the probabilities of all the possible values of X_i is 1. If X_i is known to have value x_{i_j} then $\Pr(x_{i_j}) = 1$ and if it is known not to have value x_{i_j} then $\Pr(x_{i_j}) = 0$.

While this mathematical definition of probability is rather straightforward, the same cannot be said of the semantics of probability. Indeed, there is no universal agreement on what probabilities mean. Of the various conflicting schools of thought (Shafer, 1992), there are two main positions. The first, historically, interprets a probability as a frequency of occurrence. This *frequentistic* approach takes that fact that an event a has a probability of 0.356 to mean that 0.356 of the time it will occur. This sounds like a very straightforward interpretation until you consider in more detail how such a probability can be measured[1] The second, *Bayesian*, position suggests that a probability is related to the odds that a rational person will bet on the event in question. Thus the probability of a being 0.356 means that a rational person will pay \$0.356 to bet on a happening if they win \$1 if a occurs. This approach solves some of the problems of the frequentistic interpretation (especially when estimating the probability of unique events) but has its own problems[2]. We will say no more about semantics here—all we say applies whatever semantics one chooses to use.

Given two of these variables, X_1 and X_2, then the probabilities of the various values of X_1 and X_2 may be related to one another. If they are not related, a case we distinguish be referring to X_1 and X_2 as being *independent*, then for any two values x_{1_i} and x_{2_j}, we have:

$$\Pr(x_{1_i} \wedge x_{2_j}) = \Pr(x_{1_i}) \Pr(x_{2_j})$$

If the variables are not independent, then:

$$\Pr(x_{1_i} \wedge x_{2_j}) = \Pr(x_{1_i} | x_{2_j}) \Pr(x_{2_j})$$

where $\Pr(x_{1_i} | x_{2_j})$ is the probability of X_1 having value x_{1_i} given that X_2 is known to take value x_{2_j}. Such *conditional probabilities* capture the relationship between X_1 and X_2, representing, for instance, the fact that x_{1_i} (the value "wet", say, of the variable "state of clothes") becomes

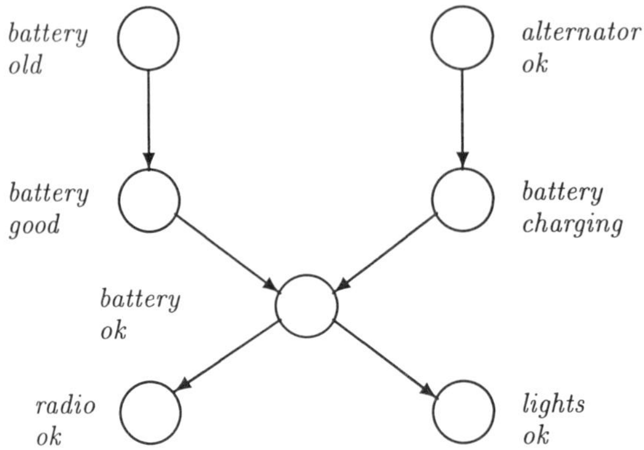

Figure 1.1. An example Bayesian network.

much more likely when x_{2_j} (the value "raining" of the variable "weather condition") is known to be true.

If we take the set of these X_i of which the agent is aware, the set $\mathbf{X}$, then for each pair of variables in $\mathbf{X}$ we can establish whether the pair are independent or not. We can then build up a graph in which each node corresponds to a variable in $\mathbf{X}$ and an arc joins two nodes if the variables represented by those nodes are not independent of each other. The resulting graph is known as a Bayesian network[3] (Pearl, 1988), and the graphical structure provides a convenient computational framework in which to calculate the probabilities of interest to the agent. In general, the agent will have some set of variables whose values it can observe, and once these observations have been taken, will want to calculate the probabilities of the various values of some other set of variables.

Figure 1.1 is an example of a fragment of a Bayesian network for diagnosing faults in cars. It represents the fact that the age of the battery (represented by the node *battery old*) has a probabilistic influence on how good the battery is, and that this in turn has an influence on whether the battery is operational (*battery ok*), the latter being affected also by whether the alternator is working and, as a result, whether the battery is recharged when the car moves. The operational state of the battery affects whether the radio and lights will work. In this network it is expected that the observations that can be carried out are those relating to the lights and the radio (and possibly the age of the battery), and that the result of these observations can be propagated through the network to establish the probability of the alternator being okay and the battery being good. In this case these latter variables are the ones which we are interested in since they relate to fixing the car.

Typically the variables an agent will be interested in are those that relate to its goals. For instance, the agent may be interested in choosing an action that will allow it to achieve a goal, and might therefore be interested in choosing that action which has the greatest chance of succeeding in achieving that goal. When the agent has many goals it could achieve, this strategy could be extended to make the agent choose to achieve the goal which has the greatest chance of being achieved, and to do this by applying the action which gives this greatest chance.

However, building an agent which follows this strategy is somewhat shortsighted since the agent will not consider the value of the goals, and will therefore choose a goal which is easy to achieve, but worthless, over a goal which is hard to achieve but very valuable. To take account of this problem, decision theory ke account of an agent's preferences. To do this, it makes use of the idea of *utility*.

2.2. UTILITY THEORY

We start from the assumption that each agent (or decision maker) has its own preferences and desires about how the world is. For the moment, we will not be concerned with where these preferences come from[4]; we will just assume that they are the preferences of the agent's user or owner. Next, we will assume that there is a set $\Omega = \{\omega_1, \omega_2, \ldots\}$ of "outcomes" or "states" that the agents have preferences over.

We will formally capture the preferences that an agent has by means of a *utility function*, which assigns to every outcome a real number, indicating how "good" the outcome is. The larger the number the better from the point of view of the agent with the utility function. Thus the preferences of an agent i will be captured by a function

$$u_i : \Omega \to \Re$$

It is not difficult to see that such a utility function leads to a *preference ordering* over outcomes. For example, if ω and ω' are both possible outcomes in Ω, and $u_i(\omega) \geq u_i(\omega')$, then outcome ω is *preferred* by agent i at least as much as ω'. We can introduce a bit more notation to capture this preference ordering. We write

$$\omega \succeq_i \omega'$$

as an abbreviation for

$$u_i(\omega) \geq u_i(\omega')$$

Similarly, if $u_i(\omega) > u_i(\omega')$, then outcome ω is *strictly preferred* by agent i over ω'. We write

$$\omega \succ_i \omega'$$

as an abbreviation for

$$u_i(\omega) > u_i(\omega').$$

In other words,

$$\omega \succ_i \omega' \text{ iff } u_i(\omega) \geq u_i(\omega') \text{ and not } u_i(\omega) = u_i(\omega').$$

We can see that the relation $\succeq_i$ really is a (partial) ordering, in that it has the following properties:

Reflexivity: For all $\omega \in \Omega$, we have that $\omega \succeq_i \omega$.

Transitivity: If $\omega \succeq_i \omega$, and $\omega' \succeq_i \omega''$, then $\omega' \succeq_i \omega''$.

Comparability: For all $\omega \in \Omega$, and $\omega' \in \Omega$ we have that either $\omega \succeq_i \omega'$ or $\omega' \succeq_i \omega$.

The strict preference relation will satisfy the second and third of these properties, but will clearly not be reflexive.

Undoubtedly the simplest way to think about utilities is as money; the more money, the better. However, it is deceptive to think that this is all that utilities are. Utility functions are a way of representing an agent's preferences. They *do not* simply equate to money.

To see why this is the case, suppose (and this really *is* a supposition) that the authors have US$500 million in the bank, while you, the reader, are absolutely penniless. A generous and rich benefactor appears, with one million dollars, which he generously wishes to donate to one or more of us. If the benefactor gives the dollar to the authors, what will the increase in the utility of our situation be? Well, we will have more money, so there will clearly be some increase in our utility. But there will not be much: after all, there is not much that you can do with US$501 million that you cannot do with US$500 million. In contrast, if the benefactor gave the money to you, the increase in your utility would be enormous; you would go from having no money at all to being a millionaire. That is a *big* difference[5].

This works the other way as well. Suppose the authors are in *debt* to the tune of US$500 million; well, there is frankly not that much difference in utility between owing US$500 million and owing US$499 million; they are both pretty bad. In contrast, there is a very big difference between being US$1 million in debt and not being in debt at all. A graph of the relationship between utility and money is shown in Figure 1.2.

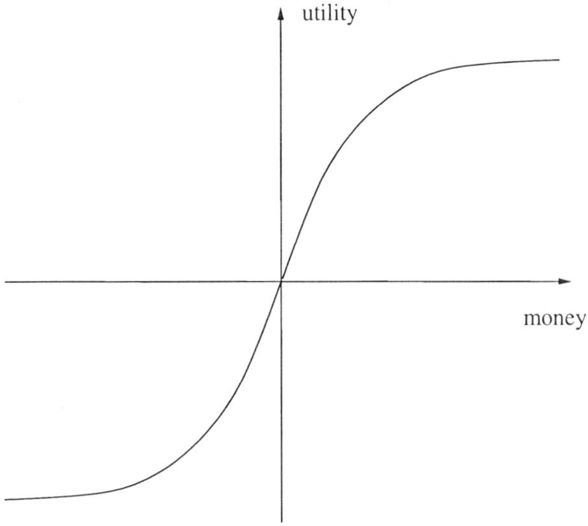

Figure 1.2. The relationship between money and utility.

So, to summarise, a utility is a value which is associated with a state of the world, and which represents the value that the agent places on that state of the world. Utilities provide a convenient means of encoding the preferences of an agent; as von Neumann and Morgenstern (von Neumann and Morgenstern, 1944) showed, it is possible to define utility functions that faithfully encode preferences such that a state S_i is preferred to S_j, if and only if it has a higher utility for the agent exactly as described above.

2.3. EXPECTED UTILITY

Now, we can consider that our agent has a set of possible actions **A**, each member A_i of which has a range of possible outcomes since the actions are not deterministic. The value of taking a particular action will depend upon what the state of the world is—it is of little value carrying a surfboard when taking a trip across the Sahara, but it is extremely valuable carrying a surfboard when taking a trip across Bondi Beach—and so in choosing which action to undertake, our agent will need to look at the value of $U(S_j)$ where S_j is the state it is in after the action. Doing this for each possible action, the agent can then choose the action which leads to the state it values most. We can certainly build an agent which works in this way, and it would unerringly choose to achieve the goal with the highest value as encoded by its utility function. However it would be just as flawed as an agent which only tried to achieve the most likely goal, trying to achieve the most valuable goal irrespective of the difficulty of that goal.

To build more sensible agents we combine probability and utility calculations for each action and calculate the *expected utility* of each. This amounts to calculating a weighted average of the utility of each outcome, where the weight is the probability of that outcome given the action being performed. Since each outcome is itself a state, we have:

$$EU(A_i) = \sum_{S_j \in \mathbf{S}} \Pr(S_j|A_i)U(S_j)$$

where $\mathbf{S}$ is the set of all states. The agent then selects action A^* where:

$$A^* = arg \max_{A_i \in \mathbf{A}} \sum_{S_j \in \mathbf{S}} \Pr(S_j|A_i)U(S_j)$$

Now, these states which are being considered here are just particular instantiations of the set of state variables $\mathbf{X}$. Thus the probabilities in this calculation are just the probabilities of the X_i having particular values given the actions.

Harking back to the discussion of Bayesian networks above, we can think of the X_i as being structured as a graph, dropping the distinction between variables and the nodes in the graph which represent them. The A_i can be brought into the graph as well, as a different kind of node (square, perhaps, in contrast to the usual round ones relating to the X_i) linked to the X_i whose values they influence. We can also incorporate utilities. This time we only require a single node (a hexagon, to keep it distinct from the others), and this is linked to those X_i which affect its value. Such a graphical structure neatly captures all the dependencies in an expected utility calculation, and is known as an influence diagram (Howard and Matheson, 1984).

Figure 1.3 is an example of a small influence diagram capturing a decision problem which a company has to make about its research and development budget. Since the budget is the thing the decision is being made about, it is represented by a square *decision node*. This is linked to the factors it directly effects, namely the technical success of the company's products and their overall profitability, that latter being captured by the hexagonal *value node*. The remaining nodes are *chance nodes* and represent the other factors which relate to the decision. These are just like nodes in a Bayesian network. Given a particular instantiation of the decision node, the relevant values can be propagated through the network, using an algorithm such as Shacter's graph reduction algorithm (Shachter, 1986) to establish the expected utility of the decision.

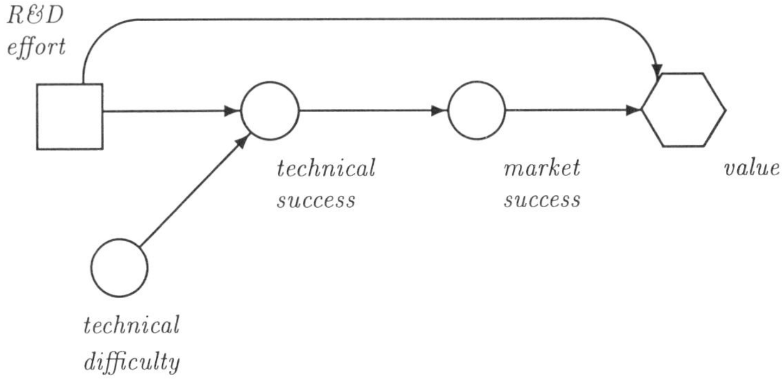

Figure 1.3. An example influence diagram.

2.4. DECISION THEORY AND AGENTS

Given that the basic mechanisms of decision theory fit so neatly into the context of intelligent agents, it is perhaps surprising that they have not been more widely employed in the field. However, agent systems which use decision theory seriously (that is adopting the notions of probability and utility) are rather scarce. One sub-area of decision theory is, however, becoming popular and that is the field of Markov decision processes (MDPs), discussed in detail in (Boutilier et al., 1999). In essence an MDP is an iterative set of classical decision problems. Consider a state of the world as a node in a graph. Carrying out an action in that state will result in a transition to one of a number of states, each connected to the first state by an arc, with some probability, and incur some cost. After a series of transitions a goal state may be reached, and the sequence of actions executed to do this is known as a *policy*. Solving an MDP amounts to finding a minimal cost policy for moving from some initial state to a goal state.

MDPs capture many of the facets of real world problems, but unrealistically assume that whatever system is solving the MDP knows at every point what state it is in. This amounts to assuming that it is possible to measure some aspect of the world and from this measurement tell precisely what state the world is in. This is rarely the case; it is far more likely is that from the measurement something can be uncertainly inferred about the world. In such a situation, the states of an MDP are replaced by beliefs about those states, and we have a partially observable Markov decision process (POMDP). Because they can capture so many real situations, POMDPs are currently a hot topic in agent research, despite the fact that they are intractable for all but the smallest problems.

3. GAME THEORY

Game theory is a branch of economics that studies interactions between self-interested agents. Like decision theory, with which it shares many concepts, game theory has its roots in the work of von Neumann and Morgenstern (von Neumann and Morgenstern, 1944). As its name suggests, the basic concepts of game theory arose from the study of games such as chess and checkers. However, it rapidly became clear that the techniques and results of game theory can equally be applied to *all* interactions that occur between self-interested agents.

The classic game theoretic question asked of any particular multi-agent encounter is: What is the best — most rational — thing an agent can do? In most multi-agent encounters, the overall outcome all depend critically on the choices made by all agents in the scenario. This implies that in order for an agent to make the choice that optimises its outcome, it must reason *strategically*. That is, it must take into account the decisions that other agent may make, and must assume that they will act so as to optimise their own outcome. Game theory gives us a way of formalising and analysing such concerns.

3.1. MULTIAGENT ENCOUNTERS

First, let us simplify things by assuming that we have just two agents; things tend to be much more complicated when we have more than two. Call these agents i and j respectively. Each of the agents is assumed to be *self-interested*. That is, each agent has its own preferences and desires about how the world is, and these are encoded as utilities. We also need to introduce a model of the environment in which these agents will act. The idea is that out two agents will simultaneously choose an action to perform in the environment, and as a result of the actions they select, an outcome in Ω will result. The *actual* outcome that will result will depend on the particular *combination* of actions performed. We will also assume that the agents have no choice about whether to perform an action — they have to simply go ahead and perform one. Further, it is assumed that they cannot see the action performed by the other agent.

To make the analysis a bit easier, we will assume that each agent has just two possible actions that it can perform. We will call these two actions C, for "cooperate", and "D", for "defect". (The rationale for this terminology will become clear below.) Let $Ac = \{C, D\}$ the set of these actions. The way the *environment* behaves is then determined by a function

$$Env : \underbrace{Ac}_{\text{agent } i\text{'s action}} \times \underbrace{Ac}_{\text{agent } j\text{'s action}} \to \Omega$$

In other words, on the basis of the action (either C or D) selected by agent i, and the action (also either C or D) chosen by agent j.

Here is an example of an environment function:

$$\begin{array}{ll} Env(D,D) = \omega_1 & Env(D,C) = \omega_2 \\ Env(C,D) = \omega_3 & Env(C,C) = \omega_4 \end{array} \tag{1}$$

This environment maps each combination of actions to a *different* outcome. This environment is thus sensitive to the actions that each agent performs. At the other extreme, we can consider an environment that maps each combination of actions to the *same* outcome.

$$\begin{array}{ll} Env(D,D) = \omega_1 & Env(D,C) = \omega_1 \\ Env(C,D) = \omega_1 & Env(C,C) = \omega_1 \end{array} \tag{2}$$

In this environment, it does not matter what the agents do: the outcome will be the same. Neither agent has any influence in such a scenario. We can also consider an environment that is only sensitive to the actions performed by one of the agents.

$$\begin{array}{ll} Env(D,D) = \omega_1 & Env(D,C) = \omega_2 \\ Env(C,D) = \omega_1 & Env(C,C) = \omega_2 \end{array} \tag{3}$$

In this environment, it does not matter what agent i does: the outcome depends solely on the action performed by j. If j chooses to defect, then outcome ω_1 will result; if j chooses to cooperate, then outcome ω_2 will result.

The interesting story begins when we put an environment together with the preferences that agents have. To see what we mean by this, suppose we have the most general case, characterised by (1), where both agents are able to exert some influence over the environment. Now let us suppose that the agents have utility functions defined as follows:

$$\begin{array}{llll} u_i(\omega_1) = 1 & u_i(\omega_2) = 1 & u_i(\omega_3) = 4 & u_i(\omega_4) = 4 \\ u_j(\omega_1) = 1 & u_j(\omega_2) = 4 & u_j(\omega_3) = 1 & u_j(\omega_4) = 4 \end{array} \tag{4}$$

Since we know that every different combination of choices by the agents are mapped to a different outcome, we can abuse notation somewhat by writing the following:

$$\begin{array}{llll} u_i(D,D)=1 & u_i(D,C)=1 & u_i(C,D)=4 & u_i(C,C)=4 \\ u_j(D,D)=1 & u_j(D,C)=4 & u_j(C,D)=1 & u_j(C,C)=4 \end{array} \quad (5)$$

We can then characterise agent i's preferences over the possible outcomes in the following way:

$$(C,C) \succeq_i (C,D) \succ_i (D,C) \succeq_i (D,D) \quad (6)$$

Now, consider the following question:

> If you were agent i in this scenario, what would you choose to do — cooperate or defect?

In this case, the answer is pretty unambiguous. Agent i prefers *all* the outcomes in which it cooperates over *all* the outcomes in which it defects. Agent i's choice is thus clear: it should cooperate. It does not matter what agent j chooses to do.

For agent j, the story is the same: we can write j's preferences as follows.

$$(C,C) \succeq_i (D,C) \succ_i (C,D) \succeq_i (D,D)$$

In just the same way, agent j prefers all the outcomes in which *it* cooperates over all the outcomes in which it defects. Notice that in this scenario, neither agent has to expend any effort worrying about what the other agent will do: the action it should perform does not depend in any way on what the other does.

If both agents in this scenario act rationally, that is, they both choose to perform the action that will lead to their preferred outcomes, then the "joint" action selected will be (C, C): both agents will cooperate.

Now suppose that, for the same environment, the agents' utility functions were as follows:

$$\begin{array}{llll} u_i(D,D)=4 & u_i(D,C)=4 & u_i(C,D)=1 & u_i(C,C)=1 \\ u_j(D,D)=4 & u_j(D,C)=1 & u_j(C,D)=4 & u_j(C,C)=1 \end{array} \quad (7)$$

Agent i's preferences over the possible outcomes are thus as follows:

$$(D,D) \succeq_i (D,C) \succ_i (C,D) \succeq_i (C,C)$$

Agent j's preferences are:

$$(D,D) \succeq_i (C,D) \succ_i (D,C) \succeq_i (C,C)$$

In this scenario, agent i can do no better than to defect. The agent prefers *all* the outcomes in which it defects over *all* the outcomes in

	i defects	i cooperates
j defects	4 4	0 5
j cooperates	1 1	3 3

Table 1.1 . A payoff matrix with dominant strategies.

which it cooperates. Similarly, agent j can do no better than defect: it also prefers all the outcomes in which it defects over all the outcomes in which it cooperates. Once again, the agents do not need to engage in *strategic* thinking (worrying about what the other agent will do): the best action to perform is entirely independent of the other agent's choice. I emphasise that in most multiagent scenarios, the choice an agent should make is not so clear cut; indeed, most are much more difficult.

We can neatly summarise the previous interaction scenario by making use of a standard game theoretic notation known as a *payoff matrix*, as in Table 1.1. The way to read such a payoff matrix is as follows. Each of the four cells in the matrix corresponds to one of the four possible outcomes. For example, the top-right cell corresponds to the outcome in which i cooperates and j defects; the bottom-left cell corresponds to the outcome in which i defects and j cooperates. The payoffs received by the two agents are written in the cell. The value in the top-right of each cell is the payoff received by player i (the *column player*), while the value in the bottom left of each cell is the payoff received by agent j (the *row playcr*). Thus in Table 1.1, if i cooperates and j defects, j gets 5 and i gets 0. As payoff matrices are standard in the literature, and are a much more succinct notation than the alternatives, we will use them as standard in the remainder of this chapter.

Before proceeding to consider any specific examples of multiagent encounter, let us introduce some of the theory that underpins the kind of analysis we have informally discussed above.

3.2. DOMINANT STRATEGIES AND NASH EQUILIBRIA

Given a particular multiagent encounter involving two agents i and j, there is one critically important question that both agents want answered: *What should I do?* We have already seen some multiagent encounters, and informally argued what the best possible outcome should be. In this section, we will define some of the concepts that are used in answering this question.

The first concept we will introduce is that of *dominance*. To understand what is meant by dominance, suppose we have two subsets of Ω, which we refer to as Ω_1 and Ω_2 respectively. We will say that Ω_1 *dominates* Ω_2 for agent i if *every* outcome in Ω_1 is preferred over *every* outcome in Ω_2. For example, suppose that:

- $\Omega = \{\omega_1, \omega_2, \omega_3, \omega_4\}$;
- $\omega_1 \succ_i \omega_2 \succ_i \omega_3 \succ_i \omega_4$;
- $\Omega_1 = \{\omega_1, \omega_2\}$; and
- $\Omega_2 = \{\omega_3, \omega_4\}$.

Then Ω_1 strongly dominates Ω_2 since $\omega_1 \succ_i \omega_3$, $\omega_1 \succ_i \omega_4$, $\omega_2 \succ_i \omega_3$, and $\omega_2 \succ_i \omega_4$. However, Ω_2 does not strongly dominate Ω_1, since (for example), it is not the case that $\omega_3 \succ_i \omega_1$. Formally, Ω_1 strongly dominates Ω_2 iff the following condition is true:

$$\forall \omega_1 \in \Omega_1, \forall \omega_2 \in \Omega_2, \text{ we have } \omega_1 \succ_i \omega_2.$$

Now, in order to bring ourselves in line with the game theory literature, we will start referring to actions (members of the set Ac) as *strategies*. Given any particular strategy s for an agent i in a multiagent interaction scenario, there will be a number of possible outcomes. Let us denote by s^* the outcomes that may arise by i playing strategy s. For example, referring to the example environment in equation (1), from agent i's point of view we have $C^* = \{\omega_3, \omega_4\}$, while $D^* = \{\omega_1, \omega_2\}$.

Now, we will say a *strategy* s_1 dominates a strategy s_2 if the set of outcomes possible by playing s_1 dominates the set possible by playing s_2, that is, if s_1^* dominates s_2^*. Again, referring back to the example of (6), it should be clear that, for agent i, "cooperate" strongly dominates "defect". Indeed, as there are only two strategies available, the cooperate strategy is *dominant*: it is not dominated by any other strategy. The presence of a dominant strategy makes the decision about what to do extremely easy: the agent guarantees its best outcome by performing the dominant strategy. In following a dominant strategy, an agent guarantees itself the best possible payoff.

Another way of looking at dominance is that if a strategy s is dominated by another strategy s', then a rational agent will not follow s (because it can guarantee to do better with s'). When considering what to do, this allows us to *delete* dominated strategies from our consideration, simplifying the analysis considerably. The idea is to iteratively consider each strategy s in turn, and if there is another remaining strategy that strongly dominates it, then to delete strategy s from consideration. If we end up with a single strategy remaining, then this will be the dominant strategy, and is clearly the rational choice. Unfortunately, for many interaction scenarios, there will not be a strongly dominant strategy; after deleting strongly dominated strategies, we may find more than one strategy remaining. What to do then? Well, we can start to delete *weakly* dominated strategies. A strategy s_1 is said to weakly dominate strategy s_2 if every outcome s_1^* is preferred at least as much as every outcome s_2^*. The problem is that if a strategy is only weakly dominated, then it is not necessarily irrational to use it; in deleting weakly dominated strategies, we may therefore "throw away" a strategy that would in fact have been useful to use.

The next notion we shall discuss is one of the most important concepts in the game theory literature, and in turn is one of the most important concepts in analysing multiagent systems. The notion is that of *equilibrium*, and more specifically, *Nash equilibrium*. The intuition behind equilibrium is perhaps best explained by example. Every time you drive a car, you need to decide which side of the road to drive on. The choice is not a very hard one: if you are in the UK, for example, you will probably choose to drive on the left; if you are in the US or continental Europe, you will drive on the right. The reason the choice is not hard is that it is a Nash equilibrium strategy. Assuming everyone else is driving on the left, you can do no better than drive on the left also. From everyone else's point of view, assuming you are driving on the left then everyone else can do no better than drive on the left also.

In general, we will say that two strategies s_1 and s_2 are in Nash equilibrium if:

1 under the assumption that agent i plays s_1, agent j can do no better than play s_2; and

2 under the assumption that agent j plays s_2, agent i can do no better than play s_1.

The *mutual* form of an Equilibrium is important because it "locks the agents in" to a pair of strategies. *Neither agent has any incentive to deviate from a Nash equilibrium.* To see why, suppose s_1, s_2 are a pair of strategies in Nash equilibrium for agents i and j respectively, and that agent j chooses to play some other strategy: s_3 say. Then by definition, i will do no better, and may possibly do worse than it would have done by playing s_1.

The presence of a Nash equilibrium pair of strategies in a game might appear to be the definitive answer to the question of what to do in any given scenario. Unfortunately, there are two important results in the game theory literature which serve to make life difficult:

1 *Not every interaction scenario has a Nash equilibrium.*

If there is no Nash equilibrium, then it may be possible to look for a *mixed* equilibrium as described in the next section.

2 *Some interaction scenarios have more than one Nash equilibrium.*

Despite these negative results, the notion of a Nash equilibrium is an extremely important concept, and will be useful in the analysis that follows.

3.3. COMPETITIVE AND ZERO-SUM INTERACTIONS

Suppose we have some scenario in which an outcome $\omega \in \Omega$ is preferred by agent i over an outcome ω' if, and only if, ω' is preferred over ω by agent j. Formally,

$$\omega \succ_i \omega' \quad \text{if and only if } \omega' \succ_j \omega.$$

The preferences of the players are thus diametrically opposed to one-another: one agent can only improve its lot (i.e., get a more preferred outcome) at the expense of the other. An interaction scenario that satisfies this property is said to be *strictly competitive*, for hopefully obvious reasons.

Zero-sum encounters are those in which, for any particular outcome, the utilities of the two agents sum to zero. Formally a scenario is said to be zero sum if the following condition is satisfied:

$$u_i(\omega) + u_j(\omega) = 0 \quad \text{for all } \omega \in \Omega.$$

It should be easy to see that any zero sum scenario is strictly competitive. Zero sum encounters are important because they are the most "vicious" types of encounter conceivable, allowing for no possibility of cooperative behaviour. If you allow your opponent positive utility, then this means that you get *negative* utility — intuitively, you are worse off than you were before the interaction.

Games such as chess and chequers are the most obvious examples of strictly competitive interactions. Indeed, any game in which the possible outcomes are win or lose will be strictly competitive. Outside these rather abstract settings, however, it is hard to think of real-world examples of zero-sum encounters. War might be cited as a zero sum interaction between nations, but even in the most extreme wars, there will usually be at least *some* common interest between the participants (e.g., in ensuring that the planet survives).

	i defects	i cooperates
j defects	-3 3	1 -1
j cooperates	0 0	-1 1

Table 1.2 . A payoff matrix for a game with no pure equilibrium.

For these reasons, some social scientists are sceptical about whether zero-sum games exist in real-world scenarios (Zagare, 1984, p.22). Interestingly, however, people interacting in many scenarios have a tendency to treat them *as if they were zero-sum.* Below, we will see that in some scenarios — where there is the possibility of mutually beneficial cooperation — this type of behaviour can be damaging.

3.4. MIXED EQUILIBRIA

All the scenarios we have considered up to now have had solutions in which the agents pick a single outcome, a single row or column in the payoff matrix[6]. As discussed above, there are scenarios in which agents that choose such *pure* strategies can arrive at Nash equilibrium or other stable solutions. However, there are some situations in which pure strategies will *not* give a stable solution. As an example, consider the game decribed by the payoff matrix in Table 1.2.

This payoff matrix describes a zero sum game. If both i and j defect, then i loses 3 and j wins 3. If i defects and j cooperates, then both get a zero payoff. If i cooperates and j defects, i wins 1 and j loses 1, while if both cooperate then the payoff is reversed. Thus neither agent has one move it can make which is definitely better than the other—it all depends on what the other agent does. The result of this arrangement of payoffs is that neither agent can make a choice which holds up if the other agent somehow finds out what the first agent is intending to do (a situation that is sometimes referred to as having a *spyproof* strategy).

To see this, consider what each agent will chose to do. If i knew that j would defect, then i would choose to cooperate. However, if j knew that i would co-operate, then j would choose to cooperate. Similarly, if i knew that j would cooperate, then it would choose to defect, while if j knew that i was planning to defect, then j would defect as well. So any piece of information about what the other is intending (which, of course, could be established by either agent by thinking about what it would do in the other agent's shoes) will cause an agent to change its strategy.

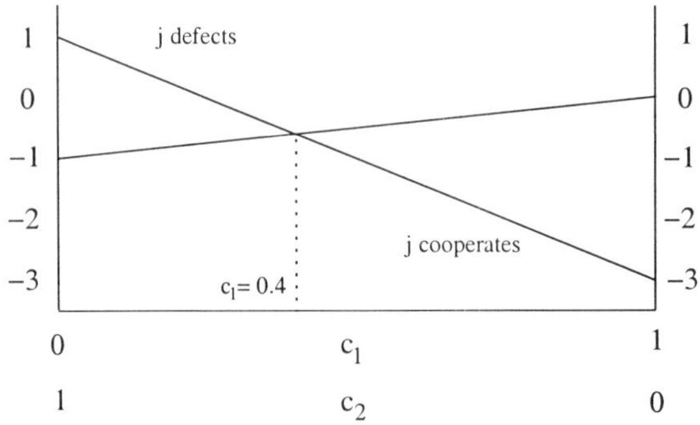

Figure 1.4. Determining a mixed strategy for Agent i.

So, rather than adopt a pure strategy, agents adopt a *mixed* strategy in which they take a random choice across a set of pure strategies. More formally, agent i picks a vector of probabilities over the columns in the payoff matrix that relate to the pure strategies it might choose. In general, where agent i has n possible strategies, it needs a vector:

$$c = (c_1, c_2, \ldots, c_n)$$

where

$$\sum_k c_k = 1$$

and

$$c_k \geq 0$$

Agent i then picks strategy k with probability c_k. Any vector of probabilities gives a mixed strategy. The question is how to obtain a good mixed strategy. Clearly, for the 2×2 game we have here, i needs to compute the best values of c_1 and c_2—the probabilities of defecting and cooperating respectively. These will be the values which give i the highest expected payoff for its mixed strategy. The diagram in Figure 1.4 is one way of thinking about solving the problem of picking a mixed strategy for agent i, which is fundamentally a problem of determining the values of c_1 and c_2 which will maximise the agent's expected utility.

The diagram plots expected utility on the vertical axis(axes), and the probabilities c_1 and c_2 across the horizontal axis. If j chooses to play the first row, in other words to defect, then i will get either -3 or 1 depending on its choice of column. When this choice is made randomly, the expected payoff to i is the weighted sum:

$$y = -3c_1 + c_2$$

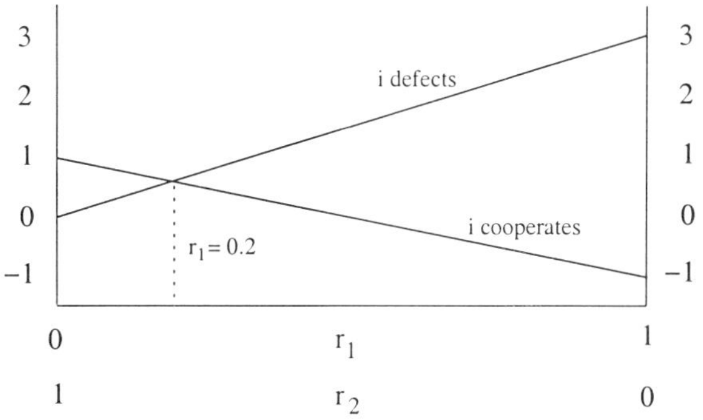

Figure 1.5. Determining a mixed strategy for Agent j.

which is the equation of the line joining 1 and -3 in Figure 1.4 (since $c_1 = 1 - c_2$). Similarly, if j chooses to play the second row, then the expected payoff to i is between 0 and -1, depending on the probability with which the two options available to i are selected, and is plotted by the other line in Figure 1.4. The intersection of the two lines gives the value of c_1 and c_2 (which is just $1 - c_1$) at which i is indifferent as to what j plays—whichever j chooses, the expected payoff to i is the same. This, then, yields a form of stability and the resulting values of c_1 and c_2 give the mixed strategy for i.

Agent j can analayse the problem in terms of a probability vector

$$r = (r_1, r_2)$$

and come up with a similar picture (Figure 1.5).

Now, let's consider the payoff's the players will expect. With i having mixed strategy (c_1, c_2) and j having (r_1, r_2), then the loss that i will expect to make will be:

$$\begin{aligned} L &= 3c_1r_1 + 0(1 - r_1)c_1 \\ &\quad - r_1(1 - c_1) + (1 - c_1)(1 - r_1) \\ &= 5c_1r_1 - 2r_1 - c_1 + 1 \end{aligned}$$

Now, assuming that i uses $c_1^* = 0.4$ as calculated above. Then:

$$\begin{aligned} L &= 5(0.4r_1) - 2r_1 - c_1 + 1 \\ &= 0.6 \end{aligned}$$

and i will have an expected loss of 0.6. Similarly, we can calculate the gain that j will expect to make as:

$$\begin{aligned} G &= 3c_1r_1 + 0(1-c_1)r_1 \\ &\quad - c_1(1-r_1) + (1-c_1)(1-r_1) \\ &= 5c_1r_1 - r_1 - 2c_1 + 1 \end{aligned}$$

If j picks $r_1^* = 0.2$ then:

$$G = 0.6$$

The neat thing is that the expected gain or loss for one agent does not depend upon the strategy of the other agent—once the correct mixed strategy has been determined, it no longer matters what the other agent is going to do, the first agent still has the same expected outcome. This result generalises to more general games, and Von Neumann's Minimax Theorem shows that you can always find a pair of mixed strategies x^* and y^* which result in i and j having the same expected value for the game.

This theory is sufficient background for the rest of the book, but before we leave the topic, it is worth considering the best-known multi-agent scenario: the *prisoner's dilemma.*

4. THE PRISONER'S DILEMMA

Consider the following scenario:

> Two men are collectively charged with a crime and held in separate cells. They have no way of communicating with each other or making any kind of agreement. The two men are told that:
>
> 1 if one of them confesses to the crime and the other does not, the confessor will be freed, and the other will be jailed for three years; and
>
> 2 if both confess to the crime, then each will be jailed for two years.
>
> Both prisoners know that if neither confesses, then they will each be jailed for one year.

We refer to confessing as defection, and not confessing as cooperating, because we are viewing the problem from the point of view of the prisoners—from the point of view of one of them, the best, most cooop-erative, thing the other can do is to not confess.

There are four possible outcomes to the prisoner's dilemma, depending on whether the agents cooperate or defect, and so the environment is of type (1). Abstracting from the scenario above, we can write down the utility functions for each agent in the payoff matrix of Table 4. Note that the numbers in the payoff matrix are *not* the length of the jail term.

	i defects	i cooperates
j defects	2 2	1 5
j cooperates	5 1	3 3

Table 1.3 . A payoff matrix for the Prisoner's Dilemma.

In other words, the utilities are as follows:

$$
\begin{array}{llll}
u_i(D,D)=2 & u_i(D,C)=5 & u_i(C,D)=1 & u_i(C,C)=3 \\
u_j(D,D)=2 & u_j(D,C)=1 & u_j(C,D)=5 & u_j(C,C)=3
\end{array}
$$

And the preferences are:

$$
\begin{array}{l}
(D,C) \succ_i (C,C) \succ_i (D,D) \succ_i (C,D) \\
(C,D) \succ_j (C,C) \succ_j (D,D) \succ_j (D,C)
\end{array}
$$

So, what should a prisoner do? The answer is not as clear cut as the previous pure strategy examples we looked at. It is not the case a prisoner prefers all the outcomes in which it cooperates over all the outcomes in which it defects. Similarly, it is not the case that a prisoner prefers all the outcomes in which it defects over all the outcomes in which it cooperates.

The "standard" approach to this problem is to put yourself in the place of a prisoner, i say, and reason as follows:

> Suppose I cooperate. Then if j cooperates, we will both get a payoff of 3. But if j defects, then I will get a payoff of one. So the best payoff I can be *guaranteed* to get if I cooperate is 1.
>
> Suppose I defect. Then if j cooperates, then I get a payoff of 5, whereas if j defects, then I will get a payoff of 2. So the best payoff I can be guaranteed to get if I defect is 2.
>
> So, if I cooperate, the worst case is I will get a payoff of 1, whereas if I defect, the worst case is that I will get 2.
>
> Since I would prefer a payoff of 2 to a payoff of 1, I should defect.

Since the scenario is symmetric (i.e., both agents reason the same way), then the outcome that will emerge — if both agents reason "rationally" — is that both agents will defect, giving them each a payoff off 2.

Notice that neither strategy dominates in this scenario, so our first route to finding a choice of strategy is not going to work. Turning to Nash equilibria, there is a single Nash equilibrium of D, D. Thus under the assumption that i will play D, j can do no better than play D, and under the assumption that j will play D, i can also do no better than play D.

But is this the best they can do? *Naive intuition says not.* Surely if they both *cooperated*, then they could do better — they would receive a payoff of 3. But if you assume the other agent will cooperate, then the rational thing to do — the thing that maximises your utility — is to defect. The conclusion seems inescapable: the rational thing to do in the prisoners dilemma is defect, even though this appears to "waste" some utility. (The fact that our naive intuition tells us that utility appears to be wasted here, and that the agents could do better by cooperating, even though the rational thing to do is to defect, is why this is referred to as a dilemma.)

The prisoners dilemma may seem an abstract problem, but it turns out to be very common indeed. In the real world, the prisoners dilemma appears in situations ranging from nuclear weapons treaty compliance to negotiating with one's children. Consider the problem of nuclear weapons treaty compliance. Two countries i and j have signed a treaty to dispose of their nuclear weapons. Each country can then either cooperate (= get rid of their weapons), or defect (= keep their weapons). But if you get rid of your weapons, you run the risk that the other side keeps theirs, making them very well off while you suffer what is called the "suckers payoff". In contrast, if you keep yours, then the possible outcomes are that you will have nuclear weapons while the other country does not (a very good outcome for you), or else at worst that you both retain your weapons. This may not be the best possible outcome, but is certainly better than you giving up your weapons while your opponent kept theirs, which is what you risk if your give up your weapons.

Many people find the conclusion of this analysis — that the rational thing to do in the prisoner's dilemma is defect — deeply upsetting. For the result *seems* to imply that cooperation can only arise as a result or *irrational* behaviour, and that cooperative behaviour can be exploited by those who behave rationally. The apparent conclusion is that nature really is "red in tooth and claw". Particularly for those who are inclined to a liberal view of the world, this is unsettling and perhaps even distasteful. As civilized beings, we tend to pride ourselves on somehow "rising above" the other animals in the world, and believe that we are capable of nobler behaviour: to argue in favour of such an analysis is therefore somehow immoral, and even demeaning to the entire human race.

Naturally enough, there have been several attempts to respond to this analysis of the prisoners dilemma, in order to "recover" cooperation (Binmore, 1992, p.355–382).

We Are Not All Machiavelli! The first approach is to argue that we are not all such "hard boiled" individuals as the prisoner's dilemma (and more generally, this kind of game theoretic analysis) implies. We are *not* seeking to constantly maximise our own welfare, possibly at the expense of others. Proponents of this kind of argument typically point to real-world examples of *altruism* and spontaneous, mutually-beneficial cooperative behaviour in order to justify their claim.

There is some strength to this argument: we do not, (or at least, most of us do not), constantly deliberate about how to maximise our welfare without any consideration for the welfare of our peers. Similarly, in many scenarios, we would be happy to trust our peers to recognise the value of a cooperative outcome without even mentioning it to them, being no more than mildly annoyed if we get the "sucker's payoff".

There are several counter responses to this. First, it is pointed out that many real-world examples of spontaneous cooperative behaviour are not really the prisoner's dilemma. Frequently, there is some built in mechanism that makes it in the interests of participants to cooperate. For example, consider the problem of giving up your seat on the bus. We will frequently give up our seat on the bus to an older person, mother with children, etc., apparently at some discomfort (= loss of utility) to ourselves. But it could be argued that in such scenarios, society has ways of punishing non-cooperative behaviour: suffering the hard and unforgiving stares of fellow passengers when we do not give up our seat, or worse, being accused in public of being uncouth!

Second, it is argued that many "counter examples" of cooperative behaviour arising do not stand up to inspection. For example, consider a public transport system, which relies on everyone cooperating and honestly paying their fare every time they travel, even though whether or not they have paid is not verified. The fact that such a system works (the buses turn up on time) would appear to be some evidence that relying on spontaneous cooperative can work. But the fact that such a system functions does not mean that the system is not exploited: it will be, and if there is no means of checking whether or not someone has paid their fare and punishing non-compliance, then all other things being equal, those individuals that do exploit the system will be better off. Unpalatable, perhaps, but true nevertheless.

The Other Prisoner is My Twin! A second line of attack is to argue that two prisoner's will "think alike", and recognise that cooperation is the best outcome. For example, suppose the two prisoners are twins, inseparable since birth; then, it is argued, if their though processes are sufficiently aligned, they will both recognise the benefits of cooperation, and behave accordingly. The answer to this is that it implies there are not actually two prisoner's playing the game. If I can make my twin select a course of action simply by "thinking it", then we are not playing the prisoner's dilemma at all.

This "fallacy of the twins" argument often takes the form "what if everyone were to behave like that" (Binmore, 1992, p311). The answer, as Yossarian pointed out in Joseph Heller's *Catch 22*, is that if everyone else behaved like that, you would be a damn fool to behave any other way.

People Are Not Rational! Some would argue[7], that we might indeed be happy to risk cooperation as opposed to defection when faced with situations where the sucker's payoff really does matter very much. For example, paying a bus fare that amounts to a few pennies does not really hurt us much, even if everybody else is defecting and hence exploiting the system. But, it is argued, when we are faced with situations where the sucker's payoff really *hurts* us — life or death situations and the like — we will choose the "rational" course of action that maximises our welfare, and defect.

The Shadow of the Future Lest the discussion so far prove too depressing, it should be emphasised that there are quite natural variants of the prisoner's dilemma in which cooperation *is* the rational thing to do. One idea is to *play the game more than once*. In the *iterated prisoner's dilemma*, the "game" of the prisoner's dilemma is played a number of times. Each play is referred to as a "round". Critically, it is assumed that each agent can see what the opponent did on the previous round: player i can see whether j defected or not, and j can see whether i defected or not.

Now, for the sake of argument, assume that the agents will continue to play the game *forever*: every round will be followed by another round. Now, under these assumptions, what is the rational thing to do? If you know that you will be meeting the same opponent in future rounds, the incentive to defect appears to be considerably diminished, for two reasons:

- If you defect now, your opponent can *punish* you by also defecting. Punishment is not possible in the one-shot prisoner's dilemma.
- If you "test the water" by cooperating initially, and receive the sucker's payoff on the first round, then because you are playing the game indefinitely, this loss of utility (one util) can be "amortized" over the future rounds. When taken into the context of an infinite

> (or at least very long) run, then the loss of a single unit of utility will represent a small percentage of the overall utility gained.

So, if you play the prisoner's dilemma game indefinitely, then cooperation is a rational outcome (Binmore, 1992, p358). The "shadow of the future" encourages us to cooperate in the infinitely repeated prisoner's dilemma game.

This seems to be very good news indeed, as truly one-shot games are comparatively scarce in real life. When we interact with someone, then there is often a good chance that we will interact with them in the future, and rational cooperation begins to look possible. However, there is a catch.

Suppose you agree to play the iterated prisoner's dilemma a *fixed* number of times, (say 100). You need to decide (presumably in advance) what your strategy for playing the game will be. Consider the last round (i.e., the 100th game). Now, on this round, you know — as does your opponent — that you will not be interacting again. In other words, the last round is in effect a one-shot prisoner's dilemma game. As we know from the analysis above, the rational thing to do in a one-shot prisoner's dilemma game is defect. Your opponent, as a rational agent, will presumably reason likewise, and will also defect. On the 100th round, therefore, you will both defect. But this means that the last "real" round, is 99. But similar reasoning leads us to the conclusion that this round will also be treated in effect like a one-shot prisoner's dilemma, and so on. Continuing this *backwards induction* leads inevitably to the conclusion that, in the iterated prisoner's dilemma with a fixed, pre-determined number of rounds, defection is the dominant strategy, as in the one-shot version (Binmore, 1992, p.354).

Whereas it seemed to be very good news that rational cooperation is possible in the iterated prisoner's dilemma with an infinite number of rounds, it seems to be very bad news that this possibility appears to evaporate if we restrict ourselves to repeating the game a pre-determined, fixed number of times. Returning to the real-world, we know that in reality, we will only interact with our opponents a finite number of times (after all, one day the world will end). We appear to be back where we started.

The story is actually better than it might at first appear, for several reasons. The first is that *actually* playing the game an infinite number of times is not necessary. As long as the "shadow of the future" looms sufficiently large, then it can encourage cooperation. So, rational cooperation can become possible if both players know, with sufficient probability, that they will meet and play the game again in the future.

The second reason is that, even though a cooperative agent can suffer when playing against a defecting opponent, it can do well overall provided it gets sufficient opportunity to interact with other cooperative agents.

5. SUMMARY

In this chapter we have discussed some of the basic concepts in decision theory and game theory. Our intention was simply to make the rest of the book comprehensible even for those who have not come across either decision theory, or game theory, or both, in the past, and this brief discussion stops short of being anything like a comprehensive introduction to the topics. To do the latter would take a book as long as this whole work (at the very least). However, for this who want to know more then there are a number of places to look.

For information on game theory, one of the best places to start is (Binmore, 1992). This is a very readable introduction from a leading expert, and, being intended as a textbook, starts at the very beginning and covers much of the game theory one is ever likely to need to know. From a more specifically agent-oriented perspective, (Rosenschein and Zlotkin, 1994) is a seminal work in the application of game theory to multi-agent encounters, and (Kraus, 2001) is a deep and rigorous investigation of similar matters. In addition, (Sandholm, 1999) gives a good general discussion of the use of game theory in designing multi-agent systems, as well as related topics such as auction theory. Initial work on the iterated prisoner's dilemma was carried out by Axelrod (Axelrod, 1984), and this is recommended as a point of departure for further reading on the topic. (Mor and Rosenschein, 1995) provides pointers into recent prisoner's dilemma literature. An non-mathematical introduction to game theory, with an emphasis on the applications of game theory in the social sciences, is (Zagare, 1984), and some of the early work in game theory is surveyed in (Schwalbe and Walker, 2001)[8].

The seminal introduction to decision theory is (Raiffa, 1968), and a good, if slightly dogmatic, alternative is (Lindley, 1975). (Smith, 1999) is not quite as detailed, but covers much of the relevant material and is very easy to read. Bayesian networks are covered in detail by (Pearl, 1988), the first book on the subject, and by (Castillo et al., 1997; Cowell et al., 1999; Jensen, 2001). The last two also cover influence diagrams and so deal with network models for the whole of decision theory (in contrast Pearl and Castillo *et al.* deal only with probability). Finally, (Puterman, 1994) is a comprehensive introduction to Markov decision processes, but for work in this area much of the most relevant material has only yet appeared in journal and conference papers.

Notes

1. You can calculate it as a ration of occurrence to the total number of events, but what counts as an occurrence? It is clear when calculating the probability of getting heads when tossing a coin, but much less obvious when, for example, estimating the probability of a particualr disease giving certain symptoms. Even trickier is the question of how many events you have to sum over. Typical definitions claim that you only truly know the probability if you sum over an infinite number of events which presents obvious practical diffculties.

2. The key step in the argument as to why betting rates are a suitable way of determing probabilities is that if the bettor misestimates the probability and so proposes a different bet—say paying $0.4 for the chance of winning $1 when a occurs—then the person they are betting with can exploit them. In particular, this second person can construct a *Dutch book*, a set of bets which can win an arbitrarily large amount of money from the bettor. Thus, the argument runs, the bettor will be motivated to get the probability right. The main problem with this argument is that it places a considerable cognitive burden on whoever is establishing the probability.

3. The notion of independence captured in the arcs of a Bayesian network is somewhat more complex than that described here, but the difference is not relevant for the purposes of this article. For full details, see (Pearl, 1988).

4. Indeed, in economics as a whole, very little thought is given to the question of where preferences come from.

5. To misquote Martin Amis, the difference is clear—a cool million dollars.

6. How might an agent choose more than one row in the kind of framework we have been considering? Well, as we will see in a very short while, it doesn't, but at the same time it doesn't choose a single row either...

7. Ken Binmore certainly did at the UKMAS workshop in December 1998.

8. They also provide the first English translation of what is generally regarded as the first paper on game theory.

References

Axelrod, R. (1984). *The Evolution of Cooperation.* Basic Books.

Binmore, K. (1992). *Fun and Games: A Text on Game Theory.* D. C. Heath and Company: Lexington, MA.

Boutilier, C., Dean, T., and Hanks, S. (1999). Decision-theoretic planning: structural assumptions and computational leverage. *Journal of Artificial Intelligence Research*, 11:1–94.

Castillo, E., Gutiérrez, J. M., and Hadi, A. S. (1997). *Expert Systems and Probabilistic Network Models.* Springer Verlag, Berlin, Germany.

Cowell, R. G., Dawid, A. P., Lauritzen, S. L., and Spiegelhalter, D. J. (1999). *Probabilistic Networks and Expert Systems.* Springer Verlag, Berlin, Germany.

Howard, R. A. and Matheson, J. E. (1984). Influence diagrams. In Howard, R. A. and Matheson, J. E., editors, *Readings on the Principles and Applications of Decision Analysis*, pages 719–762. Strategic Decisions Group, Menlo Park, CA.

Jensen, F. V. (2001). *Bayesian Networks and decision graphs.* Springer Verlag, New York, NY.

Kraus, S. (2001). *Strategic negotiation in multiagent environments.* MIT Press, Cambridge, MA.

Lindley, D. V. (1975). *Making Decisions.* John Wiley & Sons, Chichester, UK.

Mor, Y. and Rosenschein, J. S. (1995). Time and the prisoner's dilemma. In *Proceedings of the First International Conference on Multi-Agent Systems (ICMAS-95)*, pages 276–282, San Francisco, CA.

von Neumann, J. and Morgenstern, O. (1944). *Theory of Games and Economic Behaviour.* Princeton University Press.

Pearl, J. (1988). *Probabilistic Reasoning in Intelligent Systems; Networks of Plausible Inference.* Morgan Kaufmann, San Mateo, CA.

Puterman, M. L. (1994). *Markov Decision Processes.* John Wiley & Sons, New York, NY.

Raiffa, H. (1968). *Decision Analysis: Introductory Lectures on Choices under Uncertainty.* Addison Wesley, Reading, MA.

Rosenschein, J. S. and Zlotkin, G. (1994). *Rules of Encounter: Designing Conventions for Automated Negotiation among Computers.* The MIT Press: Cambridge, MA.

Sandholm, T. (1999). Distributed rational decision making. In Weiß, G., editor, *Multiagent Systems*, pages 201–258. The MIT Press: Cambridge, MA.

Schwalbe, U. and Walker, P. (2001). Zermelo and the early history of game theory. *Games and Economic Behaviour*, 34:123–137.

Shachter, R. (1986). Evaluating influence diagrams. *Operations Research*, 34:871–882.

Shafer, G. (1992). Can the various meanings of probability be reconciled? In Keren, G. and Lewis, C., editors, *A Handbook for Data Analysis in the Behavioural Sciences.* Lawrence Erblaum, Hillsdale, NJ.

Smith, J. Q. (1999). *Decision Analysis: A Bayesian Approach.* Springer Verlag, Berlin.

Zagare, F. C. (1984). *Game Theory: Concepts and Applications.* Sage Publications: Beverly Hills, CA.

Chapter 2

SELECTING PARTNERS

Bikramjit Banerje
Mathematical & Computer Sciences Department, University of Tulsa
600 South College Avenue, Tulsa, OK 74104
bikram@euler.mcs.utulsa.edu

Sandip Sen
Mathematical & Computer Sciences Department, University of Tulsa
600 South College Avenue, Tulsa, OK 74104
sandip@euler.mcs.utulsa.edu

Abstract The goal of a rational agent is to maximize utility. We consider situations where a rational agent has to choose one of several contenders to enter into a partnership. We assume that the agent has a model of the likelihood of different outcomes and corresponding utilities for each such partnership. Given a fixed, finite number of interactions, the problem is to choose a particular partner to interact with where the goal is to maximize the sum of utilities received from all the interactions. We develop a multinomial distribution based mechanism for partner selection and contrast its performance with other well-known approaches which provide exact solution to this problem for infinite interactions.

Keywords: Rational decision making, autonomous agents, expected utility.

1. INTRODUCTION

In an open environment, agents have to be extremely cautious about which other agents to interact or partner with. The goal of a self-interested agent would be to interact with or enter into partnership with those agents that will produce maximal local utility for this agent (Tesfatsion, 1998). Obviously, such an agent will also have to achieve its local goals effectively to maximize its utility. For discussions in this paper, however, we will concentrate exclusively on utilities received by interacting with other agents.

We assume that each agent interaction will ultimately generate some utility for each of the interacting agents. From a single agent X's point of view, interaction with a collection of agents can be thought of as entering a partnership (also called coalitions in game theory). We are only interested in the utility received by the agent X by interacting with a coalition. We will not concern ourselves with issues of how the coalition generates the revenue and the process by which the generated revenue is distributed among the partners.

We assume that an agent can get one of several payoffs or utilities for joining a particular coalition, and that there is a static probability distribution that governs which of the payoffs is received at any particular interaction. Our usage of the term "interaction" needs further clarification: by one interaction of an agent X with a coalition Y we refer to the entire process of X joining the coalition, the coalition generating some revenue R, and X receiving some share of that revenue r_X^Y which, as mentioned above, is determined by a probability distribution. This probability distribution of different utilities that X can receive from coalition Y will be referred to as the *payoff-structure of Y from X's viewpoint*, P_X^Y. In a later section, we will discuss a representative scenario that justifies our assumptions.

Some combination of *a priori* domain models, observation or experience-based learning schemes as well as word-of-mouth transmissions can be used to arrive at these payoff-structures. In this paper, however, we will not address the issue of how these payoff-structures are generated. Rather, we will focus on how to select a coalition given the payoff-structures for each of the coalitions an agent can choose to interact with. More precisely, we will consider choosing a single coalition to interact with repeatedly when the total number of interactions are known ahead of time. We believe, and our theoretical as well as experimental results will show, that the particular coalition chosen should vary depending on the set of payoff- structures for all the coalitions under consideration, and the total number of interactions. Our goal in this paper is to design such a coalition or partner selection mechanism that identifies the most beneficial partnership given a finite number of interactions that the selecting agent is going to partake in.

2. A SAMPLE PROBLEM

Consider a situation where an agent has to select between two partners for the next n interactions. An interaction with one of these partners, say A, yields returns of 10 and 20 units with probabilities 0.7 and 0.3 respectively. Thus the expected utility from A is $(10 * 0.7 + 20 * 0.3) =$ 13 units. On the other hand, each interaction with the other partner, B, yields payoffs of 1000 and 4 units with probabilities 0.01 and 0.99 respectively. The expected utility from B $(1000 * 0.01 + 4 * 0.99 =$ 13.96 units) is greater than that from A but if the agent takes into consideration the finite number of interactions, n, that it intends then A may be more attractive than B, since it returns higher rewards in both cases than the one that B returns almost certainly. Our contention is that to make such decisions, an agent must take into account the *number of intended interactions*, and when that is done the consequent decision may (depending on the payoff-structure as in this example) be quite unlike what risk-aversion dictates.

3. COALITION FORMATION

There has been a significant amount of work in game theory and multiagent systems field in the area of coalition formation. Representative work include:

Searching for optimal coalition structure: How should a group of agents be partitioned into subgroups or coalitions such that the sum of the revenue generated by all such coalitions is maximized (Sandholm et al., 1999)?

Decision mechanisms for forming coalitions: How should agents decide on which coalition to join (Ketchpel, 1994; Sandholm and Lesser, 1997; Shehory and Kraus, 1995; Shehory and Kraus, 1996; Zlotkin and Rosenschein, 1994)?

Payoff division in a coalition: How should the revenue generated in a coalition be divided among the partners (Ketchpel, 1994; Luce and Raiffa, 1957)?

Almost all of this body of research ignores, as we do, the issue of how the coalitions generate their revenues or the nature of problem solving adopted by individual agents after they form a coalition.

We address a slightly different problem of coalition selection under uncertainty. Perhaps Ketchpel's work (Ketchpel, 1994) is most related. That work addresses a process for negotiating who becomes the leader versus the member in a group based on how much one is willing to pay others to join a group. We, on the other hand, do not concern ourselves with our agent X building a coalition. Rather, X is selecting which coalition to join given only summary information of what payoff it can expect from that coalition. The payoff-structure encoding in the form of a probability distribution over possible payoffs for joining a coalition is the summary information on which X must base its decision. Another important distinction and the key focus of this work is the prior determination of how many interactions an agent is going to have.

Our basic hypothesis is that the most beneficial partnership to interact with can change based on how many interactions we will have. For example, consider a simple scenario of choosing from the two following partnerships. One partnership always give a steady return whereas the other offers a small return most of the times and very infrequently returns an astronomical amount. If one is only going to interact for a few times, it might be prudent to interact with the first partnership, but if one is going to have a prolonged sequence of interactions, perhaps it is worthwhile to choose the second partnership expecting that the "jackpot" is more likely to be hit at least once and will more than compensate for the smaller returns in the other interactions. Our goal is to go beyond this informal heuristic and provide a formal and well-founded decision procedure based on probability theory for this partnership selection problem.

4. PAYOFF-STRUCTURES OF PARTNERSHIPS

We now present a formal representation for *payoff-structure* of a partnership or coalition for an agent X outside the coalition. It can be represented as an n-tuple - $\{(u_1, p_1)(u_2, p_2)\dots(u_n, p_n)\}$ where p_i stands for the probability that the interaction with this partnership will yield a payoff u_i, $\forall i = 1\dots n$. X will have such a tuple for each of the other partnerships it can interact with. We note that n is not a constant but varies from one partnership to another, i.e., the number of possible payoffs depends on the particular partnership X is interacting with.

5. SELECTING THE POTENTIALLY MOST BENEFICIAL PARTNERSHIP FOR LIMITED INTERACTIONS

We are interested in devising a procedure that, given a pool of possible partnerships and the number of interactions (N) that the agent wants to make, will select a partnership that is most likely to return a maximum total utility over these N interactions. To this end, we make pairwise comparisons among the partnerships and select the most profitable one. Let p_{ij}^N be the probability that partnership i returns greater utility than j. If $p_{ij}^N > p_{ji}^N$ then we regard partnership i to be more beneficial than partnership j. Hence, we define the objective function for selection between agents i and j, as

$$S(i,j) = \begin{cases} i & \text{if } p_{ij}^N > p_{ji}^N \\ j & \text{otherwise} \end{cases}$$

We note here, that $S(i,j)$ is *not transitive* (i.e. if $S(i,j) = i$ and $S(j,k) = j$ does not imply that $S(i,k) = i$). We will illustrate this non-transitivity with an example scenario in a later section.

Our proposed decision mechanism first finds $S(i,j)$ values for all possible partnership pairs (i,j), and chooses the partnership that is returned for the maximum number of times (i.e. the partnership with the maximum number of wins against other partnerships). This mechanism trivially selects a partnership if it wins against all other partnerships. If there is a tie among a subset of these partnerships then we can use arbitration mechanism from the voting theory (Straffin, 1980), with p_{ij}^N as a relative measure of mandate between i and j, for each pair (i,j) in that subset. For example, assume that there is a tie among partnerships i, j, and k. Then the arbitration mechanism gives $p_{ij}^N + p_{ik}^N$ votes to partnership i, $p_{ji}^N + p_{jk}^N$ votes to partnership j, and $p_{ki}^N + p_{kj}^N$ votes to partnership k. If there is still a tie, we choose one of these partnerships randomly. In the rest of this section we shall derive an expression to compute p_{ij}^N, given only the payoff-structures of the prospective partnerships, i and j, and the value of N.

Let the payoff-structures of partnerships i and j be given by sequences $\{(u_k^i, p_k^i)\}$ and $\{(u_k^j, p_k^j)\}$, and their lengths be n_i and n_j respectively. We can divide N into n parts, where each part is 0 or a positive integer. The kth part in such a division represent the number of times the kth utility value was returned when interacting with a given partnership. Let any such decomposition be represented as C_n^N. There are

$$\binom{N+n-1}{n-1}$$

such distinct decompositions. If we regard such a decomposition of N interactions as actually arising in exchanges with partnership i, then the utility received as a result, is given by

$$U_i^{C_{n_i}^N} = \sum_{k=1}^{n_i} C_{n_i}^N(k) * u_k^i$$

where $C_{n_i}^N(k)$ is the *kth* component of $C_{n_i}^N$. Then the value of p_{ij}^N is computed according to the rule for conditional probability, as

$$p_{ij}^N = \sum_{C_{n_i}^N} \mathbf{Pr}[C_{n_i}^N].\mathbf{Pr}[U_i^{C_{n_i}^N} > U_j | C_{n_i}^N]$$

where $\mathbf{Pr}[U_i^{C_{n_i}^N} > U_j | C_{n_i}^N]$ (i.e. the probability that i returns greater net-utility, given a particular decomposition of its N interactions into n_i parts), is, in turn, computed as

$$\mathbf{Pr}[U_i^{C_{n_i}^N} > U_j | C_{n_i}^N] = \sum_{C_{n_j}^N} \mathbf{Pr}[C_{n_j}^N].h(C_{n_i}^N, C_{n_j}^N)$$

with $h(C_{n_i}^N, C_{n_j}^N)$ being a boolean function to decide whether or not i returns greater utility, given a pair of decompositions of N interactions into n_i and n_j parts respectively, and it is defined as

$$h(C_{n_i}^N, C_{n_j}^N) = \begin{cases} 1 & \text{if } U_i^{C_{n_i}^N} > U_j^{C_{n_j}^N} \\ 0 & \text{otherwise.} \end{cases}$$

Now, suppose $C_{n_i}^N = \langle x_1, x_2, x_3 \dots x_{n_i} \rangle$ with $x_1 + x_2 + x_3 + \dots x_{n_i} = N$. Then we compute $\mathbf{Pr}[C_{n_i}^N]$ as

$$\mathbf{Pr}[C_{n_i}^N] = \frac{N!}{x_1!x_2!x_3! \dots x_{n_i}!}(p_1^i)^{x_1}(p_2^i)^{x_2} \dots (p_{n_i}^i)^{x_{n_i}}$$

from the multinomial distribution. It can be shown that the computational complexity of this scheme for selecting a partner from among A subjects, is $O(A^2N^{2n})$ where $n = Max_i\{n_i\}$. We also note that in general,

$$p_{ij}^N + p_{ji}^N \leq 1$$

as there may be some

$$C_{n_i}^N, C_{n_j}^N$$

for which

$$U_i^{C_{n_i}^N} = U_j^{C_{n_j}^N}$$

6. AN EXAMPLE EVALUATION

Suppose the payoff-structures of three partnerships i, j and k are $\{(1, 0.4)\ (10, 0.6)\}$, $\{(5, 0.5)\ (5, 0.5)\}$, and $\{(11, 0.4)\ (3, 0.6)\}$. Here

$n_i = n_j = n_k = 2$. We choose N=1. For each partnership there are 2 decompositions of 1 trial into two parts, viz. $\langle 0,1\rangle$ and $\langle 1,0\rangle$. To compute p_{ij}^N, we see that case $\langle 0,1\rangle$ for i produces benefit = 10. This is greater than the benefits for both cases of j (benefit 5 from both $\langle 0,1\rangle$ and $\langle 1,0\rangle$). Hence $h(\langle 0,1\rangle_i, \langle 0,1\rangle_j) = h(\langle 0,1\rangle_i, \langle 1,0\rangle_j) = 1$. Again for case $\langle 1,0\rangle$ of partnership i the benefit is 1, which is lesser than that of j for both of its cases. That means, $h(\langle 1,0\rangle_i, \langle 0,1\rangle_j) = h(\langle 1,0\rangle_i, \langle 1,0\rangle_j) = 0$.

Hence

$$\begin{aligned} p_{ij}^N &= \mathbf{Pr}[\langle 0,1\rangle_i] * \{\mathbf{Pr}[\langle 0,1\rangle_j] * 1 + \mathbf{Pr}[\langle 1,0\rangle_j] * 1\} \\ &+ \mathbf{Pr}[\langle 1,0\rangle_i] * \{\mathbf{Pr}[\langle 0,1\rangle_j] * 0 + \mathbf{Pr}[\langle 1,0\rangle_j] * 0\} \end{aligned}$$

where subscript of any decomposition indicates the respective partnership with which this case arises. Now,

$$\mathbf{Pr}[\langle 0,1\rangle_i] = \frac{1!}{0!1!}(0.4)^0 * (0.6)^1 = 0.6$$

$$\mathbf{Pr}[\langle 1,0\rangle_i] = \frac{1!}{1!0!}(0.4)^1 * (0.6)^0 = 0.4$$

It is also immediately seen that

$$\mathbf{Pr}[\langle 0,1\rangle_j] = \frac{1!}{0!1!}(0.5)^0 * (0.5)^1 = 0.5$$

$$\mathbf{Pr}[\langle 1,0\rangle_j] = \frac{1!}{1!0!}(0.5)^1 * (0.5)^0 = 0.5$$

Consequently, $p_{ij}^N = 0.6 * (0.5 + 0.5) + 0.4 * (0 + 0) = 0.6$. Similarly, we can compute $p_{ji}^N = 0.4$, $p_{jk}^N = 0.6$, $p_{kj}^N = 0.4$, $p_{ik}^N = 0.36$, $p_{ki}^N = 0.64$. As a result,

$$S(i,j) = i,$$
$$S(j,k) = j,$$
$$S(i,k) = k.$$

As outlined earlier, at this situation (a non-transitive case) the arbitration by voting comes into the picture, and i gets 0.96 votes, j gets 1.0 votes, and k gets 1.04 votes. Consequently, k is chosen.

It is worthwhile to explore this example further so that we can appreciate the notion that non-transitivity is the exception rather than the rule. For $N = 2$, we have $p_{ij}^N = 0.84$, $p_{ji}^N = 0.16$, $p_{jk}^N = 0.36$, $p_{kj}^N = 0.64$, $p_{ik}^N = 0.4752$, $p_{ki}^N = 0.5248$. Hence partnership k is the obvious preferred choice, and transitivity holds. Again, for $N = 3$, the direct choice is i.

We have come up with this elaborate procedure after exploring various other alternatives that were computationally simpler, but inadequate nevertheless. The closest of these is an approximation for p_{ij}^N for all pairs (i, j), using Hoeffding Inequality (Vidyasagar, 1997). This approximation is really a crude one, and the problem called for better bounds. The direct application of Chernoff's Theorem (Billingsley, 1986) provides tighter bounds, but requires "large number of interactions" which defeats the very purpose of the problem. Lastly, the mechanism we have presented, computes p_{ij}^N accurately, using the rules for multinomial distribution of probability values, and conditional probability.

7. COMPARATIVE EVALUATION WITH A DECISION MECHANISM FOR INFINITE INTERACTIONS

To compare our decision mechanism with a standard reference we chose the *Expected Utility Maximization Principle (MEU)* (Luce and Raiffa, 1957). The MEU principle prescribes interacting with partnership i given by

$$i = \arg \max_{j \in Partners} \sum_{k=1}^{n_k} u_k^j * p_k^j,$$

where *Partners* is the set of partners the agent can interact with.

This principle is guaranteed to maximize the total utility received by the agent if the agent interacts infinitely often. In an open environment, agent relationships are often ephemeral, and *infinite interactions* are impractical. The obvious question is whether our strategy will be able to outperform a MEU choice when the assumption of infinite interactions does not hold. In particular, if we know that an agent is interested in a relationship for a *finite*, short period, a partnership with smaller expected utility may return more net utility than another partnership with greater expected utility. It would be interesting to evaluate if this conjecture is true and if so, for what range of interactions?

At this point we can also observe that the decision mechanism developed by us have the following properties:

- The strategy reduces to the MEU strategy for infinite interactions.

- It is based solely on the *payoff-structure* of the partnerships and the number of interactions to be performed.

8. EVALUATION SCENARIO

Decisions of this kind attain significant proportions in any domain where limited application of acquired knowledge is required with a fair degree of confidence in the outcomes. We can visualize a computational marketplace, where agents distribute computational tasks among service-providers (in return of payoffs, that may be determined by the time taken, quality of the service etc.), through broker agents. Such a broker has estimates of the payoff-structures for various service-providers, and is faced with the problem of choosing the potentially-most-beneficial provider (the benefit being the portion of the payoff of the provider that the broker charges as the intermediary), for a set of tasks (the number of such tasks can be a measure of N). The broker and the provider thus enter a partnership or collaboration which the broker offers to maintain for sufficient tasks (or for a sufficient period) that he is confident enough, will produce the desired payoff.

The procedure outlined in this paper enables the broker to objectively evaluate the potential of various prospective providers, as an explicit function of the intended number of interactions, and this is where the procedure gains its capability to suggest an alternative to the MEU-choice.

9. EXPERIMENTAL RESULTS

For the purpose of experimentation, we have considered only two partnerships. Of these, the payoff-structure of the MEU-partnership is $\mathcal{M}$, and that of the other partnership (non-MEU) is $\mathcal{NM}$. We can imagine, there are other partnerships in the marketplace, and the agent's choices may vary across several of these partnerships. However, for the purpose of illustration of non-MEU choices against a given MEU-partnership, only one non-MEU partnership is sufficient.

In the experiments, the payoff structure of the $\mathcal{NM}$ partnership is $\{(2, 0.1)\ (10, 0.3)\ (12, 0.6)\}$ with an expected utility of 10.4, and we vary the payoff-structure of the MEU-partnership, such that its maximum possible payoff decreases, keeping the expected utility fixed at 10.56. This means that according to the MEU strategy it is preferable to choose the second partnership. Figures 2.1 and 2.2 shows the comparative performance of the Non-MEU and MEU partnerships. For a given number of interactions, the actual payoff generation is simulated, and the percentage of cases out of 1000 simulations, in which the Non-MEU partnership yields greater payoff than the other is plotted in each graph. The shaded regions in a plot refer to the values of N for which, given the payoff-structure, the function S selects the MEU-partnership.

For all other values of N (i.e. the unshaded zones), the choice is the Non-MEU partnership. The values below the dotted-line signify cases where the MEU-partnership outperforms the other in actual simulation. The first 75 interactions have been plotted in each case.

Figures 2.1 and 2.2 show close agreement between the procedural choice and the results of the simulations. An interesting observation from these figures is that the choice function S, contrary to intuition, changes its output more than once. The choice of MEU-partnership appears in multiple distinct regions in the figures 2.1 and 2.2, roughly in the same zones where the non-MEU partnership loses out to it in actual simulations. Such change in choice depends on the probability distributions over the utilities, the multinomial coefficients, and the decomposition-pattern of the number of interactions. The summary description we can provide is that the "bands of dominance" of the MEU strategy increases in width with more interactions until after a relatively large number of interactions it becomes totally dominant. The last observation is consistent with the fact that for large number of interactions, the MEU strategy is accurate and sufficient to identify the most desirable partnership. We also note that the sample payoff-structures assumed in the experiments are representative of the type of payoff-structures where the efficacy of our procedure is clearly demonstrable. In other types of payoff-structures the results of our procedure cannot be any worse than the MEU-strategy.

It can be observed that as the highest payoff from the MEU partnership increases (and the corresponding probability of receiving that payoff is decreased to keep the expected utility constant), it takes more interactions before the first onset of dominance of the MEU partnership. This can be explained by the fact that if the expected utility of a partnership is based largely on the superlative payoff from an infrequent event (the jackpot in our earlier example), it would take more interactions to benefit from the occurrence of that infrequent event. For a more limited number of interactions, it is advisable to choose a partnership that returns a consistently high payoff even though its expected utility is less.

Another interesting observation from figure 2.2(bottom) is the unusually frequent variation of the procedural-choice in a short range of N. Because the difference in the expected utilities of $\mathcal{M}$ and $\mathcal{NM}$ is already low, the lowering of the highest utility in $\mathcal{M}$ reduces the structural difference between $\mathcal{M}$ and $\mathcal{NM}$. As a result, the choice becomes highly sensitive to N. This sensitivity, in particular, is beyond the scope of MEU-strategy.

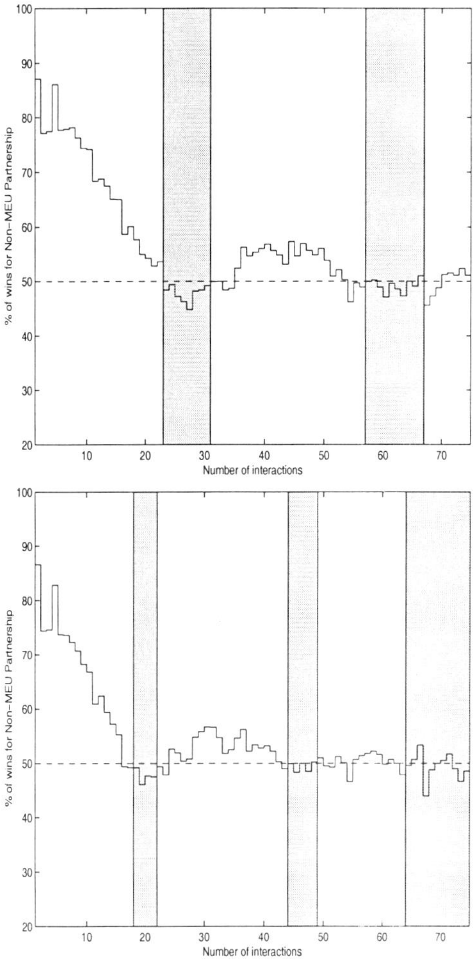

Figure 2.1. Simulation of benefit from $\mathcal{NM} = \{(2, 0.1)\ (10, 0.3)\ (12, 0.6)\}$ against $\mathcal{M} = \{(8, 0.97)\ (90, 0.02)\ (100, 0.01)\}$(top) and $\mathcal{M} = \{(8, 0.96)\ (70, 0.025)\ (75, 0.015)\}$(bottom)

10. CONCLUSIONS

In this paper, we have considered the problem of an agent deciding on which partnership to interact with given the number of interactions and a model of the environment in the form of payoff structures of each of these partnerships. We have developed a probability-theory based procedure for making the selection and compared its performance with the well known MEU principle which solves this problem exactly when the number of interactions is infinite.

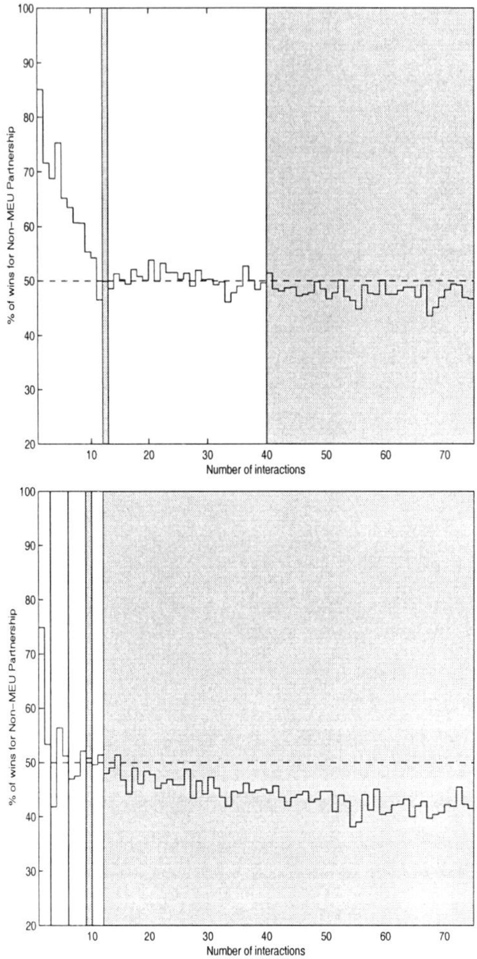

Figure 2.2. Simulation of benefit from $\mathcal{NM} = \{(2,\ 0.1)\ (10,\ 0.3)\ (12,\ 0.6)\}$ against $\mathcal{M} = \{(8,\ 0.94)\ (50,\ 0.04)\ (52,\ 0.02)\}$(top) and $\mathcal{M} = \{(8,\ 0.8)\ (20,\ 0.1)\ (21.6,\ 0.1)\}$(botom)

The agreement between the performance predicted by our selection procedure and the simulated values show that the procedure is well adapted for the sensitive differences between the payoff-structures of two coalitions. This procedure effectively captures a broad range of these subtleties and hence, is a more fine-grained measure of the relative efficacy of coalitions, than the MEU-strategy.

A related and practically important problem is to devise a strategy for interaction where the payoff-structure for coalitions is incompletely known. It would be instructive to compare the performance of Bayesian update schemes versus model-free reinforcement learning methods on these problems (Mitchell, 1997). This, admittedly more complex, problem is akin to the multi-arm bandit problem (Salganicoff and Ungar, 1995). A critical issue in this problem is that since learning is on-line, the approximation of the payoff structure from limited sampling has to be accurate to take advantage of any non-MEU strategies. Off-line learning of the payoff structure of partnerships by observing them interacting with other agents can also provide approximate models. In the latter cases, the present work can be readily used.

Acknowledgment This work has been supported in part by an NSF CAREER award IIS-9702672.

References

Billingsley, P. (1986). *Probability and Measure.* John Wiley & Sons.

Ketchpel, S. (1994). Forming coalitions in the face of uncertain rewards. In *Twelfth National Conference on Artificial Intelligence*, pages 414–419. MIT Press/AAAI Press.

Luce, R. D. and Raiffa, H. (1957). *Games and Decisions: Introduction and Critical Survey.* Dover, New York, NY.

Mitchell, T. (1997). *Machine Learning.* WCB McGraw-Hill, Boston, MA.

Salganicoff, M. and Ungar, L. H. (1995). Active exploration and learning in real-valued spaces using multi-armed bandit allocation indices. In *Proceedings of the Twelfth International Conference on Machine Learning*, pages 480–487.

Sandholm, T. and Lesser, V. (1997). Coalitions among computationally bounded agents. *Artificial Intelligence*, 94(1):99–137.

Sandholm, T. W., Larson, K., Andersson, M., Shehory, O., and Tohme, F. (1999). Coalition structure generation with worst case guarantees. *Artificial Intelligence*, 111(1-2):209–238.

Shehory, O. and Kraus, S. (1995). Task allocation via coalition formation among autonomous agents. In *Proceedings of the International Joint Conference on Artificial Intelligence*, pages 655–661.

Shehory, O. and Kraus, S. (1996). A kernel-oriented model for coalition-formation in general environments: Implementation and results. In *Proceedings of the Thirteent National Conference on Artificial Intelligence*, pages 133–140.

Straffin, P. D. J. (1980). *Topics in the theory of voting.* The UMAP expository monograph series. Birkhauser, Boston, MA.

Tesfatsion, L. (1998). *Preferential Partner Selection in Evolutionary Labor Markets: A Study in Agent-Based Computational Economics.* Springer-Verlag, Berlin.

Vidyasagar, M. (1997). *A Theory of Learning and Generalization.* Springer Verlag.

Zlotkin, G. and Rosenschein, J. S. (1994). Coalition, cryptography, and stability: Mechanisms for coalition formation in task oriented domains. In *Twelfth National Conference on Artificial Intelligence*, pages 432–437. MIT Press/AAAI Press.

Chapter 3

EVOLUTION OF AGENTS WITH MORAL SENTIMENTS IN AN ITERATED PRISONER'S DILEMMA EXERCISE

Ana L. C. Bazzan
Informatics Institute, Federal University of Rio Grande do Sul,
Av. Bento Gonçalves 9500, CP 15064, 91501-970 Porto Alegre - RS, Brazil.
bazzan@inf.ufrgs.br

Rafael H. Bordini
Informatics Institute, Federal University of Rio Grande do Sul,
Av. Bento Gonçalves 9500, CP 15064, 91501-970 Porto Alegre - RS, Brazil.
bordini@inf.ufrgs.br

John A. Campbell
Department of Computer Science, University College London,
Gower Street, London WC1E 6BT, United Kingdom.
J.Campbell@cs.ucl.ac.uk

Abstract We have been experimenting on simulations of a society of agents where some of them have "moral sentiments" towards the agents that belong to the same social group, using the Iterated Prisoner's Dilemma as a metaphor for the social interactions. We were able to observe the well-understood phenomenon of short-sighted, self-interested agents performing well in the short-term but ruining their chances of such performance in the long run in a world of reciprocators. Our results suggest that, where some agents are more generous than reciprocators, these agents have a positive impact on the social group to which they belong, without compromising too much their individual performance. In this paper, we first present the results of new simulations on these ideas based on an evolutionary setting. This initial evolutionary setting has strengthened our previous conclusions by showing the improving performance of generations of groups to which only altruistic agents belong. We also mention another evolutionary setting that we plan to use as a next step for this project.

Keywords: Multi-agent systems, iterated prisoner's dilemma, moral sentiments, evolutionary computation.

1. INTRODUCTION

The Prisoner's Dilemma (PD) has been used as a metaphor for the conflict of two individuals who engage in either mutual support (cooperation) or selfish exploitation (defection). Since there is a risk for one of the individuals to end up with the *sucker*'s payoff, to defect is always a good tactic to follow.

However, when played repeatedly, in which case it is called the Iterated Prisoner's Dilemma (IPD), mutual defection is no longer the only solution, although it meets the classic definition of rationality. This was verified in Axelrod's computer tournament (Axelrod, 1984), in which a program representing the strategy called Tit-For-Tat (TFT) won by cooperating initially, then repeating what the opponent did in the last round.

These ideas have been employed in the field of multi-agent systems in order to explain the achievement of cooperation and coordination (Rosenschein and Zlotkin, 1994). However, little work has been devoted to the IPD for social agents, particularly agents with emotions.

One reason for this may be the failure of theories based on rational choice to account for *social* action and choice as claimed by (Conte and Castelfranchi, 1995). Social agents are constantly involved in planning their actions, and they need to reason about new goals that arise in this process, which may include attempts at modifying other agents' mental states. Game theory tends to treat goals in a process of choice for each agent in isolation, and exclusively from that agent's own point of view.

It is worth mentioning that there has been marked opposition from game theorists to the widespread use of IPD and TFT as the basis for explaining complex social interactions among humans (Binmore, 1998). In our previous papers, e.g. (Bazzan et al., 1999; Bordini et al., 2000), we have extended the rules of the game so that it can relate to various issues of social agents, including moral and philosophical aspects such as why people are able to keep their promises once they agree to cooperate and why people behave altruistically. We have used the PD simply as a metaphor for social interaction within our simulations.

We have concluded that egoistic agents maximise their earnings only in the short term, failing to keep up their good performance in the long run. The results have shown clearly that in a society where agents have emotions, to behave rationally (in the classical sense of game theory) may not be the best attitude.

In the present paper, we continue to investigate how societies formed of different types of agents perform, still in the context of the IPD. However, we are now interested in comparing the performance within an evolutionary setting (through several generations of agents) to the results we have gathered so far.

The main ideas of our previous work, as well as a brief description of the results already achieved, are described in the next two sections. Sections 4 and 5 discuss the setting we have chosen for evolving populations of agents playing the IPD (some of them with Moral Sentiments), and present the results of the simulations based on that evolutionary setting. Extensive discussion of this work is given in Section 6, where we also relate it to the literature.

2. MORAL SENTIMENTS IN THE ITERATED PRISONER'S DILEMMA

2.1. A THEORY OF MORAL SENTIMENTS

In a book on the "Origins of Virtue" (Ridley, 1996, Chapter 7), Ridley makes the point that *moral sentiments* (i.e., emotions like generosity towards others and guilt for not having played fair with someone) prevent us from being *rational fools*. Rational fools act to maximise their gain in the short term, which does not pay off in the long run because people do not reciprocate with those who have proven selfish in the past. Moral sentiments lead us to sacrifice rational decisions, yet they are of fundamental importance to social relations inasmuch as they allow us to create a reputation as altruistic people. Altruism, which most people praise as a virtue, will lead an altruistic person to experience the generosity of others as well, when it is needed. However, these same emotions drive us to want those who belong to the same social group to be somewhat self-interested, which is better for the group too. We are particularly altruistic with people from the same social group or who share our genes.

In other words, moral sentiments are an important factor in the dilemma between getting the maximum out of an opportunity and being cautious about the future. They are part of our highly social nature, to favour our genes' long-term advantage. They are also a guarantee of our commitments, which makes complex social relations possible; and

Table 3.1 . A payoff matrix for the Prisoner's Dilemma.

P1/P2	C	D
C	R/R	S/T
D	T/S	P/P

- $T > R > P > S$
- $R > (S+T)/2$

$(T = 5, R = 3,$
$P = 1, S = 0)$

that too stands to our long-term advantage. Morality can be seen as a set of instincts preventing us from being selfish, which is for our own sake in the long run. When, however, people do something that is not rational and does not pay off even in the long run (let us call it *true altruism*), they could be (as Ridley puts it) "falling prey" to sentiments originally designed to obtain other people's trust, which is convenient for real life's "prisoner's dilemmas". This phenomenon, however, has beneficial consequences for groups of agents, as our simulations show.

2.2. SIMULATING AGENTS WITH MORAL SENTIMENTS IN THE IPD

We have analysed the performance of groups of agents playing the IPD in the presence of agents with moral sentiments (*altruistic agents*) or rational fools (*egoistic agents*) in that group (or a mixture of both). When possible, we also analyse the individual performance of each type of agent in the available contexts (in a group where all agents are exclusively of one of the two types just introduced or in a mixed group). The fact that two agents belong to the same group implies that they have some sort of emotional liaison (that is, they may belong to the same family or social group). Agents should be interested in the good performance of their group as a whole, as well as their own, since the social group provides also a base for support in case the agent itself is not performing well. However, that is not true for the egoistic agents, which always seek to obtain the most points, no matter with whom they are playing. We aim at demonstrating Ridley's point that the presence of agents with moral sentiments favours the well-being of the whole social group.

More details about this simulation and its theoretical motivation can be found in (Bazzan et al., 1999; Bordini et al., 2000). For the sake of completeness, we give a brief overview here. In the experiments conducted, pairs of agents are chosen randomly among agents of all groups to play the IPD. The points agents earn by playing are the standard

Table 3.2 . The tactics used by Altruistic (A) and Egoistic (E) agents.

Strategy	Op. Last M.	Ag. State	Op. State	Ag. Plays
ALLD (E)	*	*	*	D
TFT (A)	C	*	*	C
(non-partners)	D	*	*	D
MS (A)	*	S	W	D
(partners)	*	S	M,S	C
	*	W,M	*	C

amounts for the PD payoff matrix, as shown in (for player P1 and P2) Table 3.1 (where T stands for *Temptation to defect*, R stands for *Reward for mutual cooperation*, P stands for *Punishment for mutual defection*, and S stands for *Sucker's payoff*). In order to determine the *wealth state* of an altruistic agent (which influences how it will play), we compute the average number of points through the completed simulation steps or through a determined period of time. According to certain thresholds on this average, an agent's state is classified as *wealthy* (W), *medium* (M), or *straitened* (S).

In our simulations, egoistic agents defect in all interactions[1] (ALLD), regardless of which agent is chosen as an opponent for a PD interaction. Altruistic agents play Tit-For-Tat (TFT) when interacting with agents from other groups, so they play a fair game here. Then we introduce a new strategy which is used by altruistic agents when they interact with agents from the *same social group*. We call it the Moral Sentiments (MS) strategy, which is explained next.

If the altruist is in a wealthy state and the opponent (from the same group) is in a straitened one, the former cooperates despite the fact that it knows that the latter will defect; it does this to help the opponent to earn more points (remember that the agent has, so to speak, a special affection toward this opponent because they are both from the same social group). A straitened altruist cooperates if the opponent is also in a straitened (or in a medium) state, because it does not try to recover by taking advantage of a peer that is not in a wealthy state. However, an altruistic agent will defect when in a poor state if the opponent is in a wealthy one and belongs to the same group. This is a social mechanism to allow agents to recover from a bad state, which is possible through the altruism of those from the same group. Briefly, an altruist cooperates with those of the same group unless it is in a poor state and the opponent is wealthy. This can be seen clearly in Table 3.2.

In the simulations, we have considered societies of agents composed of 3 or 15 groups, each one with 4, 20, 40, 80 or 100 agents; in some simulations all agents play at a certain time, in others only 60 or 80% of agents play. In the cases where we have 3 groups, one of them has no egoistic agents (G0), another is a mixed group (G1) with equal proportions of egoists and altruists, and the third one (G2) is composed solely of egoists.

Next, we summarise the results obtained for the simulations whose main features we have just described.

3. THE UNWITTING BENEFITS OF ALTRUISM

Besides the time it takes for altruists to overtake the egoists, other measurements of performance were made using a snapshot in the simulation regarding the amount of points (per capita, so as to account for differences in the size of the groups) accumulated by the best group at that moment. This gives us a measure of the wealth of the society. Finally, we have also measured the number of times MS was selected and look at the effect this has on the wealth of the group and on the performance of the altruists. The general results can be seen in Figure 3.1. The performance of social groups and types of agents is displayed in Graphs 3.1(c) and 3.1(d), respectively. The latter depicts the performance of the different types of agents and group contexts[2]: altruistic agents in a homogeneous group (AH), i.e., those in group G0; the portion of altruistic agents in a mixed group, in this case G1, which has both altruists and egoists (AM); the egoists in a mixed group (EM), and egoists in a homogeneous group (EH), i.e., those in G2. In the graphs, the notation $\mathrm{G}_n(Ag/E)$ means that group n has Ag agents of which E are egoists.

Considering various configurations of conditions and parameters introduced in the set of simulations, we were able to confirm these major lessons:

1 The more egoists in a group, the faster the group collects points initially, but the worse is its performance after some time—which means that rational fools maximise their earnings in the short term but compromise their future performance (a consequence of the reciprocating character of TFT). Groups formed of only altruists accumulate more points than the other types of groups in the long run.

2 The more egoists in the society:

(a) the less time it takes for G0 to surpass G1, no matter what percentage of agents is playing; in most cases studied, this is also valid regarding the time for AM to surpass EM;

(b) the fewer points the society collects as a whole (regardless of the percentage of agents playing).

Thus, the presence of the egoists is harmful for all members of the group in the long run, since they would all be performing better if the egoists would stop being rational fools.

3 The generosity of MS yields a better performance for the group than mere reciprocity—recall that reciprocity is what accounts for the long-term success of individual agents; ultimately, the reason for the conclusions above. This will be further explained in the following paragraphs.

In Figure 3.1(c) and 3.1(a), we have added a fourth curve (besides the ones for the agent groups) which is the average of the performance of G0 and G2 (see `Avg(G0,G2)`, in the figure). This is to make it easier to visualise how much better G1 is doing because of the generosity of its altruists. Note that if the altruists in G1 were playing TFT instead of MS within the group, the performance of G1 would be exactly that of the average of G0 and G2's performances (given that G1 has equal proportions of altruists and egoists), rather than the curve for G1 seen in the figure[3]. This is a remarkable finding, whose message can be understood better in terms of an analogy. If one gives money to a homeless person and nobody else gets to know about it, this is clearly a case where altruism has no personal reward. It is the situation that we mention in Section 2.1 where one is falling prey to the sentiments that are an important mechanism in social life, as they bring great personal rewards in normal circumstances through trustworthiness. In circumstances where no personal reward ensues, one interesting side-effect occurs: the improvement of the social group! In the homeless analogy, it is as if one is not better off by one's altruistic act, but one's city as a whole is doing much better (compare, in Graph 3.1(c), G1 with `Avg(G0,G2)`), without reducing significantly one's own performance (compare, in Graph 3.1(d), AM with AH).

Regarding multi-agent systems, this implies a better performance of the overall system, which is important. In terms of the PD, this happens because, for the group, a joint reward of $T + S$ earned when an egoist plays against a wealthy/medium altruist partner, is better than $P + P$, received when TFT is played against ALLD. It is a similar effect to the Smithian finding that the division of labour leads to society being

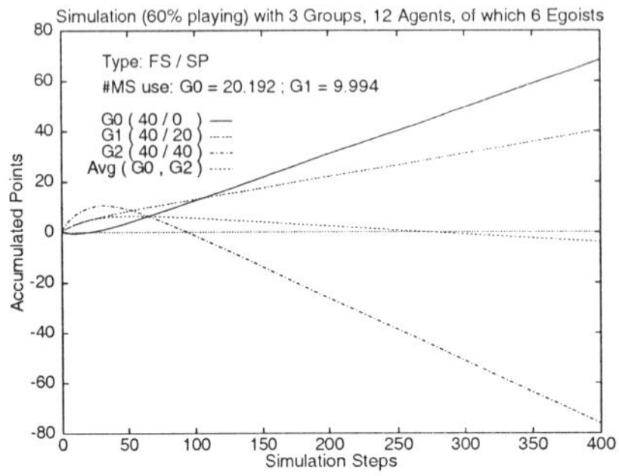

(a) Altruists using MS (Groups)

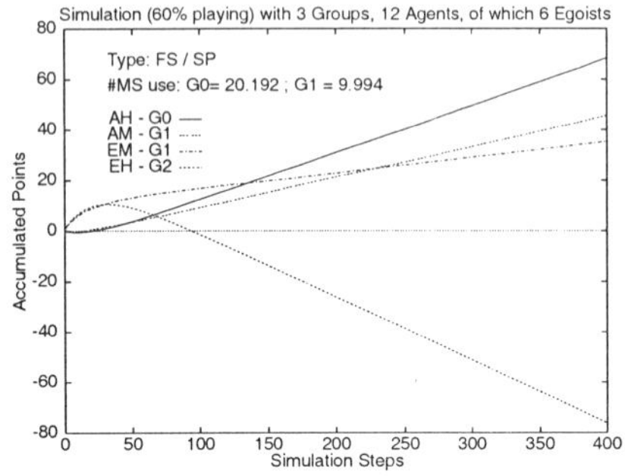

(b) Altruists using MS (Agent Types)

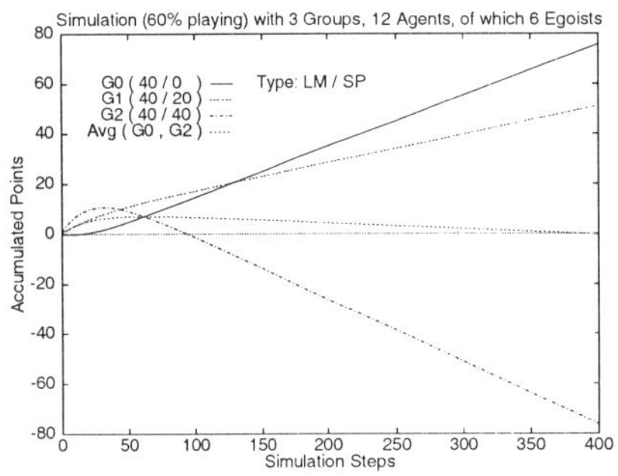

(c) Altruists as ALLC (Groups)

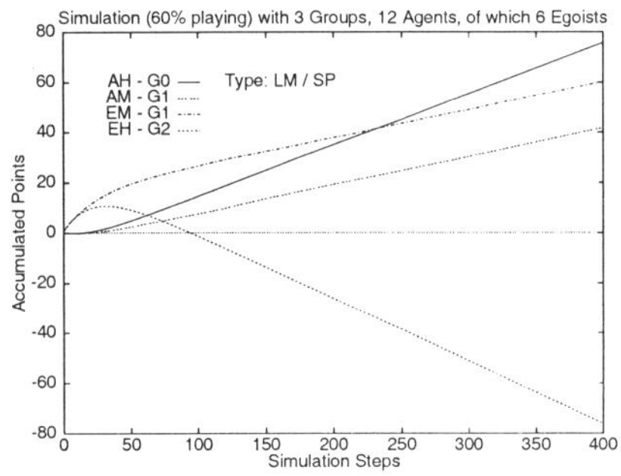

(d) Altruists as ALLC (Agent Types)

Figure 3.1. Effects of altruism and of the use of the MS strategy.

more than the sum of its parts. The lesson here is: even the individual drawback of being driven by an emotional response when it will not repay itself allows a group that is encumbered with rational fools to perform much better than it would if emotions were used only for the purpose for which they evolved in our species (i.e., the virtuous purely maximising their individual earnings in the long run).

Figure 3.1 also shows the effect of the use of the MS strategy. Graphs (a) and (b) are from a type of simulated society (called LM in the graph and in our previous papers), where MS was not used (ALLC, which means always cooperate, was used instead) by altruists playing against partners. On the other hand, graphs (c) and (d) are from a different simulated society (called FS in the graph), where the use of MS is best managed[4]. A very interesting effect is seen here. Although the graphs for the groups (3.1(a) and 3.1(c)) have very similar shapes, one can see easily that this is not so in the graphs for the types of agents (3.1(b) and 3.1(d)). While in the former type of society EM agents are doing consistently better than AM, the same does not happen in the latter society, where MS is being used! The reason is that with ALLC the altruists are being too much exploited by their selfish partners. The MS strategy is "kind" enough to keep the performance of the mixed group higher than with pure reciprocation (remember the similarity of the curves for G2 in (a) and (c)), yet being fairer to those who are actually responsible for the good performance of the group.

4. AN EVOLUTIONARY SETTING FOR THE SIMULATIONS

In our previous work, we considered the possibility that egoists might change their attitude. If egoists could reason about their performance, they would conclude that despite it being high at the beginning of the encounters, it is severely diminished in the long run. Several other changes in the simulations could also provide interesting insights.

Accordingly, we have proposed several extensions to the design of our simulations to open up such possibilities: an "artificial life" version of the simulation (where we could investigate group fitness and adaptation), egoists learning by reinforcement, altruists eventually refusing to interact with certain (non-cooperative) partners, and an evolutionary approach where population of agents (no matter what type and group) are reproduced according to their fitness.

In the present paper, we consider particularly the latter possibility. The design of this type of evolution of populations of agents was carried out according to the standard genetic algorithms proposed in (Goldberg, 1989), except that we have not allowed recombination of strings. That is, we use asexual reproduction: the k fittest individuals generate a number of offspring (as copies of themselves) proportionally to their fitness.

The main parameters of the simulation remain as described before (Section 2.2): the score of each agent is computed over the last 10 time steps, and S, P, R, and T (see Table 3.1) are set to 0, 1, 3, and 5 respectively. On the other hand, the threshold for an agent to be considered wealthy is now $T_W = R/2$. If the agent collects more than T_W points, it is considered wealthy. Less fortunate agents are classified as straitened. Thus, here we have slightly modified the previous approach, which allowed three types of wealth states.

The main characteristics of an agent are coded in a binary string: its type (egoist/altruist), its wealth state (straitened/wealthy), and the group to which it belongs (up to 16 groups are allowed in the simulation).

According to a reproduction frequency (F_R), set at the beginning of the simulation, agents "live" (play, accumulate points), reproduce, and "die" (immediately after reproducing).

In a reproduction step, the fitness of an agent (simply the accumulated points) determines the probability that the agent will reproduce. Mutation frequency (F_M) is also implemented, but not used in the simulations results reported below since the main purpose here is to study the effects of reproduction alone. Mutation will be helpful to determine to what extent a set of tactics is evolutionarily stable.

The results reported in (Bazzan et al., 1999) show that egoists either go bankrupt or perform very badly in the long run. This suggests that, taking their wealth as the fitness, they would tend to disappear, giving place to fitter individuals. We discuss this in the next section.

Note that, differently from the previous simulations (where no reproduction took place), here the number of agents in each group can increase and decrease, and groups can even disappear. This happens because only the fittest agents of the entire society generate offspring, irrespective of the group to which they belong.

5. PLAYING WITH MORAL SENTIMENTS SECURES THE SURVIVAL OF THE GROUP

For illustration, we discuss here several simulations carried out among a society (or population) of agents divided into three groups, as in the previous simulations mentioned in Sections 2.2 and 3. As for the composition of groups, G0 and G2 have only altruists and egoists, respectively. The remaining G1 is mixed, which means that it has both altruists and egoists, in different proportions. However, we discuss here the case in which G1 has 50% of each type.

Pairs of agents are selected randomly from the whole population to play the IPD. At each time step, they collect points, also as explained in Section 2.2. Hence, encounters among agents happen irrespective of the group to which they belong. The discussion here will concentrate on the results obtained with groups of 4 and 20 agents, which we call small and medium groups, respectively.

All simulations were repeated at least 500 times to account for deviations; the graphs shown here are from the average of all these repetitions. To start calibrating the simulations, we have used a configuration of three small groups within a very long time period (10,000 simulation steps), in order to be sure that it converges. Figure 3.2 shows the

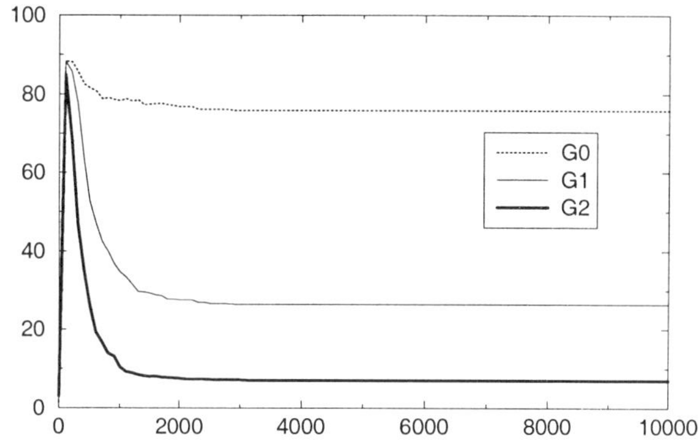

Figure 3.2. Scores (number of points) of groups in 10,000 steps.

curves for scores per time for those three groups (recall that G0 has only altruists, G1 is mixed, and G2 has only egoists).

Figure 3.3(a) also shows the performance of groups against time (simulation steps). This is some sort of snapshot of the first 1,000 steps of the simulation in Figure 3.2, as the other simulation parameters remain the same. In this way, the performance of each generation (peaks and valleys) can be better observed.

Our main result is, in all cases simulated, that the score of the groups is higher the fewer egoists are in it. This only confirms the results obtained previously as described in Section 3. Taken individually, groups formed by altruists only (G0 in the Figures), depict a very good performance without losing too many points due to the exploitation by egoists of other groups (see Figure 3.3(a)). In some situations, the maximum points collected by one generation in G0 even increase from one life span to another (Figure 3.4(a)).

In all cases we have studied, the mixed group (G1) exhibits a performance that can be described as follows: it starts by collecting many points, but eventually, due to the increasing number of egoists in it, its performance gets increasingly worse no matter how large the group is (compare Figure 3.3(a) and Figure 3.4(a)).

The group formed by egoists only (G2) performs very badly in all cases. Soon its members are recognised as defectors and the strategy TFT prevents them from exploiting other (altruistic) agents. Besides, they cannot count on their group fellows to play MS with them. Hence, once they start performing badly, they are likely to die within a life span without generating descendants. Therefore, the group as a whole tends to disappear (but not totally if groups are small).

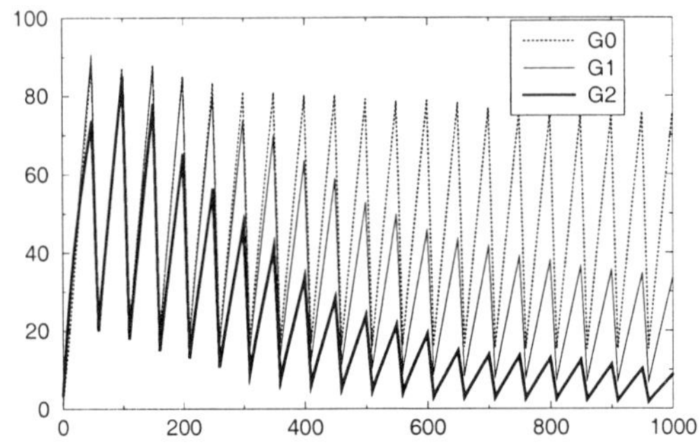

(a) Scores (number of points) of groups in 1,000 steps

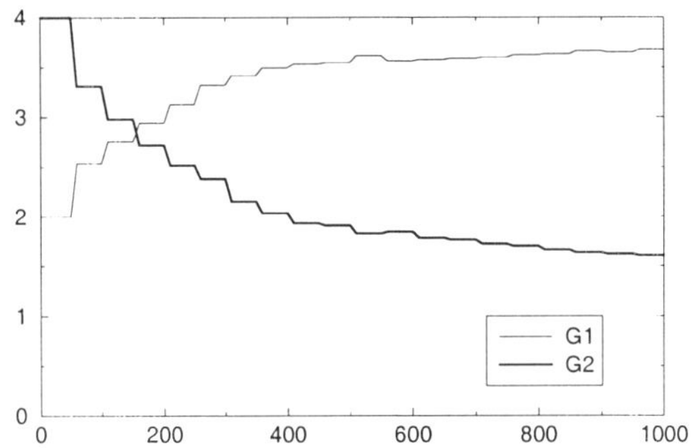

(b) Number of egoists in 1,000 steps

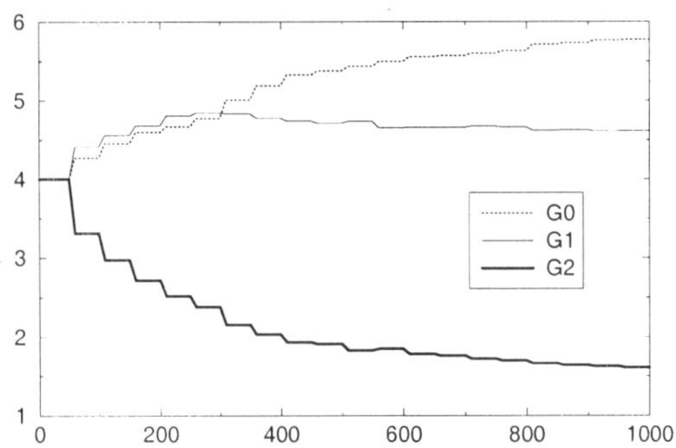

(c) Number of agents in 1,000 steps

Figure 3.3. Simulations with 3 groups, 4 agents each.

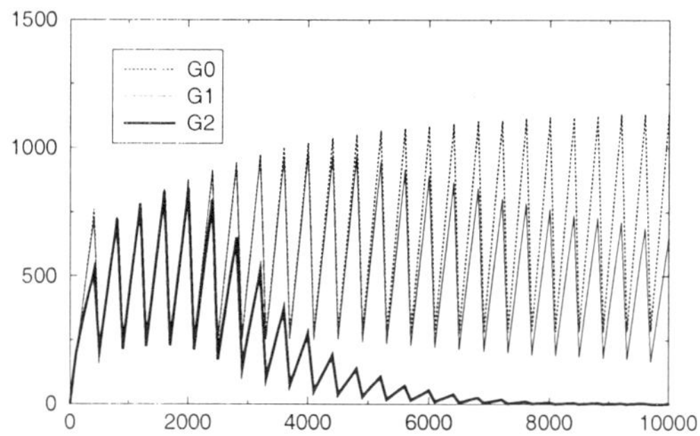

(a) Scores (number of points) of groups in 10,000 steps

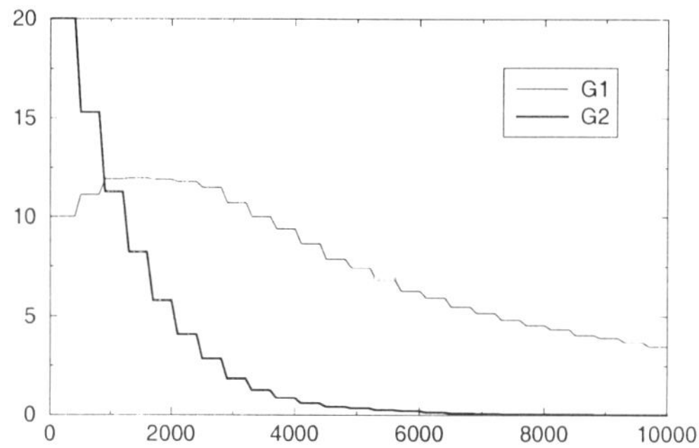

(b) Number of egoists in 10,000 steps

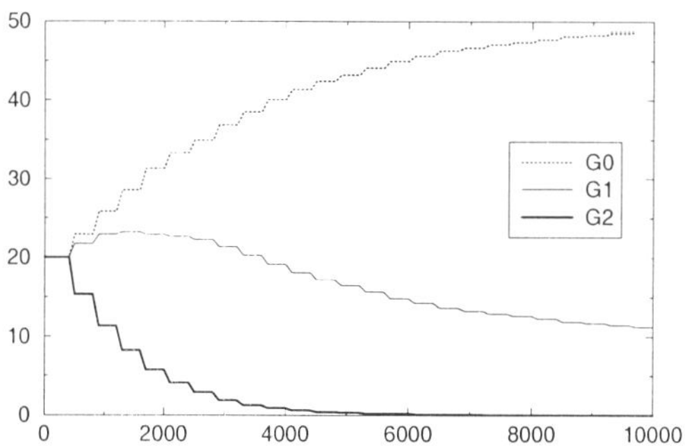

(c) Number of agents in 10,000 steps

Figure 3.4. Simulations with 3 groups, 20 agents each.

Further, the number of egoists and the number of agents in a group are important measurements we have to take into account when analysing how a group has been coping within the evolving society.

Surprisingly, the total number of egoists in the whole population does not decrease when groups are small (see Figure 3.3(b)). This happens mainly because egoists find a way into the mixed group: given that it is small, there are fewer opportunities for altruists to exercise MS among them. Therefore, the egoists can exploit the altruistic partners, which cannot recover from the exploitation. On the other hand, when G1 is medium to large, altruists there do exercise mutual support, thus egoists find it more difficult to invade. This can be seen in Figure 3.4(b). Because G0 and G1 have no place for egoists, and because those born in G2 are not fit enough to reproduce, the total number of egoists quickly decreases.

Finally, a few more words about the overall size of groups within time. In both cases, small and medium groups, G2 continually loses agents, as already said. G0 grows bigger since its agents are fitter and reproduce more. Members of the mixed group are fit in the beginning, so that the group is able to grow too. This happens until a certain point in time. After that, the presence of egoists there causes a reduction in collected points, so that agents start reproducing with decreasing rates and the group tends to shrink. Figure 3.3(c) depicts the curves for total number of agents per time for the small group, while Figure 3.4(c) shows the same for the medium group.

6. A CASE FOR BENEVOLENCE IN AGENTS

In (Bordini, 1999), an anthropological approach to interoperability of Multi-Agent Systems (MAS) was introduced. Pursuing that approach, we have been concerned with agents' emotions or sentiments. We believe that, in future, agreement on emotions (and perhaps perceptions) would avoid the need to standardise on other aspects of agency while still allowing interoperability of MAS (through such anthropological approach). We draw this lesson from social anthropology: one of the things that an anthropologist can be sure is present in all cultures is the whole spectrum of human emotions; Ridley (1996) says that they are *universally recognisable*, and even motives such as avoiding guilt are common across all cultures. An *anthropologist agent* (i.e., an agent responsible for supporting migrant agents in their adaptation to "strange" societies) needs, thus, to study the "culture" present in *target societies*. "But our cultures are not random collections of arbitrary habits. They are canalized expressions of our instincts. ... That is why, for all their superficial differences of language and custom, foreign cultures are still immediately comprehensible at the deeper level of motives, emotions and social habits." (Ridley, 1996).

Further, our view is that an *emotional stance* is yet another "missing point" (Castelfranchi, 1990) in present MAS. It should not be considered only for its role in believability, which is the point that most works that consider emotions as being relevant to agents make—see, for example, (Rizzo et al., 1997; Bates, 1994; Hayes-Roth et al., 1995). It is evident from the results shown here that to behave altruistically (even in a prisoner's dilemma context, where *rationality* has always been paramount) can prevent agents (of whatever paradigm) being *rational fools*, for it secures a good performance in the long run for the agents and in the more extreme cases for their social groups, at least (metaphorically, a social group can be seen as a complete MAS, wherever the notion of group is not available). This leads us to question the current widespread idea in MAS that autonomous agents should necessarily be selfish. Ridley wittily says that sociobiology "caught the self-interest virus" in the 1960s. It can be said that MAS has caught the same virus too.

Apart from the whole tradition of work on utility maximisation in the game-theoretic approach (Rosenschein and Genesereth, 1985; Rosenschein and Zlotkin, 1994; Gmytrasiewicz and Durfee, 1995) where agents are by definition self-interested, among the representatives of the "autonomy as selfishness" outlook in MAS are d'Inverno and Luck (1996) who have claimed:

> Cooperation will occur between two parties only when it is considered advantageous to each party to do so. Autonomous agents are thus selfish agents. A goal (whether traditionally viewed as 'selfish' or 'altruistic') will always be adopted so as to satisfy a 'selfish' motivation. (page 529)
>
> The effects of benevolence are possible, but only through self-serving motivations. (page 533)

In terms of the history of MAS, this line of thought seems to have started with the discussion on social power in (Castelfranchi, 1990). The paper was in the right direction at that stage, when benevolence was being taken for granted. It is time now to relocate benevolence, not as something taken for granted, but as an important phenomenon which may *evolve* in societies of autonomous intelligent agents from explorations of agent emotions. This issue should be of interest for those concerned with agent autonomy via the cognitive (as opposed to utilitarian) view of the field (Conte and Castelfranchi, 1995).

The misconception about autonomy and benevolence goes together with the absence of an explicit emotional component in present MAS theories. Motivations do not have to be necessarily self-serving. Consider, e.g., the idea of terminal interest adoption defined by Conte and Castelfranchi (1995). They mention the possibility of autonomous agents adopting others' interests out of affection, although they do not concentrate on the emotional aspects of agents in that book. On the other hand, they say that the usual means by which an agent can act in a selfless way, i.e. adopt a goal of another agent, is through an individual (personal) goal of being benevolent (compassionate) towards that agent. Further, they state that truly benevolent agents are those who undertake a mode of goal adoption they call *terminal*, which they claim to be the type of adoption common since the early days of MAS (i.e., it is *assumed* that agents will adopt each other's goals; essentially, these agents are not fully autonomous). This is, therefore, the perfect ground for d'Inverno and Luck's definition (quoted above). So, in this context there is a point in saying that when an agent is being benevolent, either it is not autonomous or it is pursuing an individual goal and thus its motivation is ultimately selfish. The problem is exactly in the artificial mechanism of modelling benevolence as an individual goal *per se*, due to the lack of representation and processing of "emotion". Recall that the important consequence of the emotional stance is the ability to truly exploit societal interaction in MAS, according to the discussion on the role of emotions in human social nature (see Section 2.1). In brief, emotional processing should be balanced with rational processing in the decision-making process of autonomous agents.

To round off the argument, agent autonomy is not necessarily concerned with fulfilling an agent's own selfish motivations; it has to do with its freedom to *choose* how to behave (or to have the resources to do something by itself, depending on the context)—even if that means choosing what is not best for its own goals, e.g. under the influence of emotional decisions (as we have remarked, this requires a specific part of the agent's architecture to deal with emotions). We observe in the results of our simulation that altruism prevents agents from being rational fools. As Ridley puts it, when being truly altruistic (i.e., doing something in the interests of others at one's own expense and even with no future reward at all), we are giving way to emotions which are an important mechanism behind the complex brand of social interactions that humans experience. It is certainly worth recovering this notion for the benefit of MAS. On the other hand, it is interesting to note that, to the extent that emotions are long-term utility maximisers, as opposed to a source of true altruism, they can be incorporated explicitly into the traditional decision-theoretic approach to MAS. This is precisely what has been recently done in (Gmytrasiewicz and Lisetti, 2001).

The main advantage of the simulations we present here is that it allows us to verify some of Ridley's propositions; it also yields some insights into the issues discussed above. Note that, whilst the quantitative results of the simulations presented here apply to MAS where the IPD metaphor (with the further assumptions we have made here, e.g., the existence of groups) makes sense, Ridley's line of argument, presented in Section 2.1, seems to apply quite generally; it is a high-level argument in support of our suggestion that designers of general[5] MAS architecture should consider an emotional aspect as well as the traditional rational one, as we assert in this paper. We do not aim at proposing a specific implementation of emotions in an agent architecture. There have been a few attempts to do so in general agent architectures; see, for example, (Aubé and Senteni, 1996; Sloman and Poli, 1996; Bercht and Vicari, 2000).

Emotions in agents are undoubtedly related to issues of *formation* of norms (conventions) among autonomous agents (Conte et al., 1999). It is easy to see that emotions are important in attaching agents to social norms—Doran mentioned in (Doran, 1998b) that emotions may be an important mechanism for designers of society to "manipulate" agents, e.g., to maintain their collective misbelief in "cults" (Doran, 1998a). However, in a contrary view, it is plausible that just as emotions are the right basis for seeing that autonomous agents can be *autonomous* and still behave (truly) altruistically as we have argued above, studying emotions in agent architectures can help us understand how autonomous agents form and perpetuate conventions, which are essential for social behaviour. These are interesting issues to be addressed in further research.

Similar motivations to (Conte et al., 1999) can be found in (Ossowski and García-Serrano, 1999), which relies on both a sociological and an economic approach. They use the former to define what they call structural cooperation, a coordination mechanism in which a social structure biases autonomous agent behaviour. The economic approach is then used to determine the outcome of the process of social coordination. Their main point is that the apparent "irrational" social actions of individuals in a certain society can be seen as a reflection of the society's social structure (e.g. norms, values, and behaviour rules). Respecting such social structure constrains the individuals' choices as it prescribes what one has to do as a member, while also biasing their behaviours so as to ensure the necessary conditions for the survival of that society. As in the case of (Conte et al., 1999), they are concerned with social action without compromising on agents' autonomy; this seems to be connected intrinsically to some notions of emotions or moral sentiments.

In (Castelfranchi et al., 1997), the importance of emotions is recognised in domains other than the usual believable-agents one, which seems to support our argument. When commenting on the several reasons for agents needing personalities they include *Social/Cognitive Modelling*, as follows:

> Since in nature and in society agents have personality and this seems an important construct in psychology, one might aim at modelling personality in agents (or emotions or cognitive biases) to reproduce relevant features of human interaction. (page 16)

We believe this should be a new source of guidance for the work on MAS as originally defined as the field of Distributed Artificial Intelligence concerned with coordinated intelligent behaviour among a collection of autonomous intelligent agents (Gasser, 1987). In other words, if we build agents' rationality inspired by the human counterpart but fail to provide them with other important human mechanisms like emotions, we shall find that we have built rationally foolish agents, which will be no more useful to their collaborators than rational fools are to human societies (with analogues of all the undesirable consequences to individuals too).

7. CONCLUSIONS

In our previous work, the results have strongly suggested that rational fools maximise their earnings in the short term but compromise their performance in the long run, while altruists may not have the best performance at the beginning of the simulations, but normally end up much better than the others. The results have also clearly shown that the more altruists there are in a group, the better they, and the group as a whole, perform; more importantly, their generosity, although somewhat "irrational" from an individual's point of view, implies that the group as a whole performs significantly better than when pure reciprocation is used. We have concluded that to behave rationally (in the classical sense in game theory) may not be the best attitude in the long run, or as far as the group is concerned.

Our present results using an evolutionary setting have corroborated our previous conclusions. We have seen that, in some simulations, the maximum points collected by successive generations in groups formed only of altruists increase. Also, at least for groups of medium size, egoists cannot completely invade those groups where there are altruists. Interestingly, they do invade small mixed groups, and we shall investigate this further in future papers.

It is clear that the evolutionary setting used here is a very simple one. Our plans for future work include simulation of a much more complex type of evolutionary setting based on this initial experience. Such a setting would include sexual reproduction (i.e., combining genetic material of two agents in the process of generating an offspring). Also, in that setting, the total number of agents in the society would, unlike what happens in the present set of simulations, vary according to the number of agents entitled to reproduce at each time step. For this to be possible, agents would have individually varying life spans and periods of time for sexual maturation, besides only being allowed to reproduce when they have accumulated sufficient points (representing the resources needed for bearing and raising offspring) and have mated with another agent in a similar situation.

Acknowledgements

The first two authors gratefully acknowledge the Brazilian agencies CNPq and FAPERGS for the partial support to this research project.

Notes

1. We have used the standard acronyms for the usual strategies in the IPD which we refer to: ALLD means that the agent always defects, ALLC means "always cooperate", and TFT stands for Tit-For-Tat.

2. Altruists and egoists have different performances when they are in homogeneous or mixed groups, which is why the distinction is important.

3. This is so because, if a group has half of its agents performing as well as agents in G0 and the other half performing as badly as the ones in G2, the group as a whole has the exact average of G0 and G2.

4. This best use of MS was achieved by the use of variable thresholds. In both types of societies, only some agents play (SP) at a time (60% in all four simulations shown in Figure 3.1).

5. By general we mean architecture for all sorts of agents and not only those that have a specific relation to "believability"; see references for this area given above.

References

Aubé, M. and Senteni, A. (1996). Emotions as commitements operators: a foundation for control structure in multi-agents systems. In Van de Velde, W. and Perram, J., editors, *Proceedings of the Seventh Workshop on Modelling Autonomous Agents in a Multi-Agent World (MAAMAW'96), 22–25 January, Eindhoven, The Netherlands*, number 1038 in Lecture Notes in Artificial Intelligence, pages 13–25, London. Springer-Verlag.

Axelrod, R. (1984). *The Evolution of Cooperation.* Basic Books, New York.

Bates, J. (1994). The role of emotion in believable agents. *Communications of the ACM*, 37(7):122–125.

Bazzan, A. L. C., Bordini, R. H., and Campbell, J. A. (1999). Moral sentiments in multi-agent systems. In Müller, J. P., Singh, M. P., and Rao, A. S., editors, *Intelligent Agents V—Proceedings of the Fifth International Workshop on Agent Theories, Architectures, and Languages (ATAL-98), held as part of the Agents' World, Paris, 4–7 July, 1998*, number 1555 in Lecture Notes in Artificial Intelligence, pages 113–131, Heidelberg. Springer-Verlag. UCL-CS [RN/98/29].

Bercht, M. and Vicari, R. M. (2000). Pedagogical agents with affective and cognitive dimensions. In *Fifth Ibero-American Congress on Educational Informatics (RIBIE 2000)*, Niña del Mar, Chile.

Binmore, K. (1998). Review of R. Axelrod's "The complexity of cooperation: Agent-based models of competition and collaboration". *Journal of Artificial Societies and Social Simulation*, 1(1). URL: <http://www.soc.surrey.ac.uk/JASSS/1/1/review1.html>.

Bordini, R. H. (1999). *Contributions to an Anthropological Approach to the Cultural Adaptation of Migrant Agents.* PhD thesis, University of London.

Bordini, R. H., Bazzan, A. L. C., Vicari, R. M., and Campbell, J. A. (2000). Moral sentiments in the iterated prisoner's dilemma and in multi-agent systems. *Brazilian Electronic Journal of Economics*, 3(1). URL: <http://www.beje.decon.ufpe.br/>.

Castelfranchi, C. (1990). Social power: a point missed in multi-agent DAI and HCI. In Demazeau, Y. and Müller, J.-P., editors, *Decentralized A.I.—Proceedings of the First European Workshop on Modelling Autonomous Agents in a Multi-Agent World (MAAMAW'89), 16–18 August, Cambridge, 1989*, pages 49–62. Elsevier Science B.V., Amsterdam.

Castelfranchi, C., de Rosis, F., and Falcone, R. (1997). Social attitudes and personalities in agents. In *Proceedings of the AAAI Fall Symposium on Socially Intelligent Agents*, Cambridge, Massachusetts.

Conte, R. and Castelfranchi, C. (1995). *Cognitive and Social Action.* UCL Press, London.

Conte, R., Castelfranchi, C., and Dignum, F. (1999). Autonomous norm acceptance. In Müller, J. P., Singh, M. P., and Rao, A. S., editors, *Intelligent Agents V—Proceedings of the Fifth International Workshop on Agent Theories, Architectures, and Languages (ATAL-98), held as part of the Agents' World, Paris, 4–7 July, 1998*, number 1555 in Lecture Notes in Artificial Intelligence, pages 99–112, Heidelberg. Springer-Verlag.

d'Inverno, M. and Luck, M. (1996). Understanding autonomous interaction. In Wahlster, W., editor, *Proceedings of the 12th European Conference on Artificial Intelligence (ECAI'96).*

Doran, J. (1998a). Simulating collective misbelief. *Journal of Artificial Societies and Social Simulation*, 1(1).
URL: <http://www.soc.surrey.ac.uk/JASSS/1/1/3.html>.

Doran, J. (1998b). Social simulation, agents and artificial societies. In Demazeau, Y., editor, *Proceedings of the Third International Conference on Multi-Agent Systems (ICMAS'98), Agents' World, 4–7 July, Paris*, pages 4–5, Washington. IEEE Computer Society Press. Extended Abstract for an Invited Talk.

Gasser, L. (1987). Distribution and coordination of tasks among intelligent agents. In *Proceedings of the First Scandinavian Conference on Artificial Intelligence*, Tromsø, Norway.

Gmytrasiewicz, P. J. and Durfee, E. H. (1995). A rigorous, operational formalization of recursive modeling. In Lesser, V. and Gasser, L., editors, *Proceedings of the First International Conference on Multi-Agent Systems (ICMAS'95), 12–14 June, San Francisco, CA*, pages 73–80, Menlo Park, CA. AAAI Press / MIT Press.

Gmytrasiewicz, P. J. and Lisetti, C. L. (2001). Using decision theory to formalize emotions. In Parsons, S., Gmytrasiewicz, P., and Wooldridge, M., editors, *Game Theory and Decision Theory in Agent-Based Systems*. Kluwer Academic Publishers. (In this volume).

Goldberg, D. E. (1989). *Genetic algorithms in search, optimization, and machine learning*. Addison-Wesley, Reading, Massachusetts, reprinted with corrections edition.

Hayes-Roth, B., Brownston, L., and van Gent, R. (1995). Multiagent collaboration in directed improvisation. In Lesser, V. and Gasser, L., editors, *Proceedings of the First International Conference on Multi-Agent Systems (ICMAS'95), 12–14 June, San Francisco, CA*, pages 148–154, Menlo Park, CA. AAAI Press / MIT Press.

Ossowski, S. and García-Serrano, A. (1999). Social structure in artificial agent societies: Implications for autonomous problem-solving agents. In Müller, J. P., Singh, M. P., and Rao, A. S., editors, *Intelligent Agents V—Proceedings of the Fifth International Workshop on Agent Theories, Architectures, and Languages (ATAL-98), held as part of the Agents' World, Paris, 4–7 July, 1998*, number 1555 in Lecture Notes in Artificial Intelligence, pages 133–148, Heidelberg. Springer-Verlag.

Ridley, M. (1996). *The Origins of Virtue*. Viking Press, London.

Rizzo, P., Veloso, M., Miceli, M., and Cesta, A. (1997). Personality-driven social behaviors in believable agents. In *Proceedings of the AAAI Fall Symposium on Socially Intelligent Agents*, Cambridge, Massachusetts.

Rosenschein, J. and Genesereth, M. (1985). Deals among rational agents. In Joshi, A., editor, *Proceedings of the Ninth International Joint Con-*

ference on Artificial Intelligence (IJCAI-85), pages 91–99. AAAI Press / Morgan Kaufmann, Los Angeles, CA.

Rosenschein, J. and Zlotkin, G. (1994). *Rules of Encounter*. MIT Press, Cambridge, MA.

Sloman, A. and Poli, R. (1996). SIM_AGENT: A toolkit for exploring agent designs. In Wooldridge, M., Müller, J. P., and Tambe, M., editors, *Intelligent Agents II—Proceedings of the Second International Workshop on Agent Theories, Architectures, and Languages (ATAL'95), held as part of IJCAI'95, Montréal, Canada, August 1995*, number 1037 in Lecture Notes In Artificial Intelligence, pages 392–407, Berlin. Springer-Verlag.

Chapter 4

DYNAMIC DESIRES

Mehdi Dastani
Institute of Information and Computing Sciences, Utrecht University
P.O. Box 80.089, 3508 TB Untrecht, The Netherlands
mehdi@cs.uu.nl

Zhisheng Huang
Division of Mathematics and Computer Science, Vrije Universiteit Amsterdam
De boelelaan 1081a, 1081 HV Amsterdam, The Netherlands
huang@cs.vu.nl

Leendert van der Torre
Division of Mathematics and Computer Science, Vrije Universiteit Amsterdam
De boelelaan 1081a, 1081 HV Amsterdam, The Netherlands
torre@cs.vu.nl

Abstract Logics of desires, preferences and goals have recently been proposed in planning and agent theory. In this paper we use a dynamic logic with utilitarian goals and desires to discuss the relation between goals, desires and utilities. From a static point of view goals are defined using desires and beliefs, and desires are defined using the utility function and a context. The consequences for the dynamics are that utilities are the most stable, then desires, and goals change most often. We illustrate the use of the logic to formalize certain aspects of negotiation. In particular, we show how one agent can influence the behavior of another agent by influencing his desires.

Keywords: Utility functions, changing desires, dynamics of agent decisions.

1. INTRODUCTION

Logics for desires and goals have been studied in several areas (see e.g. (Rao and Georgeff, 1991; Cohen and Levesque, 1990; Doyle and Thomason, 1999; Keeney and Raiffa, 1976)). Both desires and goals play important roles in practical decision making. Here we understand desires as motivational forces inside agents, and goals as desirable states that the agents try to reach. They are understood as in the BDI process: desires are used to determine goals. A typical distinction is that an agent can choose his goals, whereas he cannot choose his desires. In this paper we formalize desires and goals in a decision-theoretic setting. An important issue in the logics of desires and goals is the relation between utility functions, desires, and goals. In decision theories as well as in BDI theories desires are usually formalized in terms of utility functions which implies that desires becomes as unchangeable as the utility function. This perspective is in line with the traditional perspective in economic and game theory, and has been questioned by e.g. Shoham (Shoham, 1997), where utility is interpreted as one which has close relationship with probabilities, and Castelfranchi (Castelfranchi, 1998) who suggests that one can manipulate others' utilities. We are interested in formal models of *changing* desires and goals.

In this paper we look for a formal model in which the utility functions are typically constant, desires are relatively stable, whereas goals change much more frequently. For example, a person may want to go to a vacation island by airplane this week, but hears that the airplanes are in strike for the coming weeks. Since he can only go on holiday in the coming week, he changes his goal and now wants to stay at home and do house renovation. However, desires are much more stable. If an agent desires to go to the beach, then even if he cannot go because he has to work in the office, his desire does not disappear or change. Nevertheless, there are examples where desires change, and the typical example is the agent who stops to desire smoking once he falls in love with a girl that has lung cancer. It is not the case that he still likes to smoke but does not do it anymore; the cigarette no longer tastes. The change of the desire may be encoded by various factors. The issue has been discussed in the social sciences when researchers try to explain the effect of putting a nice looking girl on top of a car. This does not give us any new information about the car, but it nevertheless influences desires. The point is that agents can be influenced in many different ways, which do not operate only through their beliefs. This fundamental insight has already been part of the BDI model: goals are products of beliefs and desires.

Utility function

——————— Lifting problem: Context

desires

——————— Realizability problem: Beliefs

goals

Figure 4.1. Utilities, desires and goals.

Our representation is based on the simple model in figure 4.1. This figure should be read as follows. From decision theory we use the notion of a utility function. To define desire in terms of the utility function we have to solve the so-called lifting problem (to be explained later), for which we use a notion of *context*. To select desires which can function as goals we get into a second problem, the realizability problem: which goals can be realized? To solve this second problem we refer to the agent's beliefs. Since this second step has already been discussed extensively in the literature, for example in BDI logic, we focus in this paper on the first step.

Our aim is to develop a formal model which explains that the utility function is the most stable, then the desires and finally the goals. If the utility function changes then the other two change too, and if the desire changes then the goals change too. However, the inverse is not true. For example, goals can change when desires do not change because the agent's beliefs change. Moreover, the utility function may be constant whereas desires change, simply because the context changes.

This paper is organized as follows. In Section 2 we discuss the logic of utilities, desires and goals. In Section 3 we introduce the dynamic logic of utilitarian desires, and we show by an example with reasoning about desire in negotiation that utilities are more stable than desires. In Section 4, we discuss some properties of the logic and we discuss some future work. Finally, we conclude the paper in Section 5.

2. UTILITIES, DESIRES AND GOALS

> "The basic problem of social life among cognitive agents lies beyond mere coordination: *how to change the mind of the other agent? How to induce the other to believe and even to want something?* How to obtain that *y* does or does not do something? Of course, normally – but not necessarily – by communicating." (Castelfranchi, 1998, p.170)

In this paper, desires of rational agents are assumed to reflect their utility function which in turn reflects their preferences. This relation

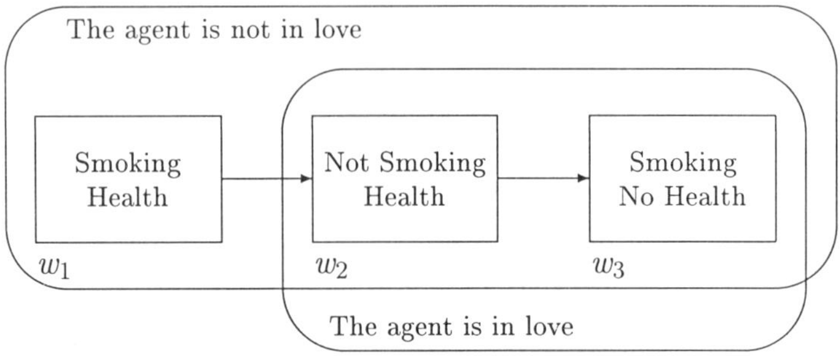

Figure 4.2. The smoking example.

formalizes the intuition that *rational agents desire to achieve states with the highest utility* when making decisions. This type of desire is called *utilitarian desire*; see (Lang et al., 2002; Pearl, 1993; Boutilier, 1994) for a further discussion on the relation between desires and decision making. Here utilitarian desire is understood as in classical decision theory: a rational agent acts as if he is maximizing expected utility based on an imaginary (or subjective) utility function and probability distribution. However, although we agree that desire and utility function are closely related concepts, we believe that they are distinct concepts since a utility function is assumed to remain constant while desires can change. Below, we try to elucidate this distinction by giving two examples. These examples illustrate how agents act to influence each other's desires and consequently affect the choice of actions to be taken by them.

2.1. CHANGING DESIRES: AN EXAMPLE

Consider again the agent who used to desire to smoke, but who no longer desires to do so since he has fallen in love with a girl that has lung cancer. If we consider the two propositions 'healthy' and 'smoking', then the following three states can be distinguished: 1) w_1: healthy and smoking, 2) w_2: healthy and not smoking, and 3) w_3: unhealthy and smoking. His utility function obeys the constraint $u(w_1) > u(w_2) > u(w_3)$. When the agent is not in love, then he focuses his attention to all states and he desires to smoke. But, when he falls in love with that girl, he focuses his attention to states w_2 and w_3 and he desires not to smoke.

Note that in this example, the utility function is defined on complete specifications (states) and desires on incomplete specifications (sets of states). To infer the desirability of a proposition the utility function defined on states has to be *lifted* to a desirability function on sets of states (called the lifting problem (Thomason and Horty, 1996)). The core of our idea is that although the utility function itself does not change, the lifting condition can change. Summarizing, the lifting condition in the first case (i.e. he is not in love) lifts the utility of the states to the desirability to smoke, and the lifting condition in the other case (i.e. he is in love) lifts the utility of the states to the desirability not to smoke.

2.2. THE LIFTING PROBLEM

The main problem of defining utilitarian desires is that utility is defined on states, whereas desirability is defined on sets of states or propositions. This is called the *lifting problem*: the utility function on states has to be lifted to a desirability function on sets of states. For example, we can simply define the desirability of a set of states as the maximum of the utilities of its elements, or the lowest one, or some kind of average. As discussed in (van der Torre and Weydert, 2001), the maximum reflects a kind of optimism that the best state will obtain, and the lowest a kind of pessimism that the worst state will obtain. The first is justified if we can determine or influence which of these states will obtain, and the latter is justified when an opponent can decide it (or when nature is seen as an opponent, like in Wald's criterion). In this paper we do not discuss the details of this particular choice and we therefore use the most popular choice, the maximum. Thus, a desire for p is satisfied by a utility function if the *relevant* state with the highest utility satisfies p, i.e. when for all *relevant* states w satisfying $\neg p$ there is a *relevant* state w' satisfying p such that the utility of w' is higher than the utility of w (the advantage of the latter definition is that it is more intuitive for infinite sets of states).

In this paper we are in particular interested in the question *which* states should be considered *relevant*. That is, we explicitly distinguish the following two stages in the lifting problem, of which in this paper we are primarily interested in the first one:

1 Select a relevant subset of all states. One constraint on the states is that the condition of the desire is true. Thus, for a desire "p if q" we select all q states. However, this is not all. For example, in the smoking example we restricted the focus first to states in which the agent is not in love, and thereafter to states in which the agent is in love.

2 Of these states, select the ones with the highest utility. A proposition p is desired if it is true in all these maximal utility states.

To solve the problem of state selection we introduce a rather abstract restriction of the set of states, inspired by recent work on modeling and using context (van der Torre and Tan, 2000). We therefore distinguish between two types of utilitarian desires,

ideal utilitarian desire, written $Id(p|q)$ for ideally p if q. It is defined in terms of utilities of all possible states, and

feasible utilitarian desire, written $D(p|q)$ for desirable p if q. It is defined in terms of utilities of all possible states in the agent's context.

2.3. LOGIC OF DESIRES AND GOALS

In this section we formalize both types of utilitarian desires in Kripke models and in terms of possible states, their accessibilities, and their utilities. The starting point of our formalization is that the utility function is the same in all states. The following definition formalizes the pure utilitarian desire. An agent finds something ideal if and only if for all states in which it is false there is a state with a higher utility in which it is true. In this definition we use a mapping to the non-negative integers to ensure that there are no infinite descending chains.

Definition 1 *Let $\mathcal{M} = \langle W, U_i, C_i, N_i \rangle$ be a possible states model that consists of a set W of all possible states, a function U_i from W to the non-negative integers indicating the utility of states for agent i, an accessibility relation C_i that indicates which states the agent i considers relevant for his desires, and an accessibility relation N_i indicating the normality of states for agent i. A formula "ϕ is ideal if ψ" holds in state w for agent i (i.e. $\mathcal{M}, w \models Id_i(\phi|\psi)$) if and only if for all $\neg\phi \wedge \psi$ states there exists a $\phi \wedge \psi$ state with a higher utility, i.e.*

$$\mathcal{M}, w \models Id_i(\phi|\psi) \iff \forall w' \in W(\mathcal{M}, w' \models \neg\phi \wedge \psi)\exists w'' \in W \; (\mathcal{M}, w'' \models \phi \wedge \psi) \wedge (U_i(w'') > U_i(w'))$$

$$\mathcal{M}, w \models D_i(\phi|\psi) \iff \forall w' \in C_i(w)(\mathcal{M}, w' \models \neg\phi \wedge \psi)\exists w'' \in C_i(w) \; (\mathcal{M}, w'' \models \phi \wedge \psi) \wedge (U_i(w'') > U_i(w'))$$

$$\mathcal{M}, w \models G_i(\phi|\psi) \iff \mathcal{M}, w \models D_i(\phi|\psi), \exists w' \in N(w) M, w' \models \phi \wedge \psi$$

Although desire and goal are defined as binary operator $D_i(\phi|\psi)$ and $G_i(\phi|\psi)$, the unary desire can be defined in terms of its binary counterpart as follows:

$$D_i\phi =_{def} D_i(\phi|\top) \qquad G_i\phi =_{def} G_i(\phi|\top)$$

If we allow conflicting desires (i.e. if we allow $D(p) \wedge D(\neg p)$ to be consistent), which is intuitive, [1] then we can derive $D(q)$ for any q. This is a consequence of the fact that $D(p)$ is a normal modal operator defined on normal possible state semantics. This means that the K-axiom and its derived conjunction rule (i.e. $\mathcal{M}, w \models (D_i p \wedge D_i q) \rightarrow D_i(p \wedge q)$) hold if $C_i(w)$ is always assumed to be a non-empty set. For simplicity, we do not allow conflicting desires and thus cannot satisfactorily reason with them.

Applying the conjunction rule to conflicting desires $D_i p \wedge D_i \neg p$ implies $D_i(p \wedge \neg p)$. However, it is possible to have conflicting desires if we do not require that $C_i(w)$ is always a non-empty set. Furthermore, note that the following weak version of conflicting desires is satisfiable:

$$D_i(\phi|\psi) \wedge D_i(\neg\phi|\rho).$$

The notion of context in the definition of feasible utilitarian desire is crucial. In fact, the context, which distinguishes the two types of desires, specifies which states are relevant (or feasible) for the agent's desires. It is this context that functions as a filter on possible states and determines the desirable states. This filtering function combined with the belief accessible states determine the goals accessible states. In general, given a set of formula specifying the set of possible accessible states, the context determines a set of desire formulae that specifies the desirable states. The set of desirable states is considered to be a subset of possible states. The set of desire formulae is then combined with the set of belief formulae, which result the set of formulae that specify the goal states.

Taking again the smoking example, it is now obvious that the change of desire from smoking to not smoking can be modeled by considering love as a determinating factor of the context: before love the context consisted of all three states and after falling in love the context contains only the state w_2 and w_3.

The reader may have noticed the close relation between our work and the BDI system. In order to illustrate this relation, consider the diagram illustrated in figure 4.3. A part of this diagram shows the general scheme of the BDI system consisting of beliefs, goals, and intentions related by dash arrows. According to this part of the diagram, desires determines goals from beliefs (this is indicated by the thick arrow) and commitments determines intentions from goals. Another part of this diagram shows a different path (characterized by solid arrows) starting from ideal

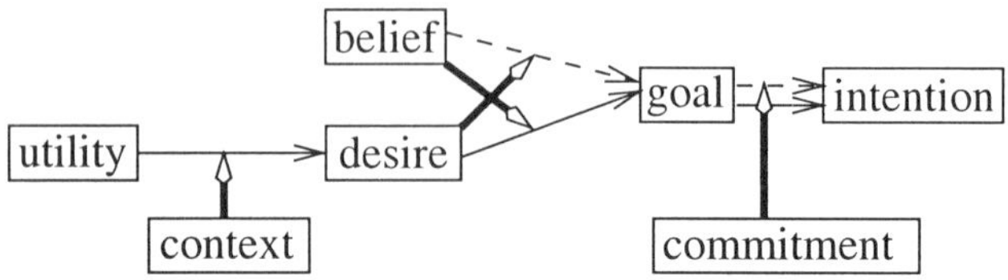

Figure 4.3. Enriching BDI with utilitarian desires.

utilitarian desire and ending at intentions. According to this part of the diagram, context determines (feasible utilitarian) desires from ideal utilitarian desires, beliefs determine goals from (feasible utilitarian) desires, and like in the BDI system, commitments determines intentions from goals.

3. AN EXAMPLE FROM NEGOTIATION

In the remainder of this paper we are interested in modeling the agent's desires in negotiation. Rational agents in negotiation decide what action to take based on their desires which reflect their utility functions. In this section we give an example to illustrate that the agent's desires are subject of change during negotiation. Moreover, since negotiation is usually modeled by game theory and because in game theory the utility function is assumed to remain constant during a game, we have to solve an apparent contradiction: on the one hand, it is reasonable to assume that the agent's desires can be changed during negotiation and on the other hand the utility function which is reflected by the rational agent's desire have to remain constant during negotiation. This apparent contradiction can be solved by *lifting* the utility function to a desirability function and allow the lifting condition to change on the basis of some context parameters.

Imagine a washing machine that wants to wash and thereby it must buy electricity from an electricity company. Figure 4.4 illustrates three possible washing states, w_1, w_2 and w_3 for the washing machine. In the state w_1, the washing machine washes in the afternoon and the electricity delivery is certain. Other states should be interpreted in similar way. Let the arrows between these three states denote the decrease of the utility of the states for the washing machine. Since w_1 provides the washing machine the highest utility, he desires to wash in the afternoon. However, the electricity provider desires to deliver electricity at night since they have usually more customers during the day. In this scenario, the electricity provider tries to influence the desires of the washing machine and informs him about the high prices of (certain deliverage) electricity

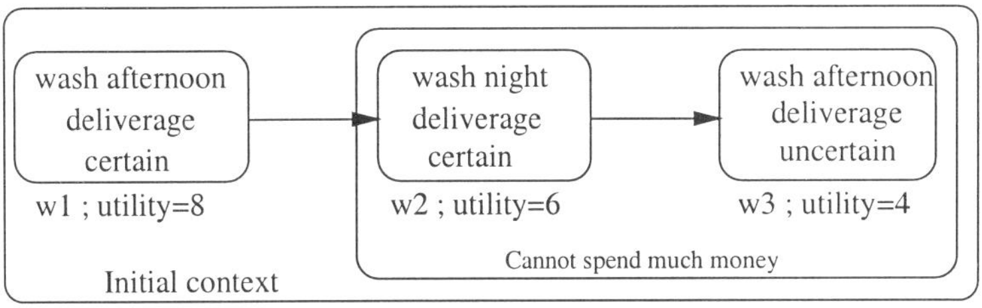

Figure 4.4. The washing machine example.

for day consumption. The washing machine may not be able to spend much money for electricity. Consequently, the washing machine concentrates on states in which washing is cheap (i.e. w_2 and w_3) and because w_2 provides him more utility than w_3, he desires to wash at night. Note that the electricity provider may also change the focus of the washing machine in various ways such as making advertisements or refusing to sell electricity for the day consumption. In this paper, we do not consider what causes desires to change, and concentrate on modeling the dynamics of desires.

Note that the agent cannot decide to change his own desire to wash at night, like an agent can decide to change his adopted goal to wash at night. Desires are not consequences of the decision making process, they are their input. It is due to the influence from outside that desires are changed. Analogous examples can be found in pricing airline tickets, telephone calls, internet connection bandwidth, data storage, insurance contracts, and many other areas.

In order to model this type of negotiations, we represent the space of possible negotiation states and the accessibility relation defined on them as a Kripke frame. Note that the accessibility relation defined on negotiation states can be specified by negotiation actions such as propose, inform, and require actions. Moreover, we use dynamic logic to represent and reason about agent's negotiation strategies (actions). Since agents should reason about both their own desires as well as the desires of other agents we define agent's desire formally as a modal operator. This modal operator will be combined with dynamic operators resulting in a multi-modal logic. In this way, we can express the effects of actions on desires and enable agents to reason about desires and actions, and to decide on negotiation strategies.

We are now in position to show how agents can change their desires without changing their utility functions. This is because an agent's desires are now defined relative to the current agent's context and the fact that the current agent's context may change. Let UA, CN, and CA

denote three washing states indicating washing in the afternoon with uncertain delivery of electricity, washing at night with certain delivery of electricity, and washing in the afternoon with certain delivery of electricity, respectively. Let also the utility of these states for the washing machine be: $UA = 4$, $CN = 6$, and $CA = 8$. Suppose that the initial context of the washing machine consists of all states. We assume that the washing machine can wash at one and only one period of the day in each state. In this initial context, the washing machine desires to wash in the afternoon since the only accessible state with highest utility is the one in which he washes in the afternoon. The washing machine receives now either the prices of the electricity company or become influenced by the advertisement announced by the electricity company. Consequently, the context of the washing machine may be changed such that it now consists of two states UA and CN. In this new context, the washing machine's desire has changed to wash at night since the only accessible state with the highest utility is the one in which he washes at night.

4. DYNAMIC LOGIC

The machinery of dynamic logic, encompassing *choice* ($\cup$), *sequence* (;) and *iteration* (*), has been proposed in (Broersen et al., 2000) to reason in the context of negotiation protocols. Protocols are usually represented by finite state machines, which are equivalent in expressive power to the regular actions that are central to the ontology of dynamic logic. The *concurrency* operator $\cap$ is, for instance, used to model the reasoning of an agent that offers some service to different agents simultaneously. Action *negation* ($\sim$) is used in the formalization of the deontic notions 'obligation', 'permission' and 'prohibition', according to an approach first taken by Meyer (Meyer, 1988). The *skip* denotes an action whose effect is void: doing a *skip* does not change the state we are in. It can be used in an expression like $T_{ij}(\alpha \leq skip)$ to express that by performing α our trust is decreased in comparison to the situation where we do nothing (we skip). The *any*-action refrains from action labels completely, which is useful if we want to reason about reachability of states, in a more temporal fashion. The *none*-action is the empty action. The most important elements of the logic introduced in (Broersen et al., 2000) are the modalities $C_{ij}(\alpha \leq \beta)$, whose intended meaning is 'agent i, is more *committed*, towards agent j, to perform α than to perform β', and $T_{ij}(\alpha \leq \beta)$ with the intended meaning 'the *trust* of agent i in agent j after performance of α is less than, or equal to the trust after performance of β'.

In this paper we extend dynamic logic with an operator D for desires and an operator B for beliefs, and we do not discuss commitments or trust. We (implicitly) assume that the utility function is global, i.e. the same in every state of the model. Thus, given utility function U_i for agent i, the fact that the utility function is global can be formally expressed by $U_i(\phi) \rightarrow [\alpha]U_i(\phi)$ which states that the utility function is not changed by any action. Since we are not interested in the desire operator by itself, but only in the interaction between actions and desires, we use a simple normal modal logic to formalize them. The belief operator is introduced in the logic, so that we define the goal operator G in terms of desires and beliefs, by which we can discuss the relations between desires and goals.

Given a set $\mathcal{A}$ of action symbols and $a \in \mathcal{A}$, a set $\mathcal{P}$ of proposition symbols and $p \in \mathcal{P}$, a ***well formed formula*** ϕ is defined through the following BNF:

$$\begin{array}{lll} \phi & ::= & p \mid \neg\phi \mid \phi \vee \psi \mid \langle\alpha\rangle\phi \mid Id_i(\phi|\psi) \mid D_i(\phi|\psi) \mid B_i\phi \\ \alpha & ::= & a \mid any \mid none \mid skip \mid \sim\alpha \mid \alpha \cup \alpha' \mid \alpha \cap \alpha' \mid \alpha;\alpha' \mid \alpha^* \end{array}$$

The semantics is defined in a fairly standard way.

Definition 2 *Given a set $\mathcal{A}$ of action symbols, a set $\mathcal{P}$ of proposition symbols, a structure $\mathcal{S} = (S, R_{\mathcal{A}}, \pi, U_i, C_i, B_i)$ is defined as:*

- *S is a nonempty set of possible states*
- *$R_{\mathcal{A}}$ is an $\mathcal{A}$-indexed collection of (reachability) relations over $S \times S$*
- *π is a valuation function $\pi : \mathcal{P} \rightarrow 2^S$ that interprets propositions $p \in \mathcal{P}$*
- *U_i is a total ordering over S that indicates the utility of states for agent i*
- *C_i is an accessibility relation that indicates which states the agent i considers relevant for his desires.*
- *B_i is a serial accessibility relation that indicates which states the agent i considers possible.*

Definition 3 *The meaning of well-formed formulas in a state s of a structure $\mathcal{S}$ is given by:*

$$
\begin{array}{lcl}
R_\alpha & = & R_a \; for \; \alpha = a \; for \; a \in \mathcal{A} \\
R_{none} & = & \emptyset \\
R_{skip} & = & \{\langle s, s\rangle : \; for \; all \; s \in S\} \\
R_{\alpha \cap \alpha'} & = & R_\alpha \cap R_{\alpha'} \\
R_{\alpha \cup \alpha'} & = & R_\alpha \cup R_{\alpha'} \\
R_{any} & = & (R_{a \cup b \cup c \ldots})^* \; with \; \{a, b, c, \ldots\} = \mathcal{A} \\
R_{\sim\alpha} & = & R_{any} \setminus R_\alpha \\
R_{\alpha;\alpha'} & = & R_\alpha \circ R_{\alpha'} \\
R_{\alpha^*} & = & (R_\alpha)^*
\end{array}
$$

$$
\begin{array}{lll}
\mathcal{S}, s \models P & iff & s \in \pi(P) \\
\mathcal{S}, s \models \neg\phi & iff & not \; \mathcal{S}, s \models \phi \\
\mathcal{S}, s \models \phi \wedge \psi & iff & \mathcal{S}, s \models \phi \wedge \mathcal{S}, s \models \psi \\
\mathcal{S}, s \models \langle\alpha\rangle\phi & iff & \exists s' \in S, \; s.t. \; \langle s, s'\rangle \in R_\alpha \wedge \mathcal{S}, s' \models \phi \\
\mathcal{S}, s \models Id_i(\phi|\psi) & iff & \forall s' \in S(\mathcal{S}, s' \models \neg\phi \wedge \psi) \\
 & & \exists s'' \in S(\mathcal{S}, s'' \models \phi \wedge \psi) \wedge (U_i(s'') > U_i(s')) \\
\mathcal{S}, s \models D_i(\phi|\psi) & iff & \forall s' \in C_i(s)(\mathcal{S}, s' \models \neg\phi \wedge \psi) \\
 & & \exists s'' \in C_i(s)(\mathcal{S}, s'' \models \phi \wedge \psi) \wedge (U_i(s'') > U_i(s')) \\
\mathcal{S}, s \models B_i\phi & iff & \forall s' \in S, s.t. \langle s, s', \rangle \in B_i \wedge S, s' \models \phi
\end{array}
$$

For the discussion in this paper, in which we are interested in the interaction between the two types of modal operators, the exact definition of the operators is not relevant. We define $[\alpha]\phi$ as $\neg\langle\alpha\rangle\neg\phi$. Moreover, the goal operator G is defined in terms of desires and beliefs as follows. ϕ is a goal for agent i (under the condition ψ) if and only if ϕ is a desire for i and the agent i does not believe that ϕ is impossible (under the condition ψ).

$$G_i(\phi|\psi) \Leftrightarrow D_i(\phi|\psi) \wedge \neg B_i\neg(\phi \wedge \psi).$$

The unary operator of the goal is defined as:

$$G_i\phi \Leftrightarrow G_i(\phi|\top).$$

Thus, we have the following relation between goals and desires:

$$G_i\phi \leftrightarrow D_i\phi \wedge \neg B_i\neg\phi.$$

4.1. THE EXAMPLE

The following example illustrates how desires change in a dynamic situation such as the negotiation setting. Consider the following Kripke model $\mathcal{S}$, illustrated in Figure 4.5, representing one possible step in the negotiation between the washing machine and the electricity provider. In this model there are two types of relations defined on states. One relation

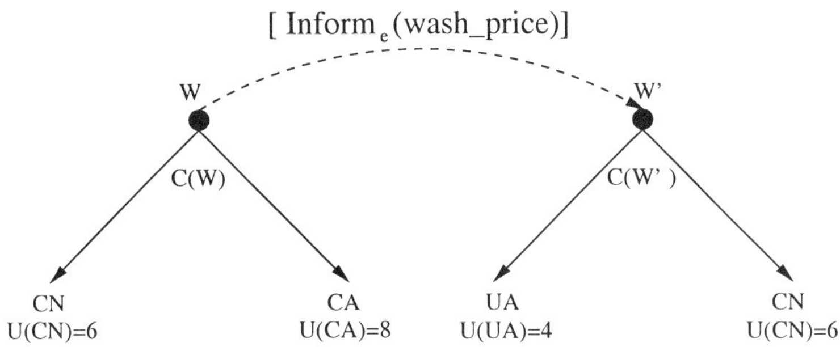

Figure 4.5. A model $\mathcal{S}$ for the washing machine.

is represented by dashed lines and they indicate possible negotiation actions, i.e. the relation specifies the space of possible negotiation actions (extensive negotiation game). The second relation is represented by solid lines and they indicate accessible states of an agent, i.e. the relation expresses the context of an agent.

In this model, the dash line indicates a proposal initiated by the electricity provider, i.e. $Propose_e(Wash(night))$, and the solid lines from each state indicates the context of an agent in that state. It is this context which determines the desires of agents. For instance, in the state w of model $\mathcal{S}$, the washing machine desires to wash in the afternoon since the only accessible state with the highest utility in the context of w for the washing machine is CA, i.e. $\mathcal{S}, w \models D_{wm}(Wash(afternoon)|\top)$. Suppose, after the proposal done by the electricity provider the context for the washing machine changes such that he desires now to wash at night, i.e. $\mathcal{S}, w' \models D_{wm}(Wash(night)|\top)$. Note that although the utility of the washing machine for states are not changed, his desire to wash in the afternoon is changed.

4.2. THEOREMS

The following formulas are theorems of the logic.

Stability

$Id_i(\phi|\psi) \rightarrow [\alpha]Id_i(\phi|\psi)$.

(Actions do not change pure utilitarian desires at all.)

Left Expansion

$Id_i(\phi|\psi) \rightarrow Id_i(\phi \wedge \psi|\psi)$.

$D_i(\phi|\psi) \rightarrow D_i(\phi \wedge \psi|\psi)$.

$G_i(\phi|\psi) \rightarrow G_i(\phi \wedge \psi|\psi)$.

Right Expansion

$Id_i(\phi|\psi) \rightarrow Id_i(\phi|\phi \wedge \psi)$.
$D_i(\phi|\psi) \rightarrow D_i(\phi|\phi \wedge \psi)$.
$G_i(\phi|\psi) \rightarrow G_i(\phi|\phi \wedge \psi)$.

Consistency

$\neg Id_i(\phi|\neg\phi)$ if $\neg\phi$ satisfiable.
$\neg D_i(\phi|\neg\phi)$ if $\neg\phi$ satisfiable.
$\neg G_i(\phi|\neg\phi)$ if $\neg\phi$ satisfiable.
$\neg B_i\bot$.
$\neg G_i\bot$.

Conjunction Decomposition

$Id_i(\phi \wedge \rho|\psi) \leftrightarrow Id_i(\phi|\psi) \wedge Id_i(\rho|\psi)$.
$D_i(\phi \wedge \rho|\psi) \leftrightarrow D_i(\phi|\psi) \wedge D_i(\rho|\psi)$.
$G_i(\phi \wedge \rho|\psi) \leftrightarrow G_i(\phi|\psi) \wedge G_i(\rho|\psi)$.
$B_i(\phi \wedge \rho) \leftrightarrow B_i(\phi) \wedge B_i(\rho)$.

Dynamics

$G_i(\phi|\psi) \wedge [a]B_i\neg(\phi \wedge \psi) \rightarrow \neg\langle a\rangle G_i(\phi|\psi)$.
$\langle a\rangle B_i(\phi \wedge \psi) \wedge [a]\neg G_i(\phi|\psi) \rightarrow \langle a\rangle\neg D_i(\phi|\psi)$.

4.3. CLASSIFICATION OF ACTIONS

The formal language combines dynamic action operators with the desire operator. This combined language is useful to characterize several properties. First we distinguish between several types of formulas:

objective formulas do not contain any modality

static formulas do not contain a dynamic modality

desire atom starts with the operator D or the operator Id

desire formula is a combination of conjunction, disjunction and desire atoms

Some formulas hold for certain types of actions, and thus *characterize* these actions. For example:

pure actions α do not change pure utilitarian desires at all and are characterized by $Id_i(\phi|\psi) \rightarrow [\alpha]Id_i(\phi|\psi)$

influence actions α only change desires and are characterized by $\phi \rightarrow [\alpha]\phi$ if ϕ is objective

fulfillment actions α only change facts if this is desirable and are characterized by $\phi \rightarrow ([\alpha_i](\neg\phi \wedge \psi) \rightarrow D_i(\neg\phi|\psi))$

5. FUTURE RESEARCH

In this paper we focused on motivation and semantics. However, if agents reason about negotiation they may prefer to use the logical proof theory. For example, the agent can reason about the other agent's desires and decide how to influence them and which negotiation action to take. Thus there is a need for an axiomatization.

Besides, the logical language may be useful to distinguish different concepts. For example, we can classify actions in actions that only change facts, actions that only change desires, and actions that change both. Moreover, actions can be partitioned in actions that fulfill desires and actions that do not. This can be the basis of agent profiles: an agent that often does what you desire may become one of your friends.

6. CONCLUDING REMARKS

In this paper we studied utilitarian desires in a dynamic environment. We proposed a semantics for a dynamic logic of context dependent desires. The context formalizes the intuition that desires can change whereas the utility function remains constant. We have shown that the notion of context involved is different from notions of belief or knowledge which have been used in a decision-theoretic setting. Technically, the work is related to dynamic epistemic logic. In these studies, a distinction is made between general actions and observation actions. In the former the agents state changes, whereas in the latter only the agent's beliefs change.

This work is presented in the context of practical negotiations where agents can reach agreements by influencing other agent's desires. In the negotiation context, the interaction between dynamic operators and the desire operator is illustrated by the effect of negotiation actions (dynamic operators) on an agent's desire (desire operator).

Notes

1. Intuitively, one can for example desire to be rich and not-working although not-working implies not-rich.

References

Boutilier, C. (1994). Toward a logic for qualitative decision theory. In *Proceedings of the 4th International Conference on Principles of Knowledge Representation and Reasoning (KR'94)*, pages 75–86.

Broersen, J., Dastani, M., and van der Torre, L. (2000). Levelled commitment and trust in negotiation. In *Proceedings of the Autonomous*

Agents 2000 Workshop on Deception, Fraud and Trust in Agent Societies, Barcelona.

Castelfranchi, C. (1998). Modelling social action for ai agents. In *Artifical Intelligence 103(1-2)*, pages 157–182.

Cohen, P. and Levesque, H. (1990). Intention is choice with commitment. In *Artificial Intelligence, 42*, pages 213–261.

Doyle, J. and Thomason, R. (1999). Background to qualitative decision theory. In *AI magazine 20(2)*, pages 55–68.

Keeney, R. and Raiffa, H. (1976). Decisions with multiple objectives: Preferences and value trade-offs. In *Wiley and Sons*. New York.

Lang, J., van der Torre, L., and Weydert, E. Utilitarian desires. In *Autonomous Agents and Multi Agent Systems*. To appear.

Meyer, J.-J. C. (1988). A different approach to deontic logic: deontic logic viewed as a variant of dynamic logic. In *Notre Dame Journal of Formal Logic, 29*, pages 109–136.

Pearl, J. (1993). From conditional oughts to qualitative decision theory. In *Proceedings of the Ninth Conference on Uncertainty in Artificial Intelligence (UAI'93)*, pages 12–20.

Rao, A. and Georgeff, M. (1991). Modeling rational agents within a bdi architecture. In *Proceedings of the 1st International Conference on Principles of Knowledge Representation and Reasoning (KR'91)*.

Shoham, Y. (1997). Conditional utility, utility independence, and utility networks. In *Proceedings of the Thirteenth Conference on Uncertainty in Artificial Intelligence (UAI'97)*, pages 429–436.

Thomason, R. and Horty, R. (1996). Nondeterministic action and dominance: foundations for planning and qualitative decision. In *Proceedings of the Sixth Conference on Theoretical Aspects of Rationality and Knowledge*, pages 229–250, De Zeeuwse Stromen, The Netherlands. Morgan Kaufmann Publishers.

van der Torre, L. and Tan, Y. (2000). Contextual deontic logic: violation contexts and factual defeasibility. In P. Bonzon, M. C. and Nossum, R., editors, *Formal Aspects of Context, Applied Logic Series*, pages 143–166. Kluwer.

van der Torre, L. and Weydert, E. (2001). Parameters for utilitarian desires in a qualitative decision theory. In *Applied Intelligence, 14*, pages 285–301.

Chapter 5

EMOTIONS AND PERSONALITY IN AGENT DESIGN AND MODELING

Piotr J. Gmytrasiewicz
Department of Computer Science, University of Illinois at Chicago, 851 S. Morgan M/C 154 Chicago, IL 60607.
piotr@cs.uic.edu

Christine L. Lisetti
Department of Computer Science, University of Central Florida, Orlando, FL 32816-5419.
lisetti@cs.ucf.edu

Abstract

Our research combines two diverse strands of work in AI and cognitive science. We start from the principled paradigm of rational agent design based on decision theory. We then use this paradigm to formally define the emotional states and personality of an artificial intelligent agent. We view the emotional states as the agent's decision-making modes, predisposing the agent to make its choices in a specific, yet rational, way. Change of the emotional state, say due to an external stimulus, invokes a transformation of the agent's decision-making behavior. We define personality as consisting of the agent's emotional states together with the specification of transitions taking place among the states. To enable an artificial agent to model the personalities and emotional states of agents and humans that it interacts with, we additionally provide a precise definition of a personality models of other agents. Our definition allows the personality models to be learned over the course of multiple interactions with the users and other agents.

Keywords: Decision theory, emotions, personality, multi-agent systems.

1. INTRODUCTION

The objective of this research is to develop a fundamental understanding of the role and usefulness of the notions of emotions and personality in designing rational artificial agents. The main hypothesis of our work is that notions of personality and emotions are useful in designing competent artificial agents that are to operate within complex uncertain environments populated by other artificial and human agents. Our work draws on and combines an emerging technology of rational agent design from artificial intelligence on the one hand (Dennett, 1986; Doyle, 1992; Russell and Norvig, 1995; Wooldridge and Anand Rao, 1999), with research on human emotions in cognitive science and psychology on the other hand (Dimasio, 1994; Fridja, 1986; James, 1884; James, 1894; Johnson-Laird and Oatley, 1992; Leventhal and Scherer, 1987; Ortony et al., 1988; Simon, 1967; Sloman and Croucher, 1981; Sloman, 1990).

Our work accepts the decision-theoretic paradigm of rationality, according to which a rational agent should behave so as to maximize the expected utility of its actions (see (Boutilier et al., 1999; Doyle, 1992; Russell and Norvig, 1995) and other references therein). The expected utilities of the alternative courses of action are computed based on their possible consequences, the desirability of these consequences to the agent,[1] and the probabilities with which these consequences are thought by the agent to obtain.[2] We aim to examine ways in which components of the decision-theoretic model, i.e., the utility functions, the set of behavioral alternatives, and the probabilities of consequences, can be transformed in ways that has been recognized in cognitive science as interactions between emotional states and decision-making. Further, we want to make the personality and the emotional components useful to model other agents, including humans, during interactions.

As research in cognitive science shows, some of the most important functions of emotions is to manage the individual's cognitive resources, allow him to understand the other agents' internal states, and to effectively communicate his internal state to others. The importance of managing the cognitive resources is clear when one considers the cognitive limitations of animals, humans and machines on the one hand, and the computational demands imposed by complex dynamical environments on the other hand. Given these limitations, biological systems are equipped with simplified ways of arriving at the appropriate action to be executed. Lower-level mechanisms, such as instincts, are essentially condition-action rules. Higher-level mechanism is provided by emotions, which dictate not actions but action tendencies (Johnson-Laird and Oatley, 1992; Simon, 1967; Sloman and Croucher, 1981). For example, if an external event threatens the safety of an individual, the emotional state of fear restricts the alternative behaviors to a small repertoire of possible appropriate actions such as: interrupt current activity, monitor environment, flee, or fight. In this context, emotional states are modifying the parameters of deliberative rationality to control its complexity under time pressure. The significance of our work lies in the formal definitions of these functions and their use for the benefit of the artificial interactive agents we are designing.

As we mentioned, the other aspect in which emotions and personality are important is the ability to understand and appropriately model the internal states of other agents. For example, during human-computer interactions the users' emotional state, such as anger, fear, boredom, panic, surprise, joy, or excitation, can be assessed using measurable factors (facial expression recognition, vocal intonation, prosody, galvanic skin response, heart rate and breathing patterns, haptic and tactile feedback, body posture (Cottrell and Metcalfe, 1991; Hayes-Roth et al., 1998; Lisetti and Rumelhart, 1998; Lisetti and Schiano, 1999; Picard, 1997)). In spite of availability of these technologies the HCI literature abounds with cases in which the machine is hopelessly out-of-step with the emotional state of the human user. In such cases, the interaction may turn out to be all together damaging rather then helpful. Our definitions of emotional states allow the system to correctly model the user in such cases. Further, our definition of personality introduces a principled dynamic model of the user's emotional states. Using it enables a prediction that, for example, a user already annoyed will not be calmed down by another system response that is not along the user's wishes. Even if the machine is interacting with an artificial agent, the ability to understand and model the agent's internal state can be of essence. For example, interacting with a bounded agent that is under a time pressure requires an understanding that the agent probably does not have the time to evaluate all of the available options in great detail.

Closely related to the above is the need for reliable and universally understood vocabulary using which the agents can communicate their internal states to each other. For example, it may be more expeditious to communicate that an agent is "panicked" than to explain at length that the time horizon of the alternative plans being considered had to be shortened due the computational limitations and time pressure of the situation at hand. Also, the lengthy explanation may be meaningful only to agents of similar designs. Thus, a principled understanding of what emotions are in terms of already known theories of behavior serves as semantics of communicative terms that describe the agents' internal states.

As we mentioned, our approach complements and builds on the existing approaches to designing rational and socially competent agents (Bicchieri, 1993; Boutilier et al., 1999; Cohen and Levesque, 1990; Doyle, 1992; Durfee et al., 1993; Gmytrasiewicz and Durfee, 2000; Gmytrasiewicz and Durfee, 2001; Kraus and Sycara, 1989; Rosenschein and Zlotkin, 1994; Russell and Norvig, 1995; Sycara, 1998; Wooldridge and Anand Rao, 1999). Such agents should be able to function efficiently under time and other environmental pressures, and be able to interact and communicate with other agents. This includes informing each other about details of the external environment and about the agents' own internal states, as well as the ability to model and predict the internal states of other agents. Apart from the area of multi-agent systems, our approach has applications in Human-Computer Interaction (HCI) that range from intelligent tutoring systems and distance learning support systems (with recognition of expressions signaling interest, boredom, confusion), to stress and lie detectors, to monitors of pilots' and drivers' state of alertness, to software product support systems (with recognition of users being dis/pleased with software products), to entertainment and computer games (enjoyment, confusion), and to ubiquitous computing and smart houses (Lesser et al., 1999).

2. DECISION-THEORETIC PRELIMINARIES

As we mentioned, the objective of our research is to develop a fundamental understanding of the role and usefulness of emotional states and personality in designing intelligent artificial systems. Our approach draws on and combines an emerging technology of rational agent design of Artificial Intelligence on the one hand (Boutilier et al., 1999; Dennett, 1986; Doyle, 1992; Russell and Norvig, 1995; Wooldridge and Anand Rao, 1999), with research on human emotions in cognitive science and psychology on the other hand (Dimasio, 1994; James, 1884; James, 1894; Johnson-Laird and Oatley, 1992; Leventhal and Scherer, 1987; Ortony et al., 1988; Simon, 1967; Sloman and Croucher, 1981; Sloman, 1990).

We use the decision-theoretic paradigm of rationality, according to which a rational agent should behave so as to maximize the expected utility of its actions (see (Boutilier et al., 1999; Doyle, 1992; Russell and Norvig, 1995) and references therein). The expected utilities of the alternative courses of action are computed based on their possible consequences, the desirability of these consequences to the agent, and the probabilities with which these consequences are thought by the agent to obtain.

Thus, a rational agent formulates its decision-making situation in terms of a finite set, A, of the alternative courses of action, or behaviors, it can execute, which we will call the agent's *action space*. An alternative behavior, say a_i, is a plan consisting of consecutive actions extending into the future time t_{a_i}, which we will call the time horizon of this particular plan. Alternative courses of action in set A can stand for abstract actions as well as for detailed elaborations; increasing the level of abstraction facilitates keeping the size of A down to manageable proportions. We demand that the actions be distinct and that the set A be exhaustive, i.e., that all of the possible behaviors be accounted for. Sometimes an "all-else" behavioral alternative is used for compactness, and represents all other possible behaviors except the ones explicitly enumerated.

At any point, an agent finds itself in some state of the world, but due to the fact that the environment may not be fully observable the agent may be uncertain about the state. The fact that the actual state may be unknown to the agent can be formalized by specifying the set of all possible states of the world, S, together with a family of probability distributions, $\mathbf{P}(S)$, over these states. One of these distributions, say $P_c(S)(\in \mathbf{P})$, specifies which of these states are currently possible and how likely they are. Thus $P_c(S)$ fully describes the information the agent has about the present state of the world.

The agent can ponder the consequences of its alternative actions. Due to possible nondeterminism, each action, $a_i \in A$, may lead to many resulting possible states. The likelihoods of the resulting states can be specified by another probability distribution, $P_i(S)(\in \mathbf{P})$, also over S. The process of determining the probabilities of different results, i.e., the distribution P_i has been called a probabilistic temporal projection. The projection is a function $Proj : \mathbf{P}(S) \times A \rightarrow \mathbf{P}(S)$, so that the result of projecting the results of action a_i, given the current information about the state $P_c(S)$, results in the projected information about the resulting state, $P_i(S)$: $Proj(P_c(S), a_i) = P_i(S)$. The above formulation does not preclude state changes due to actions of other agents or erogenous events; here these effects are implicit and folded into the projection function (Boutilier et al., 1999).

The desirabilities of the states of the world to the agent are encoded using a utility function $U : S \rightarrow \Re$, which maps states of the world to real numbers. Intuitively, the higher the utility value of a state the more desirable this state is to the agent. The agent decision problem involves choosing which of the alternative actions in the set A it should execute. One of the central theorems of decision theory states that if the agent's utility function is properly formed, and the agent expresses its uncertain beliefs using probabilities, then the agent should execute an action, a^*, that maximizes the expected utility of the result.

$$a^* = ArgMax_{a_i \in A} \sum_{s \in S} p_i^j \, U(s^j), \tag{1}$$

where the p_i^j is the probability the projected distribution $P_i(S)$ assigns to a state $s^j \in S$. Frequently, it is convenient to represent the utility function, U, as depending on a small number of attributes of the states of the world, as opposed to depending on the states themselves. This is intuitive; humans may prefer, say, all of the states in which they have more money, are more famous, and are healthier. The attributes of wealth, fame, and health are then convenient factors in terms of which the utility function can be expressed. Multi-attribute utility theory postulates that, in some simple cases, the utility of a state be a weighted sum of the utilities, $U(X_l(s))$ of individual attributes:

$$U(s) = \sum_{X_l \in Attributes} W_{X_l} U(X_l(s)), \tag{2}$$

where the W_{X_l} is the weight or, intuitively, the importance, of the attribute X_l. Having the weights of the attributes explicitly represented is convenient since it enables the trade offs among the attributes the agent may have to make. For example, the agent may have to give up some of its wealth to improve its health, and so on.

The elements defined above are sufficient to formally define a decision-making situation of an agent:

Definition 1: A decision-making situation of an agent is a quadruple: $D =< P_c(S), A, Proj, U >$, where S, $P_c(S)$, A, $Proj$ and U are as defined above.

The above quadruple fully specifies the agent's knowledge about the environment, the agent's assessment as to its possible courses of action, the possible results of the actions, and desirability of these results. Our definition here is a version of partially observable Markov decision process (see (Boutilier et al., 1999; Russell and Norvig, 1995) and references therein), but it makes explicit the time horizons of the alternative action sequences the agent is choosing among.

Given its decision-making situation, an agent can compute its best action, a^*, as specified in Equation 1. It is clear that this computation can be fairly complex. In a multi-agent environment, for example, all of the information the agent has about the physical environment and about the other agents could be relevant and impact the expected utilities of alternative courses of action. Sometimes the agent may have

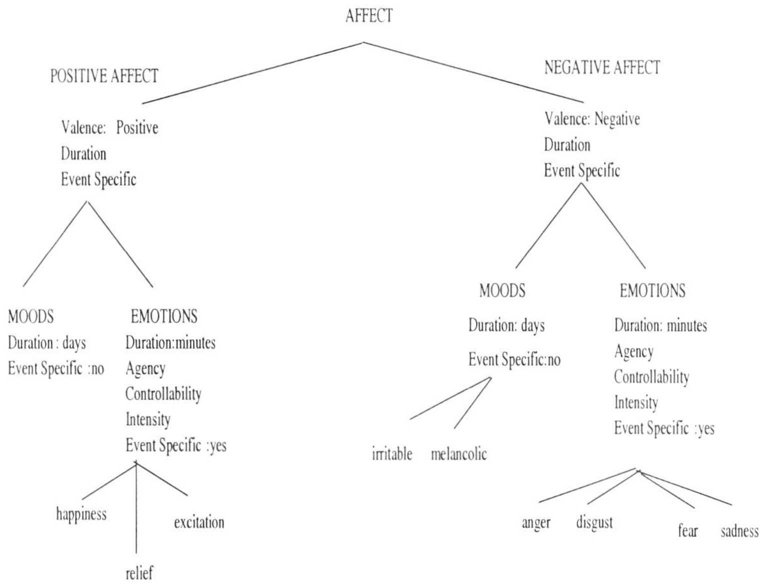

Figure 5.1. An example taxonomy of emotional states.

information about the other agents' state of knowledge, which is also potentially relevant. Given these complexities it is clear that a mechanism for managing the agent's computational resources is needed. In our work, we exploit emotional states and personality, as defined below, as a tool providing for such ability.

3. EMOTIONAL STATES AND PERSONALITY

As we mentioned, we view emotional states as different modes of decision-making, defined above. The set of emotional states and transitions between them comprise the agent's personality, which we define further below.

A simple taxonomy (Ortony et al., 1988) of emotional states in Figure 5.1 is aimed at differentiating among emotional states by using values of well-defined attributes. In its leaves, the taxonomy includes the basic, also called primitive (O'Neill, 2000), emotions of happiness, sadness, fear and anger. Each of these predisposes the agent to make decisions in a different way, and we associate each of them with a different decision-making situation:

Definition 2: An emotional state of an agent is associated with its decision-making situation $D =< P_c(S), A, Proj, U >$, defined above.

The above definition *associates* an emotional state with a decision-making situation, but it does not *identify* the two to be the same. The reason is that emotions, apart from being decision-making modes, may involve physical and other changes (say breathing rate, muscle tension, dilation of the pupil or rate of power consumption.) However, this definition does allow us to describe different emotions in terms of the elements of the decision-making situation, and naturally illustrates the impact emotions have on an agent's behavior. For example, an emotional state characterized by a low utility U assigned to all states, which could be called sadness, will have a well-defined effect on the behavior the agent will arrive at. We detail several other ways of modifying the elements of the decision-making situation that can be associated with various emotions later. Now we go on to define personality.

We conceptualize an agent's personality as the set of emotional states the agent is capable of being in, the transformations that are possible between these states, and the events that trigger the transformations.

Definition 3: An agent's personality is a finite state machine $P =< \mathbf{D}, \mathbf{IN}, \Delta, N >$, where

- $\mathbf{D}$ is a finite set of emotional states, defined above,
- $\mathbf{IN}$ is a set of environmental inputs,
- Δ is an emotional transformation function, $\Delta : \mathbf{D} \times \mathbf{IN}^* \rightarrow \mathbf{D}$,
- $N \in \mathbf{D}$ is an initial (or neutral) emotional state.

The above definition specifies that the agent's emotional state can change due to a, possibly empty, sequence of environmental inputs. This means that emotions are passive - they happen to the agent without the agent's control. As we mentioned, in biological systems they are built-in and are of essential survival value in emergency situations that are too important to rely on cognition (O'Neill, 2000). In artificial systems, the personality construct we defined above is a method controlling deliberative decision-making imposed by the agent's designer. The major challenge is to establish what kinds of personalities are most beneficial, given the range of environments the agent is to operate in.

The definition of personality above formalizes this notion from the individual agent's point of view, i.e., it describes the personality of the agent itself. To equip the agent with a tool with which to model emotional states of other agents we define a related notion of a personality model.

Definition 4: A personality model of agent R is a probabilistic finite state machine $P_R =< \mathbf{D}, \mathbf{IN}, \Delta, N >$, where

Figure 5.2. Simple personality model of a tit-for-two-tats agent.

- **D** is a finite set of emotional states of agent R,
- **IN** is a set of environmental inputs,
- Δ is a probabilistic transformation function, $\Delta : \mathbf{D} \times \mathbf{IN}^* \times \mathbf{D} \rightarrow [0, 1]$,
- $N \in \mathbf{D}$ is an initial (or neutral) emotional state of agent R.

An agent that has a personality model of another agent, R, can use this model to probabilistically predict R's emotional state, given an initial state and an environmental input. The transformation function is probabilistic to allow for uncertainty as to the next emotional state of the modeled agent. It assigns a probability $\Delta(N, IN, D)$ to a possible transformation between emotional states N and D due to an environmental input IN. The main advantage of using this approach is that a personality model can be learned, given limited amount of observations of the other agent's behavior, using an unsupervised US-L learning algorithm (see, for example (Carmel and Markovitch, 1996; Littman, 1994; Sandholm and Crites, 1995).)

In Figure 5.2 we present a very simple example of a personality model. In it, the transformation function happens to be Boolean, and there are only three emotional states: COOPERATIVE, SLIGHTLY ANNOYED, and ANGRY. The transitions among the states are caused by environmental inputs which are divided into Cooperative and Uncooperative ones. Using this dynamic model one can predict that an agent that is in COOPERATIVE emotional state will become SLIGHTLY ANNOYED given Uncooperative input. Further, the emotional state of SLIGHTLY ANNOYED will evolve into ANGRY if another Uncooperative response follows.

In Figure 5.2, if the COOPERATIVE state is the initial (neutral) state, then this personality model corresponds to the Tit-for-Two-Tats strategy widely investigated in the literature on repetitive prisoner's dilemma game (Axelrod, 1984; Carmel and Markovitch, 1996; Sandholm and Crites, 1995). Here, all three emotional states radically change the agent's decision-making model by limiting the alternative behaviors to cooperative (executed in the COOPERATIVE and SLIGHTLY ANNOYED states) to uncooperative (executed in the ANGRY state.)

We now present an initial approach we have taken to defining classes of emotional states in terms of the elements of the agent's decision-making situation. We find it intuitive to describe them as differing from, or transforming, the elements defining the "neutral" emotional state: $N =< P_c(S), A, Proj, U >$.

3.1. TRANSFORMATIONS OF THE ACTION SPACE A

Transformation of the action space A, for example by narrowing the set of alternative actions considered to encompass only a small subset of all of the actions, predisposes the agent to take action from this smaller set. This constitutes the *action tendency* that the emotion is invoking in the agent, as postulated, for example, by Fridja in (Fridja, 1986). In the extreme, narrowing the set A to a single action implements a behavioral condition-response rule, as in the emotional states COOPERATIVE, SLIGHTLY ANNOYED, and ANGRY of the Tit-for-Two-Tats agent above.

Formally, these are emotional transformations $\Delta(N, IN) = D$ such that $N =< P_c(S), A, Proj, U >$, $D =< P_c(S), A', Proj, U >$. The new emotional state, D, corresponds to an action tendency if $A' \subset A$. For example, an agent becoming angry may result in it considering only a subset of its behavioral alternatives – the ones of aggressive nature. A special case is when A' is a singleton set, as in the Tit-for-Two-Tats above.

Another intuitive special case of such transformation is one that results in the agent's deliberating in a more short-term fashion, such as it being rushed or panicked under time pressure. Formally we have: $\forall a_i^{'} \in A' : t_{a_i^{'}} \leq t_{a_i}$, which states that the time horizon of alternative plans considered has diminished. This is characteristic of human decision-makers; people frequently become more short-sighted when they are rushed or panicked, since they have no time to consider long-term effects of their alternative behaviors.

3.2. TRANSFORMATIONS OF THE UTILITY FUNCTIONS U

Intuition behind this transformation is that emotions both implement U, as well as modify it. Humans evaluate desirability of states by having positive or negative feelings about them. Positive or negative emotions or moods may alter these evaluations by, say, decreasing them, as in melancholic or depressed moods (when everything looks bleak), or increasing them, as in elated or happy moods. Other emotional states can change the weights of the factors contributing to the utility ratings (Equation 2). The agent's behavior resulting from these transformations may be one of inaction, when all of the behaviors available lead to seemingly worthless results, or recklessness, when some factor regarded as important previously in the utility function has its weighting reduced.

Formally, these are transformations $\Delta(N, IN) = D$ such that $N =< P_c(S), A, Proj, U >$, and $D =< P_c(S), A, Proj, U' >$. A special case of sadness is when the desirability of every state diminishes: $\forall s \in S : U'(s) \leq U(s)$.

3.3. TRANSFORMATIONS OF THE PROBABILITIES OF STATES

The intuition behind this transformation is that changing these probabilities, for instance by simplifying them, can be helpful and save time under time pressure. The most radical simplification is one that makes the most likely state to be the only possible state. This corresponds to considering only the most likely result of action and neglecting all less likely states, which is often observed in human decision-makers.

Formally, these are transformations $\Delta(N, IN) = D'$ such that $N =< P_c(S), A, Proj, U >$, $D' =< P_c^{'}(S), A, Proj^{'}, U >$. The special case described above obtains is when the probability distribution $P_c^{'}$, as well as every projected distribution $P_i^{'}$ returned by the projection function $Proj^{'}$ are deterministic.

4. CONCLUSIONS AND FUTURE WORK

In this paper, we intended to show how two diverse strands of work in AI and cognitive science can be combined to create a fundamental theory of emotions and personality. We started from the principled paradigm of rational agent design based on decision theory. We then used this paradigm to formally define the emotional states and personality of an artificial intelligent agent. We view the emotional states as the agent's decision-making modes, predisposing the agent to make its choices in a specific, yet rational, way. Change of the emotional state, say due to an external stimulus, invokes a transformation of the agent's decision-making behavior. We defined personality as consisting of the agent's emotional states and specification of transitions taking place among the states. We also defined a notion of a personality model, which is a probabilistic and learnable version of the personality construct.

Having the formal definitions of personality and emotions allows us to show how, and why, they are useful to a rational artificial agent. First, by modifying the decision-theoretic model used for deliberative rationality, emotions allow the agent to control the allocation of its cognitive resources and the complexity of its deliberations under time pressure in an uncertain environment. For example, limiting the number of alternative behaviors considered, or shortening the time horizon of these alternatives allows for more rapid decision-making. Second, emotions and terms related to personality are valuable when the agent finds it useful to inform other agent(s) about its own internal state. Instead of having to describe the details of its internal state, and running the risk of being misunderstood if the other agents are engineered differently, the agent can use more abstract and universal terms. For example, notions of stress or panic may be convenient to express the fact that the urgency of the situation forced the agent to look at only short-term effects of its actions. Or, say, anger, may be a good way to summarize the agent's belief that only hostile behavioral alternatives are worth considering and putting into action. Thus, in the context of communication with other agents, the formal definitions of emotions provide semantics of terms with which the agents can express their own internal states and understand the states the other agents are in. Third, well-defined emotional states of self and others are crucial in the agent's interaction with humans. Frequently, human-computer interaction is impeded by the machine being hopelessly out-of-step with the emotional state of the human user. However, since the user's emotional state can be now assessed using measurable and inferred factors, it is important for the machine to understand the user's emotional state and model its effects on the user's decision-making and his/her tendency to action.

The result of our work is a fundamental and principled understanding of the notions of personality and emotions in artificial intelligent agents. This, we believe, will significantly enhance the applications of intelligent systems in dynamic and unpredictable environments, multi-agent systems, and in human-computer interaction.

Our future work will provide clearer definitions of emotional states as decision-making modes, use these definitions to design useful personalities of agents operating in various environments, and to investigate robust methods of learning of personality models used during interaction and communication with other agents. We will investigate the effects of various modifications of the basic decision-making model, as defined above, on the abilities of the agents to accomplish goals under conditions of emergency and time pressure. Further, we will go beyond the simple transformations exemplified above, and create a comprehensive catalogue of transformations and their mixtures. We will experiment with agents implemented in the simulated environments of Tile-World, Wumpus Environment, and MICE. The agents will be made to satisfy given goals but act under time pressure and face emergency situations.

Also needed are extensive experiments with agents endowed with simple personalities, acting under moderately variable simulated environments. We will concentrate on agents under time pressure and attempt to determine what emotional states and decision-making modes are most useful, given parameters of a moderately variable environment. We expect that more complex personalities, with more five and more emotional states, will be needed to deal with complex environments, calling for the agents to be, for example, curious in their explorations, but also ready to go into panic mode under time pressure, and exhibiting angry and aggressive behavior when threatened.

5. ACKNOWLEDGEMENTS

This research has been supported by the Office of Naval Research grant N00014-95-1-0775, and by the National Science Foundation CAREER award IRI-9702132.

Notes

1. Such agents are sometimes called self-interested.
2. The probabilities are, therefore, subjective.

References

Axelrod, R. (1984). *The Evolution of Cooperation.* Basic Books.

Bicchieri, C. (1993). *Rationality and Coordination.* Cambridge University Press.

Boutilier, C., Dean, T., and Hanks, S. (1999). Decision-theoretic planning: Structural assumptions and computational leverage. *Journal of Artificial intelligence Research*, 11:1–94.

Carmel, D. and Markovitch, S. (1996). Learning models of intelligent agents. In *Proceedings of the National Conference on Artificial Intelligence*, pages 62–67, Portland, OR.

Cohen, P. R. and Levesque, H. J. (1990). Rational interaction as the basis for communication. In Cohen, P. R., Morgan, J., and Pollack, M. E., editors, *Intentions in Communication.* MIT Press.

Cottrell, G. and Metcalfe (1991). Empath: Face, emotion and gender recognition using holons. In *Advances in Neural Information Processing.* Morgan Kaufman Publishers.

Dennett, D. (1986). Intentional systems. In Dennett, D., editor, *Brainstorms.* MIT Press.

Dimasio, A. R. (1994). *Descartes' Error.* Grosset/Putnam.

Doyle, J. (1992). Rationality and its role in reasoning. *Computational Intelligence*, 8:376–409.

Durfee, E. H., Lee, J., and Gmytrasiewicz, P. (1993). Overeager rationality and mixed strategy equilibria. In *Proceedings of the National Conference on Artificial Intelligence.*

Fridja, N. H. (1986). *The Emotions.* Cambridge University Press.

Gmytrasiewicz, P. J. and Durfee, E. H. (2000). Rational coordination in multi-agent environments. *Autonomous Agents and Multiagent Systems Journal.*

Gmytrasiewicz, P. J. and Durfee, E. H. (2001). Rational communication in multi-agent environments. *Autonomous Agents and Multiagent Systems Journal, in print.*

Hayes-Roth, B., Ball, B., Lisetti, C., and Picard, R. (1998). Panel on affect and emotion in the user interface. In *Proceedings of the 1998 International Conference on Intelligent User Interfaces*, pages 91–94.

James, W. (1884). What is an Emotion? *Mind*, 9:188–205.

James, W. (1894). The Physical Basis of Emotion. *Psychological Review*, 1:516–529.

Johnson-Laird, P. N. and Oatley, K. (1992). Basic Emotions, Rationality, and Folk Theory. *Cognition and Emotion*, 6(3/4):201–223.

Kraus, S. and Sycara, K. (1989). Argumentation in negotiation: A formal model and implementation. *Artificial Intelligence*, 104(1-2):1–69.

Lesser, V., Atighetchi, M., Benyo, B., Bryan Horling, R., Anita, V., Regis, W., Thomas, P., Xuan, S., and Zhang, Z. (1999). The intelligent home testbed. In *Proceedings of the Autonomy Control Software Workshop (Autonomous Agent Workshop).*

Leventhal, H. and Scherer, K. (1987). The relationship of emotion to cognition:a functional approach to semantic controversy. *Cognition and Emotion*, 1(1):3 – 28.

Lisetti, C. L. and Rumelhart, D. E. (1998). Facial expression recognition using a neural network. In *Proceedings of the 1998 Florida Artificial Intelligence Research Symposium (FLAIRS'98).*

Lisetti, C. L. and Schiano, D. J. (1999). Automatic facial expression interpretation: Where human interaction, artificial intelligence and cognitive science intersect. *Pragmatics and Cognition, Special Issue on Facial Information Precessing and Multidisciplinary Perpective.*

Littman, M. L. (1994). Markov games as a framework for multi-agent reinforcement learning. In *Proceedings of the International Conference on Machine Learning.*

O'Neill, B. (2000). Approaches to modelling emotions in game theory. Technical report, Department of Political Science, Stanford University, http://www.stanford.edu/~boneill/emotions.html.

Ortony, A., Clore, G., and Collins, A. (1988). *Cognitive Structure of Emotions.* Cambridge University Press.

Picard, R. (1997). *Affective Computing.* MIT Press.

Rosenschein, J. S. and Zlotkin, G. (1994). *Rules of Encounter.* MIT Press.

Russell, S. and Norvig, P. (1995). *Artificial Intelligence: A Modern Approach.* Prentice Hall.

Sandholm, T. and Crites, R. H. (1995). Multiagent reinforcement learning and iterated prisoner's dilemma. *Biosystems Journal*, 37:147–166.

Simon, H. (1967). Motivational and Emotional Controls of Cognition. *Psychological Review*, 74:29–39.

Sloman, A. (1990). Motives, Mechanisms, and Emotions. In Boden, M., editor, *The Philosophy of Artificial Intelligence.* Oxford: Oxford University Press.

Sloman, A. and Croucher, M. (1981). Why robots will have emotions. In *Proceedings of the Seventh IJCAI Vancouver, B.C.*, pages 197–202. San Mateo, CA: Morgan-Kaufmann.

Sycara, K. (1998). Multiagent systems. *AI Magazine*, 10(2):79–93.

Wooldridge, M. and Anand Rao, E. (1999). *Foundations of Rational Agency.* Kluwer Academin Publishers.

Chapter 6

IMPLEMENTING A DECISION-THEORETIC APPROACH TO GAME THEORY FOR SOCIALLY COMPETENT AGENTS

Piotr J. Gmytrasiewicz

Department of Computer Science, University of Illinois at Chicago, 851 S. Morgan M/C 154 Chicago, IL 60607.

piotr@cs.uic.edu

Sanguk Noh

Department of Computer Science, University of Missouri–Rolla, 1870 Miner Circle, Rolla, MO 65409.

noh@umr.edu

Abstract We describe an implementation of decision-theoretic paradigm of expected utility maximization applied to design rational socially competent agents. Our implementation uses a frame-based knowledge base which explicitly represents what the agent knows about the world, what it knows about the other agents, what it knows about what they know, and so on. We argue that this representation has to include uncertainty, since an agent does not have a direct access to the other agents' knowledge states. We concentrate on realistic cases of finitely nested knowledge states, and, for these states, we briefly illustrate a dynamic programming solution method for our representation. The solution allows the agent to process the representation of its state of knowledge and arrive at an assignment of expected utilities to physical and communicative actions it can execute. The agent's executing actions that maximize its expected utility leads to it being rational while coordinating and communicating with the other agents. We briefly summarize results we obtained of coordination and communication in interactions of our agents among themselves and with human subjects.

Keywords: Game theory, decision theory, multi-agent systems, coordination, communication, air defense.

1. BACKGROUND ON DECISION-THEORETIC APPROACH TO GAME THEORY

The goal of our research is to implement artificial agents that exhibit social competence. We define social competence as an agent's ability to coordinate its actions with actions of other agents sharing its environment, and its ability to effectively communicate with them if needed or desirable. We define coordination as a choice of action that takes into account the anticipated actions of the other agents. By effective communication we understand the agent's ability to choose, among possibly large number of alternatives, the proper communicative action – one with the content best suited to the situation at hand, and one that is transmitted using the most appropriate language encoding and communication medium. The artificial agents we are interested in should be capable of competent social interactions not only within a homogeneous group of agents of same design; they should be able to interact with artificial agents of radically different designs and with human beings. Further, they should not need to rely on the assumption that the other agents are cooperatively predisposed; they should cooperate with others when it is appropriate, but be able to act in their own best interest if the situation demands it.

We believe that the paradigm of decision theory (Chernoff and Moses, 1959; Coles et al., 1975; Doyle, 1992; Feldman and Sproull, 1977) is able to provide the set of principles guiding the development of such socially competent agents. Our reason for turning toward decision theory is that it offers an answer to one of the central issues in designing intelligent agents, as articulated by Newell (Newell, 1981) within his knowledge-level view: How should an agent represent knowledge, and how should it operate on it to arrive at purposeful, rational action? Decision theory provides an answer by postulating that probability (intuitively, what an agent knows) (Cheeseman, 1985; Pearl, 1988) and utility (i.e., what an agent prefers) (Fishburn, 1981; Wellman, 1991) be combined to define rationality as the agent's behavior that maximizes its expected utility (Rasmusen, 1989; Russell and Norvig, 1995). It can be shown (see (Rasmusen, 1989) and (Russell and Norvig, 1995)) that, if the agent's beliefs are consistent with the axioms of probability theory and its preferences are consistent with the axioms of utility theory, then the agent should make its decisions so as to maximize its expected utility.

The issue of a rational agent interacting with other rational agents is one that is central to game theory. The main stream of research in game theory assumes, and attempts to refine, the central game-theoretic solution concept of Nash equilibrium. We, on the other hand, are taking the knowledge level perspective, understood in decision-theoretic terms, which leads toward a less known area of game theory, called a decision-theoretic approach to game theory (Kadane and Larkey, 1982; Raiffa, 1982). The difference between the two is illustrated by the following quote, due to Myerson (Myerson, 1991):

> The decision-analytic approach to player i's decision problem is first to assess some subjective probability distribution to summarize i's beliefs about what strategies will be used by the other players and then to select a strategy for i that maximizes his expected payoff with respect to these beliefs. [...] In contrast, the usual game-theoretic approach is to analyze and solve the decision problems of all players together, like a system of simultaneous equations in several unknowns.

Clearly, the strategy of first assessing a probability of other agents' alternative actions, and then selecting the agent's own best strategy can be applied recursively to the other agents as well since their own actions depend on what they, in turn, expect of others, including the original agent. While, as Myerson points out, it seems that this recursion leads to circularity, we do not believe this is the case. The reason is that the solution of the agent's own decision problem, and what the agent thinks another agent thinks is solution to the original agent's decision problem, are arrived at based on different bodies of information. The first problem is solved based on the information the original agent has about the environment and about other agents, i.e., based on its own decision-making situation. The second problem, on the other hand, is solved based on the information the original agent has about what the other agent knows about the decision-making situation of the original agent. Clearly, these two do not have to coincide. In particular, it may be that the original agent, while knowing its own decision-making situation quite well, has less information about what another agent knows about the original agent's decision-making situation. The postulate that the agents do run out of information at some level of nesting is the basis of our implementation of the solution using dynamic programming, as we describe briefly below.

From another perspective, the difference between the traditional game theory and decision-theoretic approach to multi-agent interactions stems from the different questions being asked. As an example, consider the task of mechanism design, for which traditional game theory is perfectly suited. The mechanism (for example auction, mediation, or negotiation) is to be imposed on individuals while interacting with each other, but these individuals are not being designed or otherwise controlled by the game theorists (usually, they are humans). A mechanism is judged as good, typically, when it it is expected to result in desirable properties of the multi-agent system as a whole, when, and if, the system reaches equilibrium (see (Fudenberg and Tirole, 1991) Chapter 7). Here, the equilibrium is defined without any reference to the state of knowledge of the agents involved. In the decision-theoretic approach, on the other hand, the central question is framed from the individual agent's point of view: Given my state of knowledge of the multi-agent environment, what is my rational decision? Thus, while the AI knowledge-level problem of how an agent is to represent and process its available knowledge is peripheral to traditional game theory, it is central to decision-theoretic approach.

Our interest in decision-theoretic approach is justified not only for those taking the AI knowledge-level perspective, however. Over the years if has become increasingly clear that the assumption of common knowledge, which underlies the concept of Nash equilibrium, is problematic (Aumann and Brandenburger, 1995; Bicchieri, 1993; Binmore, 1982; Brandenburger, 1992; Geanakoplos, 1992; Kadane and Larkey, 1982; Reny, 1988; Tan and Werlang, 1988). Similar results have been arrived at in the field of distributed systems. Halpern and Moses (Halpern and Moses, 1990) have proven that arriving at the state of common knowledge in a realistic distributed environment is impossible. Take a simple realistic example situation of agents that have interacted with other agents for a finite amount of time, and therefore could have exchanged only a finite number, say K, of messages. Clearly, the information they have about each other cannot possibly be nested deeper then level K, unless they jumped into unjustified conclusions (see the proof in (Halpern and Moses, 1990) for details). Thus, in such realistic situations, there is a level of nesting in the agents' beliefs about each other below which the agents have *no information* about each other.

Finally, it could be postulated that while it is not strictly justifiable for the agents to assume common knowledge and use the game-theoretic equilibrium solution concept, it may be convenient for them to do so and rely on the rich body of research in traditional game theory. While it may be the case in some domains, we found that this approach encounters problems with multiple equilibria. In the domains we studied, for example in the pursuit game, the cooperative assembly, and in the air defense domain, there are usually many equilibria and no clear way for the agents to choose among them. In particular, in the air defense domain there are as many equilibria as there are alternative targets that can be intercepted, rendering this approach useless as the strategy to select the best target to shoot at next.

In the following section, we summarize our work in implementing the knowledge bases of agents as a hierarchy of concepts/frames. Our design accommodates information the agents have about the environment and about the other agents, including the content of their KB's. Further, our implementation includes uncertainty and lack of information the agents are inevitably faced with during realistic finite interactions. We then go on to the decision-theoretic method that uses the information in the KB, implemented as our Recursive Modeling Method (RMM) (Gmytrasiewicz and Durfee, 2000; Gmytrasiewicz and Durfee, 1995). We motivate the dynamic programming solution of the agent's decision problem. We briefly mention results of coordination with and without communication among our agents, and compare them to performance of coordinating human teams.[1]

2. KNOWLEDGE BASE DESIGN

Our design of the knowledge base (KB) is based on work on frame-based (Brachman and Schmoltze, 1985; Karp et al., 1995) and object-oriented (Patel-Schneider, 1990; Yelland, 1993) knowledge representation formalisms (see also (Brachman and Levesque, 1985) and references therein). These formalisms postulate that the KB be organized as a set of interrelated frames representing classes, i.e., sets of entities, and instances, i.e., the individual entities themselves. The frames representing the classes form a superclass/subclass hierarchy (ontology) allowing for usual inheritance of properties, while the leaves of the hierarchy are occupied by instances of classes identified in the agent's environment. An example of such KB for the air defense domain, depicted in Figure 6.1, is presented in Figure 6.2.

A fundamental limitation of the frame formalisms is that they do not support uncertainty, which, as we mentioned, is ubiquitous and particularly important for socially competent agents. First, the properties of objects in the environment may be not known with certainty. Second, to effectively interact, the agents have to be able to represent and reason about limits of knowledge of other agents, uncertainty of others' knowledge, their own uncertainty as to others' states of knowledge, and so forth.

The framework widely used for uncertainty reasoning is that of Bayesian networks (Pearl, 1988), which represent probabilistic relations among uncertain variables describing the domain at hand. Our KB design builds on an effort that overcomes some of the limitations of frame-based sys-

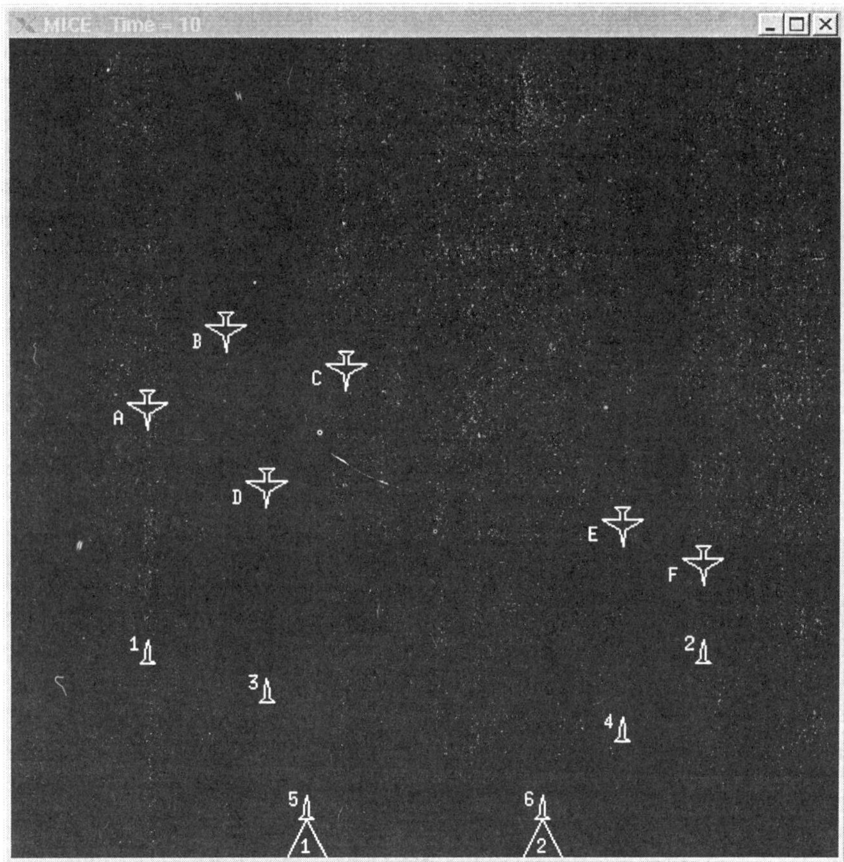

Figure 6.1. A simulation of the air defense domain.

tems by combining them with Bayesian networks (Koller and Pfeffer, 1995; Koller and Pfeffer, 1998). The basic idea is to treat the slots of frames (or attributes of objects) describing the properties that may not be known with certainty as nodes of a Bayesian network, or its extension, called an influence diagram (see, for example (Neapolitan, 1990) for introduction to Bayesian nets and influence diagrams.) Such probabilistic slots allow values in form of probability distributions, and further contain information about the slots' parents in the influence diagram, as well as the conditional probability tables that allow the probabilities to be updated in response to change in the parents' probabilities.

For example, in KB used by the Battery1 in scenario depicted in Figure 6.1, the speed and the range of the missile objects may be uncertain, subject to inaccuracies of the radar systems. Further, an interceptor

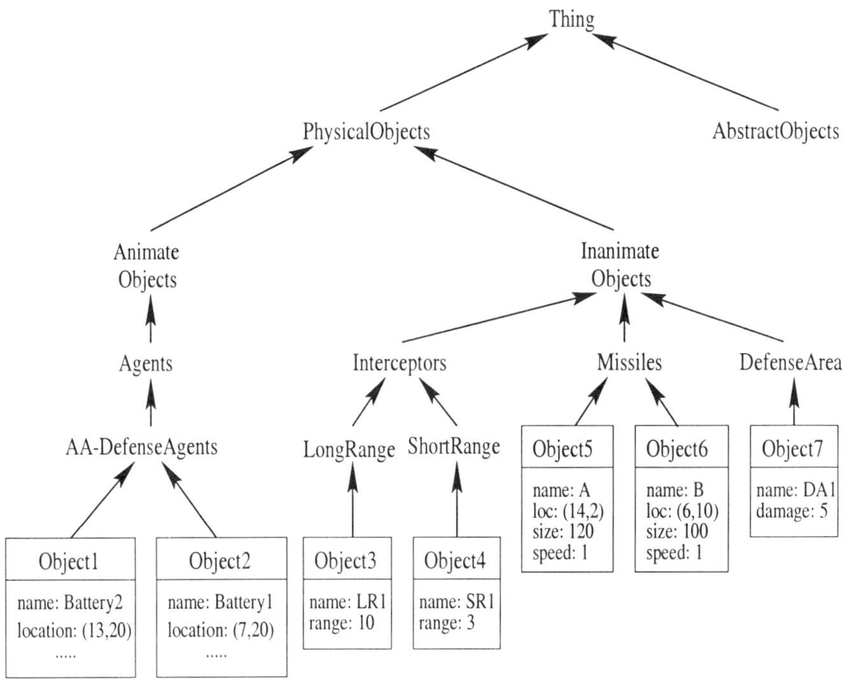

Figure 6.2. A high-level view of a KB design for the air defense domain scenario.

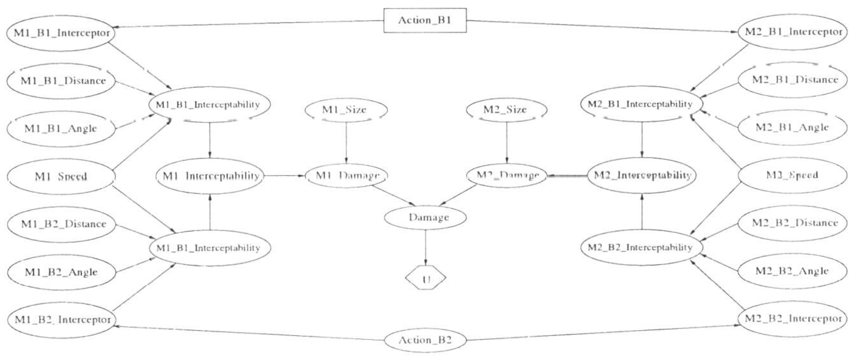

Figure 6.3. Influence diagram spanning the KB of Battery1.

launched against a missile may destroy it only with some probability, which, in turn can cause the missile to penetrate the defenses, strike its intended target, and cause damages, magnitude of which is also uncertain. The relations among these and other uncertain attributes are depicted in a network in Figure 6.3.

While it is clearly useful to combine the frame-based and Bayesian representations to overcome the limitation of traditional frames in representing uncertainty, the other important advantage is in making it possible to generate the Bayesian representation of decision-making situation on-the-fly. In our implementation, for example, as new objects are identified by the agent's sensory systems, they are automatically represented as objects belonging to appropriate classes in the frame-based KB, and automatically become part of the influence diagram representation of the agent's decision-making situation.

The following theorem establishes a correspondence between influence diagrams, contained in the agent's probabilistic frame-based KB, and payoff matrices used by RMM for reasoning in multi-agent situations:

Theorem 1: Each influence diagram has a corresponding and unique payoff matrix representing the same decision-making situation.

Proof (Outline): The proof proceeds by repeated absorption (following (Shachter, 1989)) of the nature nodes[2] other than the node(s) representing the actions of the other agent(s) in the influence diagram. The information contained in the absorbed nodes is compiled into the probabilities and values of the remaining nodes. When the only nodes remaining are the decision node, the utility node, and the nodes representing the actions of the other agent(s), it is straightforward to tabulate the values of the utility node depending on the values of the agents' action nodes and to form a payoff matrix.

An example matrix for the air defense scenario in Figure 6.1 is depicted on top of Figure 6.5. The rows and columns are labeled with the alternative actions of the agents, and the entries in the matrix are payoffs (i.e., the expected utilities) resulting from the agents' executing their corresponding alternatives. For example, the entry 65.1 in the row labeled B and column labeled F means that the expected payoff (in terms of the threat reduction, as we explain in (Noh and Gmytrasiewicz, 1997; Noh and Gmytrasiewicz, 1998), also on http://dali.uta.edu) to Battery1 resulting from Battery1's attempting to intercept missile B and Battery2's attempting to intercept missile F is 65.1.

As we mentioned, the agent's information about the other agent's decision-making situation is also represented in the KB. In the scenario above we assumed that Battery1 is uncertain whether Battery2 has short range interceptors, or whether it has been incapacitated. Thus, Battery1 has three alternative models of the state of knowledge of Battery2, depicted in Figure 6.4. The models representing different decision-making situations of Battery2 can also be compiled, and these, together with the Battery1's own payoff matrix for RMM's recursive model structure, are depicted in Figure 6.5.

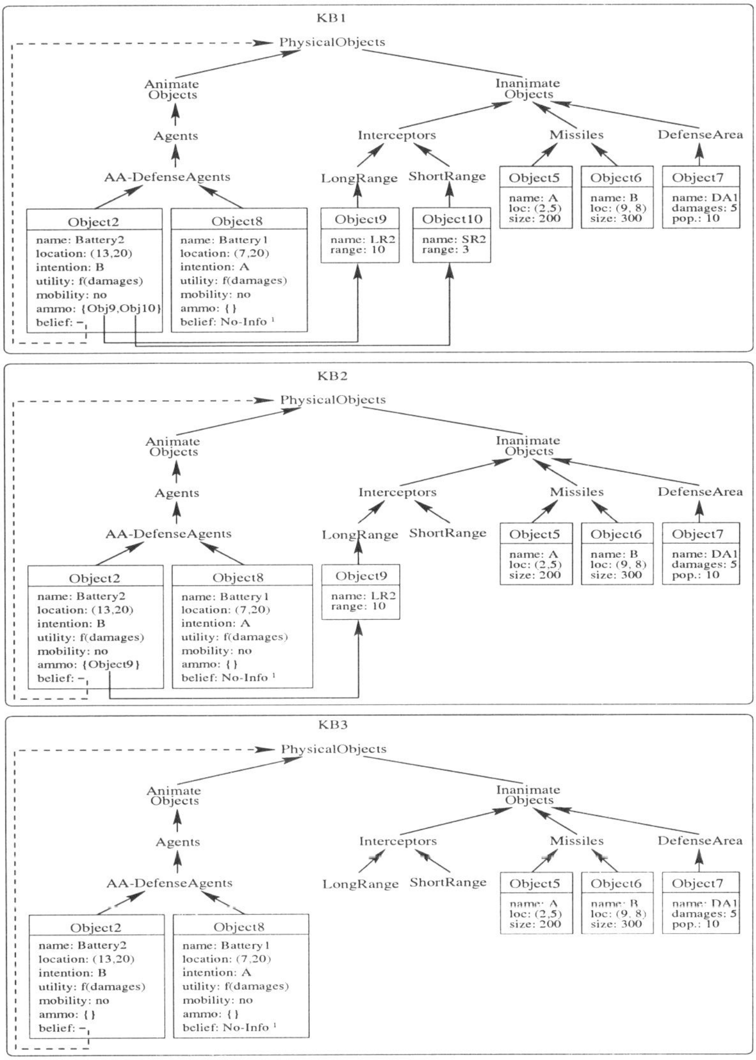

Figure 6.4. Three alternative models of Battery2's state of knowledge Battery1 maintains.

In Figure 6.5, the matrix corresponding to the possibility that Battery2 has both long and short range interceptors has seven rows, representing the Battery2's ability to intercept any of the six missiles visible, in addition to its possibly remaining still. The matrix representing the possibility that Battery2 does not have any short range interceptors has less rows since Battery2 cannot target the missiles that are too close to it, i.e., D, E and F in Figure 6.1. The third model on the second level in

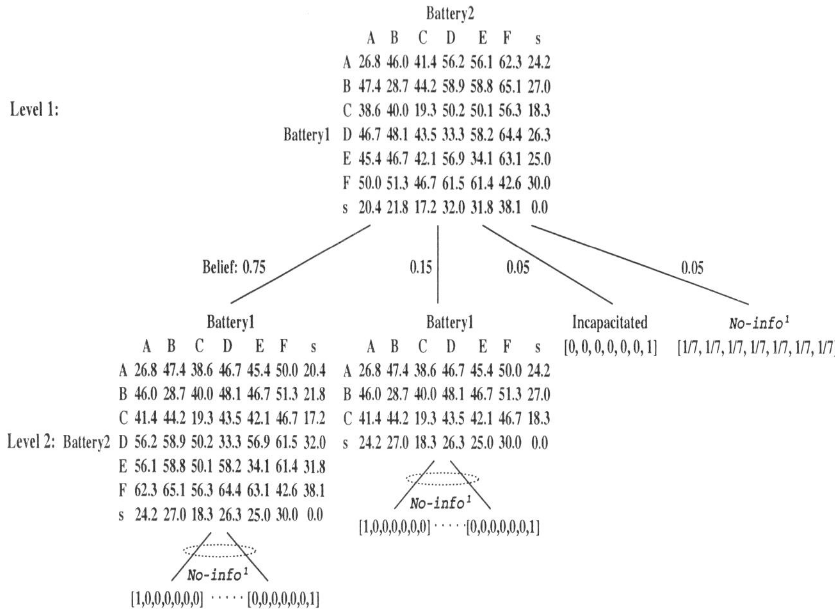

Figure 6.5. Hierarchy of payoff matrices corresponding to nested models of agents.

Figure 6.5 corresponds to the possibility that Battery2 is incapacitated. Since, in this case, Battery2 cannot attempt to intercept any of the missiles and has to remain stationary, this model is simply a probability distribution over Batter2's possible actions that assigns 100% probability that Battery2 will execute it's "s" option, i.e., stay put. The fourth model on the second level in Figure 6.5 is the No-Info^1 model. Since it represents the fact that Battery1 has no information it can use to model Battery2, this model assigns equal probability to each of the Battery2's seven alternative actions. The uniform probability distribution has a maximum entropy and contains zero information, correctly representing the state of Battery1's knowledge in this case.

3. RMM'S SOLUTION METHOD

We now briefly describe how dynamic programming method that can be used to solve recursive model structures, like the one in Figure 6.5; the reader is referred to (Gmytrasiewicz and Durfee, 1995; Gmytrasiewicz and Durfee, 2000) for details.

Dynamic programming is an appropriate implementation of decision-theoretic approach to games since the decision-making problem an interacting agent has to solve exhibits the properties of optimal substructure and overlapping subproblems, which are characteristic for dynamic programming (Cormen et al., 1990). Optimal substructure reveals itself when we notice that an agent making the optimal decision in presence of other agents has to solve their decision-making situations first. In other words, the optimal solution of the agent's decision-making problem contains, as subproblems, the optimal solutions to the decision-making problems the agent thinks the other agents are facing. The fact that other agents may have information about others, including the original agent, gives rise to recursive nesting of models. The dynamic programming solution explores this recursion and, after the recursion bottoms out, propagates the results of solving the models, such as the one in Figure 6.5, bottom-up.

The above procedure is applicable, of course, only if the recursive model structures, like the one in Figure 6.5, are finite and actually do bottom out. We argue that this is always the case. Since the agents have finite amount of information about others, and what others know about others, and so on, the nesting of models has to terminate when the agents run out of information. We represent it by using the No-Info models at the leaves of the recursive model structures.

The property of overlapping subproblems reveals itself when we realize that some of the alternative models the agents have of other agents may be identical to the models other agents have. While this is not the case in the structure in Figure 6.5, it is easy to see if we consider three interacting agents. Among the three agents, the first agent may think that some its models of the second agent are the same as some of the models the third agent uses to model the second agent.[3] Once the overlapping models are identified, they can be solved once and their solution can be cached and reused in other parts of the recursive model structure. In domains in which such overlap does exist, it can substantially reduce the complexity and run time.[4]

We now describe the solution of recursive model structures using dynamic programming using the structure in Figure 6.5 as an example. As we mentioned the solution proceeds bottom-up. Starting with the model of Battery2 on the left consisting of the 7×7 payoff matrix and a No-Info^1 model below it, we use sampling over the space of the alternative distributions in the No-Info^1 model, and, for each distribution compute the best action for Battery2.[5] For example, if the sampled distribution is $[0.9, 0.1, 0, 0, 0, 0, 0]$, then the expected utilities of Battery2's alternative actions A through F and s, are 28.72 ($= 26.8 \times 0.9 + 46.0 \times 0.1$), 45.53, 38.74, 46.84, 45.53, 50.13 and 20.27, respectively. This shows that is Battery2 thinks that Battery1 will execute its action A with probability 0.9 and its action B with probability 0.1, then the best action for Battery2 would be to intercept missile F, since it has the highest expected payoff in this case.

To summarize the results of processing all of the distributions in the No-Info[1] model, our sampling algorithm counts the proportion of distributions that favor each particular action of Battery2. It turned out that 78% of them favored Battery2's action F (as the distribution above), 10% favored action E, and 12% favored D. This means that, given that it is unknown what Battery2 expects Battery1 to do, one can assign the following probability distribution to alternative behaviors of Battery2: $[0, 0, 0, 0.12, 0.10, 0.78, 0]$. This is the case in which Battery2 has both kinds of ammunition. Similarly, processing the next model of Battery2 in Figure 6.5 according to which it has no short range interceptor, results in the distribution $[0.37, 0.58, 0.05, 0, 0, 0, 0]$, which means that, if Battery2 has no short interceptors, we can expect it to shoot at missile A with probability 37%, at missile B with probability 58%, and at missile C with probability 5%.

Now, using the four alternative models of Battery2, Battery1 can form an overall probability distribution describing Battery2's likely actions. The four probability distributions, describing Battery2's expected behavior in each of four models, are combined with weights representing the likelihood that each of these models is correct, i.e., 0.75, 0.15, 0.05 and 0.05, respectively. This procedure is called Bayesian model averaging (Heckerman and Chickering, 1996). The result is

$$[0.063, 0.094, 0.015, 0.097, 0.082, 0.592, 0.057]$$

which summarizes Battery1's overall expectation as to Battery2's alternative actions. Now, Battery1 can compute the expected utilities of its own alternative actions using its payoff matrix (on top on Figure 6.5) and the probabilities of Battery2's actions, as:

$$\begin{aligned} U_A = \; & 0.063 \times 26.8 + 0.094 \times 46.0 + 0.015 \times 41.4 + 0.097 \times 56.2 + \\ & 0.082 \times 56.1 + 0.592 \times 62.3 + 0.057 \times 24.2 = 54.9460 \end{aligned}$$

Using the same probability distribution we can similarly compute: $U_B = 56.9601$, $U_C = 49.8316$, $U_D = 55.7424$, $U_E = 54.9772$, $U_F = 46.6022$ and $U_s = 31.8592$.

The final result shows that Battery1's best action is to attempt to intercept missile B, and to expect the payoff of 56.96. This conclusion is derived from processing all of the information Battery1 has about the environment and about Battery2 contained in the knowledge base. Based on the information it has, Battery1 formed expectation as to Battery2's likely actions, and chose its own action appropriately. The resulting action is, therefore, *coordinated* with action expected of Battery2.

4. RATIONAL COMMUNICATION

Our approach to rational communication associates a communicative act that can be executed with its *decision-theoretic (DT) pragmatics*, defined as the transformation of the state of knowledge about the decision-making situation (i.e., the recursive model structure described above) the act brings about (see (Gmytrasiewicz and Durfee, 2001) for more detailed discussion).[6] The transformation in the agent's decision-making situation may change the expected utilities of alternative actions. It is natural to identify the change of the expected utility brought about by a communicative action as the expected utility of this action itself.[7]

Formally, the expected utility of the communicative act M is defined as the difference between the payoff the agent expects before and after executing the act:

$$U(M) = U_{p^M}(Y) - U_p(X). \tag{1}$$

where $U_p(X)$ is the utility of the best physical action, X, expected before sending the message (in the example solved in previous section the best action was B and its expected utility was 56.96), and $U_{p^M}(Y)$ is the utility of the best physical action, Y, expected if a message were to be sent.

In considering the set of messages available to an automated agent, we classify the messages into types: intentional messages, modeling messages, questions, imperatives, etc. (Gmytrasiewicz et al., 1991; Gmytrasiewicz and Durfee, 2001). Intentional messages describe the intended actions of the agent, and modeling messages describe the environment or the status of the agent relevant to the achievement of its goal. Both types of messages are intended to remove some of the uncertainty present in the absence of communication.

The language we use is similar to KQML, and performatives provided in KQML, which include *Attempt* and *Assert* (Labrou and Finin, 1994). To send an intentional message our agents use the *Attempt* performatives, and to send a modeling messages, they use *Assert*. For simplicity, we assume here that the agents can mutually understand communication primitives, but it turns out that this assumption can be relaxed.

5. EXPERIMENTS IN AIR DEFENSE DOMAIN

In our implementation, the anti-air defense simulator with communication was written in Common LISP and built on top of the MICE simulator (Durfee and Montgomery, 1989). During the experiments two coordinating defense units were faced with an attack by seven incoming missiles. Therefore, the RMM agents used an 8×8 payoff matrix to represent the agents' decision-making situations. Each of the two defense units was assumed to be equipped with three interceptors each, if they were not incapacitated. Thus, they could launch one interceptor at a time, and did it three times during a course of one defense episode.

5.1. COORDINATION WITHOUT COMMUNICATION

Our experiments was aimed at determining the quality of modeling and coordination achieved by the RMM agents in a team, when paired with human agents, and when compared to other strategies. To evaluate the quality of the agents' performance, the results were expressed in terms of the total expected damage to friendly forces after all six interceptors were launched. The total expected damage is defined as a sum of the residual warhead sizes of the attacking missiles. Thus, if a missile was targeted for interception, then it contributed $(1 - P(H)) \times Size$ to the total damage. If a missile was not targeted, it contributed all of its warhead size value, T_j, to the damage.

The target selection strategies are as follows:

- Random: selection randomly generated.
- Independent, no modeling: selection of arg $\max_j\{P(H_{ij}) \times T_j\}$ for agent i.
- Human:[8] selection by human.
- RMM: selection by RMM.

The random agents were included to provide the worst-case base line of the system performance in our experiments. We included the independent agents to show what coordination can be expected when agents maximize but do not model each other in making their choices. It turned out that the ways that human agents choose a missile were different and sometimes quite arbitrary. For example, some of our human subjects shot only the 3 missiles coming to their own side of the grid world. This, however, was difficult when the missiles were clustered at the center area, which led to much duplicated effort. Others tended to choose

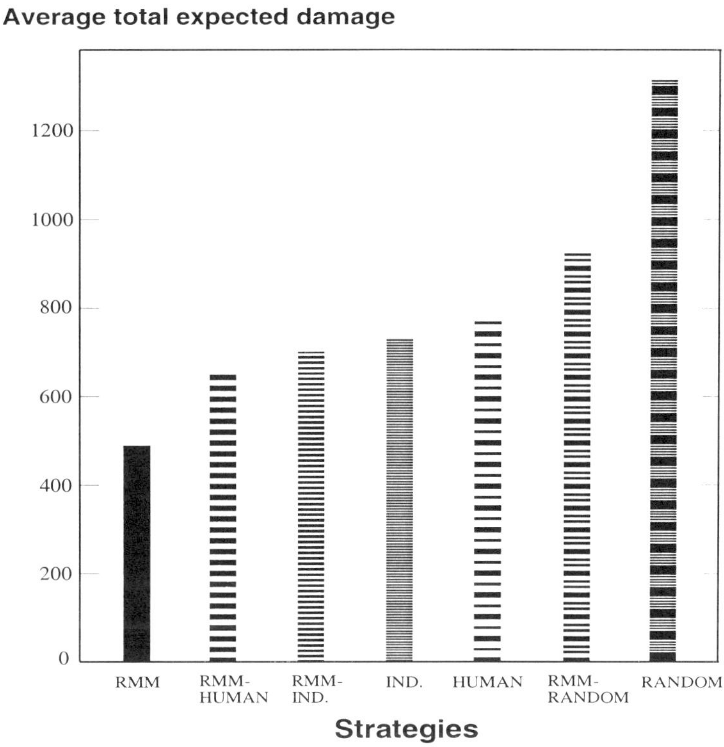

Figure 6.6. Average total expected damage (over 100 runs).

missiles with the largest missile size. Still others tried to consider the multiplication of the missile size and the hit probability, but did not model the other agent appropriately.

We experimented with the above policies to understand the agent interactions in two groups: heterogeneous teams of agents with the same policy and the mixed agent teams with different policies.

As shown in Figure 6.6, we found that the all-RMM team outperformed the human and independent teams. The human performance was very similar to the performance of independent agents. The most obvious reason for this is that humans tend to depend on their intuitive strategies for coordination, and, in this case, found it hard to engage in deeper, normative, decision-theoretic reasoning. The common reason for the lower score of human teams was the choice of the greatest threat simultaneously–they made redundant efforts to attack the same missile. This, again, suggests that human agents attempted to minimize their expected damage but did not model the decision making of the other human very well, while RMM agents were rationally optimizing given what they expected of the other agent. The performance of the RMM team was not perfect, however, since the agents were equipped with limited and uncertain knowledge of each other.

The above results show that the decision-theoretic approach to multi-agent interactions allows the automated agents to achieve high quality coordination in this unpredictable environment without relying on predefined protocols. As we argued, methods using traditional game-theoretic equilibria would not be sufficient to coordinate agents in this domain. A particularly promising facet of our results is that our implementation seems to provide a robust mechanism for modeling and coordination not only among RMM agents, but also with the human-controlled agents.

5.2. COORDINATION WITH COMMUNICATION

In the experiments with communication we ran, two defense units were faced with an attack by seven incoming missiles. Therefore, the RMM agents used an 8×8 payoff matrix to represent the agents' decision-making situations. For all settings, each defense unit was initially assumed to have the following uncertainties (beliefs) in its knowledge base: the other battery is fully functional and has both long and short range interceptors with probability 60%; the other battery is operational and has only long range interceptors with probability 20% (in this case, it can shoot down only distant missiles, which are higher than a specific altitude.); the other battery has been incapacitated by enemy fire with probability 10%; the other battery is unknown with probability 10%. Further, each battery has no deeper nested knowledge about the other agent's beliefs.

The warhead sizes of missiles were $470, 410, 350, 370, 420, 450$, and 430 unit for missiles A through G, respectively. In these experiments each of the two defense units was assumed to be equipped with three interceptors, if they were not incapacitated. Thus, they could launch one interceptor at a time, and did it three times during a course of one defense episode.

We set up 100 scenarios for RMM team and 20 scenarios for human team. We allowed for one-way communication between defense units before each salvo. If both agents wanted to send messages, the speaker was randomly picked in the RMM team, and the human team flipped a coin to determine who would be allowed to talk. The listener was silent and could only receive messages. For uniformity, in all of the anti-air defense scenarios, each battery was assumed to have a choice of the

Table 6.1 . The performances of RMM, human, and mixed teams.

Cases	Team (B1-B2)	No of selected targets	Total expected damage	ANOVA
Case I (B2:both ammo, w/o comm.)	RMM-RMM	5.95 ± 0.21	717.01 ± 110.71	f=3.45
	RMM-Human	5.70 ± 0.47	797.39 ± 188.35	
	Human-Human	5.75 ± 0.44	800.45 ± 147.69	
Case II (B2:both ammo, w/ comm.)	RMM-RMM	6.00 ± 0.00	652.33 ± 58.97	f=3.43
	RMM-Human	6.00 ± 0.00	717.93 ± 94.45	
	Human-Human	6.00 ± 0.00	710.20 ± 100.92	
Case III (B2:only long, w/o comm.)	RMM-RMM	5.83 ± 0.37	852.01 ± 160.79	f=6.96
	RMM-Human	5.75 ± 0.44	862.70 ± 120.19	
	Human-RMM	5.40 ± 0.50	895.92 ± 127.32	
	Human-Human	5.30 ± 0.47	997.32 ± 145.15	
Case IV (B2:only long, w/ comm.)	RMM-RMM	5.88 ± 0.32	787.42 ± 110.26	f=4.58
	RMM-Human	5.85 ± 0.36	842.50 ± 131.26	
	Human-RMM	5.75 ± 0.44	815.67 ± 133.60	
	Human-Human	5.80 ± 0.41	908.08 ± 103.25	
Case V (B2:incap. or random)	RMM-Incap.	3.00 ± 0.00	1742.22 ± 64.45	f=122.01
	Human-Incap.	3.00 ± 0.00	1786.86 ± 87.94	
	RMM-Random	4.86 ± 0.58	1079.52 ± 210.64	
	Human-Random	4.85 ± 0.59	1115.57 ± 228.94	

Note) For all of cases, Battery1 is fully functional. $f_{.05,2,57} = 3.15$, $f_{.01,3,76} = 4.13$.

following communicative behaviors: "No communication," "I'll intercept Missile A." through "I'll intercept Missile G.", "I have both long and short range interceptors.", "I have only long range interceptors.", or "I'm incapacitated.". As human subjects, we used 20 CSE and EE graduate students.

The summary of the experimental runs we performed is shown in Table 1. It presents the average number of selected targets and the average total expected damage by RMM agents after 100 trials and by human agents after 20 trials. We focus on the performances of three different teams: RMM-RMM, RMM-human, and human-human team. To see whether the differences in results obtained by the teams are not due to chance, we performed an analysis of variance (ANOVA) on the basis of total expected damage. For this purpose we randomly chose 20 trials of the RMM team among the 100 trials available.

In Case I of Table 1, both defense units are fully functional and there is no communication. Since the computed value $f = 3.45$ in ANOVA exceeds $3.15 (= f_{.05,2,57})$, we know that the three teams are not all equally effective at the 0.05 level of significance, i.e., the differences in their performance are not due to chance with probability 0.95. Results in Case II include communication between fully functional defense units. ANOVA test shows that the differences in performance are not due to chance with probability 0.95 again. When the communication was available, the performances achieved by three teams were improved, with the RMM team performing slightly better than the other teams. The experiments in Case III and IV intended to examine how fully functional battery can cope with the situation in which the other battery has only long range interceptors. Like Case I and II, the experimental results of Case III and IV proved the effectiveness of communication. The ANOVA test indicates that the observed differences in the performance of four teams for the target selection are significant at the 0.01 level of significance, and are not due to chance with probability 99%. In Case V, the battery controlled by human or RMM cannot take advantage of communication, because the other battery is incapacitated or behaves randomly. The huge value of $f = 122.01$ tells that the performances of four different teams are clearly distinct.

6. CONCLUSIONS

We have outlined the implementation of our research on decision-theoretic approach to decision-making in multi-agent situations. We argued that this approach to agent design complements the traditional game-theoretic approaches, mainly used for mechanism design. We presented the implementation of the knowledge base of the interacting agents that accommodates uncertainty and nested information agents may have about other agents. We then showed how our KB implementation interfaces with the decision-theoretic computations performed within the framework of Recursive Modeling Method. We briefly summarized some of our experimental results, which we believe, show the viability of our approach in our chosen domain.

Acknowledgements

This research has been supported by the Office of Naval Research grant N00014-95-1-0775, and by the National Science Foundation CAREER award IRI-9702132.

Notes

1. These results are available in more detailed version in (Gmytrasiewicz and Durfee, 2000; Gmytrasiewicz and Durfee, 2001).
2. The absorption operation is available in many of the Bayesian network tools, including Netica.
3. See our implementation of the pursuit coordination game for an example of such overlap, at http://dali.uta.edu.
4. In the pursuit game, taking advantage of overlapping subproblems allows us to reduce the running time of decision-making from exponential to polynomial.
5. As we mentioned, the alternative distributions represent different ways in which Battery2 may be modeling Battery1's intended action.
6. Since our formalism is probabilistic, it naturally handles cases when the meaning of a message is itself uncertain.
7. The notion of the utility of a message we use here differs from the notion of the value of information considered in decision theory (Pearl, 1988; Russell and Norvig, 1995). The latter expresses the value of information to its recipient. We, on the other hand, consider the value of a message to its sender, since, of course, it is the sender that makes the decision of whether and what, to communicate. The two notions coincide in two special cases: when the preferences of the agents perfectly coincide, and when a speaker requests information from the hearer by asking a question.
8. We should remark that our human subjects were simply CSE and EE graduate students who were informed about the criteria for target selection. We would expect that anti-air specialists, equipped with a modern defense doctrine, could perform better than our subjects. However, the defense doctrine remains classified and was not available to us at this point.

References

Aumann, R. J. and Brandenburger, A. (1995). Epistemic conditions for Nash equilibrium. *Econometrica.*

Bicchieri, C. (1993). *Rationality and Coordination.* Cambridge University Press.

Binmore, K. (1982). *Essays on Foundations of Game Theory.* Pitman.

Brachman, R. and Levesque, H. (1985). *Readings in Knowledge Representation.* Morgan Kaufman.

Brachman, R. and Schmoltze, J. (1985). An overview of kl-one knowledge representation system. *Cognitive Science*, 9:171–216.

Brandenburger, A. (1992). Knowledge and equilibrium in games. *Journal of Economic Perspectives*, 6:83–101.

Cheeseman, P. (1985). In defense of probability. In *Proceedings of the Ninth International Joint Conference on Artificial Intelligence*, pages 1002–1009, Los Angeles, California.

Chernoff, H. and Moses, L. E. (1959). *Elementary Decision Theory.* John Wiley, New York.

Coles, L. S., Robb, A. M., Sinclar, P. L., Smith, M. H., and Sobek, R. R. (1975). Decision analysis for an experimental robot with unreliable sensors. In *Proceedings of the Fourth International Joint Conference on Artificial Intelligence*, Stanford, California.

Cormen, T. H., Leiserson, C. E., and Rivest, R. L. (1990). *Introduction to Algorithms.* The MIT Press.

Doyle, J. (1992). Rationality and its role in reasoning. *Computational Intelligence*, 8:376–409.

Durfee, E. H. and Montgomery, T. A. (1989). MICE: A flexible testbed for intelligent coordination experiments. In *Proceedings of the 1989 Distributed AI Workshop*, pages 25–40.

Feldman, J. A. and Sproull, R. F. (1977). Decision theory and Artificial Intelligence II: The hungry monkey. *Cognitive Science*, 1(2):158–192.

Fishburn, P. (1981). *The Foundations of Expected Utility.* D. Reidel Publishing Company.

Fudenberg, D. and Tirole, J. (1991). *Game Theory.* MIT Press.

Geanakoplos, J. (1992). Common knowledge. In *Proceedings of the Conference on Theoretical Aspects of Reasoning about Knowladge*, pages 255–315. Morgan Kaufman.

Gmytrasiewicz, P. J. and Durfee, E. H. (1995). A rigorous, operational formalization of recursive modeling. In *Proceedings of the First International Conference on Multiagent Systems, ICMAS'95*, pages 125–132.

Gmytrasiewicz, P. J. and Durfee, E. H. (2000). Rational coordination in multi-agent environments. *Autonomous Agents and Multiagent Systems Journal.*

Gmytrasiewicz, P. J. and Durfee, E. H. (2001). Rational communication in multi-agent environments. *Autonomous Agents and Multiagent Systems Journal, in print.*

Gmytrasiewicz, P. J., Durfee, E. H., and Wehe, D. K. (1991). The utility of communication in coordinating intelligent agents. In *Proceedings of the National Conference on Artificial Intelligence*, pages 166–172.

Halpern, J. Y. and Moses, Y. (1990). Knowledge and common knowledge in a distributed environment. *Journal of the ACM*, 37(3):549–587.

Heckerman, D. and Chickering, D. M. (1996). A comparison of scientific and engineering criteria for bayesian model selection. Technical Report MSR-TR-96-12, Microsoft Research, Microsoft Corporation, Redmond, WA.

Kadane, J. B. and Larkey, P. D. (1982). Subjective probability and the theory of games. *Management Science*, 28(2):113–120.

Karp, P. D., Myers, K., and Gruber, T. (1995). The generic frame protocol. In *Proceedings of the Fourteenth International Joint Conference on Artificial Intelligence*, pages 768–774.

Koller, D. and Pfeffer, A. (1995). Object oriented bayesian networks. In *Proceedings of The Conference on Uncertainty in Artificial Intelligence.*

Koller, D. and Pfeffer, A. (1998). Probabilisitc frame based systems. In *Proceedings of the National Conference on Artificial Intelligence*, pages 580–587, Madison, WI.

Labrou, Y. and Finin, T. (1994). A semantics approach for KQML a general purpose communication language for software agents. In *Proceedings of the Third International Conference on Information and Knowledge Management.*

Myerson, R. B. (1991). *Game Theory: Analysis of Conflict.* Harvard University Press.

Neapolitan, R. E. (1990). *Probabilistic Reasoning in Expert Systems.* John Wiley and Sons.

Newell, A. (1981). The knowledge level. *AI Magazine*, 2(2):1–20.

Noh, S. and Gmytrasiewicz, P. J. (1997). Agent modeling in antiair defense. In *Proceedings of the Sixth International Conference on User Modeling*, pages 389–400.

Noh, S. and Gmytrasiewicz, P. J. (1998). Rational communicative behavior in anti-air defense. In *Proceedings of the Third International Conference on Multiagent Systems*, pages 214–221.

Patel-Schneider, P. (1990). Practical, object-oriented knowledge representation for knowledge-based systems. *Information Systems*, 15.

Pearl, J. (1988). *Probabilistic Reasoning in Intelligent Systems: Networks of Plausible Inference.* Morgan Kaufman.

Raiffa, H. (1982). *The Art and Science of Negotiation.* Harvard University Press.

Rasmusen, E. (1989). *Games and Information.* B. Blackwel.

Reny, P. J. (1988). Extensive games and common knowledge. In *Proceedings of the Conference on Theoretical Aspects of Reasoning about Knowladge*, page 395. Morgan Kaufman.

Russell, S. and Norvig, P. (1995). *Artificial Intelligence: A Modern Approach.* Prentice Hall.

Shachter, R. D. (1989). Evidence absorption and propagation through evidence reversals. In *Proceedings of the Fifth Workshop on Uncertainty in Artificial Intelligence*, pages 303–310, University of Windsor, Ontario, CA.

Tan, T. C. and Werlang, S. R. (1988). A guide to knowledge and games. In *Proceedings of the Second Conference on Theoretical Aspects of Reasoning about Knowledge.*

Wellman, M. P. (1991). The preferential semantics for goals. In *Proceedings of the National Conference on Artificial Intelligence*, pages 698–703.

Yelland, P. M. (1993). Object oriented knowledge representation. *British Telecom Technology Journal*, 11:41–51.

Chapter 7

SHOPBOT ECONOMICS

Jeffrey O. Kephart
IBM Institute for Advanced Commerce, IBM Thomas J. Watson Research Center
Yorktown Heights, NY 10598 USA
kephart@us.ibm.com

Amy R. Greenwald
Department of Computer Science, Brown University, Box 1910
Providence, RI 02912 USA
amygreen@cs.brown.edu

Abstract Shopbots are Internet agents that automatically search for information pertaining to the price and quality of goods and services. As the prevalence and usage of shopbots continues to increase, one might expect the resultant reduction in search costs to alter market behavior significantly. We explore the potential impact of shopbots upon market dynamics by proposing, analyzing, and simulating a model that is similar in form to some that have been studied by economists investigating the phenomenon of price dispersion. However, the underlying assumptions and methodology of our approach are different, since our ultimate goal is not to explain human economic behavior, but rather to design economic software agents and study their behavior. We study markets consisting of shopbots and other agents representing buyers and sellers in which (i) search costs are nonlinear, (ii) some portion of the buyer population makes no use of search mechanisms, and (iii) shopbots are economically motivated, strategically pricing their information services so as to maximize their own profits. Under these conditions, we find that the market can exhibit a variety of hitherto unobserved dynamical behaviors, including complex limit cycles and the co-existence of several buyer search strategies. We also demonstrate that a shopbot that charges buyers for price information can manipulate markets to its own advantage, sometimes inadvertently benefitting buyers and sellers.

Keywords: Shopbots, Economic software agents, Price dispersion, Search costs

1. INTRODUCTION

Shopbots—software agents that automatically query multiple on-line vendors to gather information about prices and other attributes of consumer goods and services—herald a future in which autonomous agents are an essential component of nearly every facet of electronic commerce (Chavez and Maes, 1996; Kephart et al., 1998; Tsvetovatyy et al., 1997; Eriksson et al., 1996). In response to a buyer's expressed interest in an item, a shopbot can query several dozen sellers' web sites, and return sorted information within seconds. Shopbots outperform and out-inform humans by providing extensive product coverage in just a few seconds, far more than a patient, determined human shopper could achieve after hours of manual search.

Since the launch of the BargainFinder (Krulwich, 1996), a CD shopbot, in 1995, scores of shopbots have emerged. One popular shopbot, `mySimon.com`, provides price and product information for groceries, apparel, office supplies, consumer electronics, toys, and other goods, allowing shoppers to sort by price or by merchant reputation as provided by `Gomez.com`. Another shopbot, `DealTime.com`, reports prices and expected delivery times on a variety of products offered for sale on line, including books, CDs, and movies. It computes the partition of a set of goods across multiple vendors that minimizes the total price, taking shipping charges into account.

Shopbots deliver on one of the great promises of the Internet and electronic commerce: a radical reduction in the economic friction that is associated with the cost of obtaining and distributing information. Transportation costs, menu costs (the costs to sellers of evaluating, updating, and advertising prices), and search costs (the costs to buyers of seeking out optimal price and quality) are all expected to decrease, as a consequence of the digital nature of information as well as the presence of autonomous agents that collect, process, and disseminate that information at little cost. The fluent manipulation of freely flowing information by agents is predicted to have a profound effect on the efficiency and dynamics of markets (Bakos, 1997; Green, 1998; Lewis, 1997; DeLong and Froomkin, 1998).

In this paper, we explore how shopbots might influence the behavior of electronic markets of the future. More specifically, we study how shopbot-induced changes in the cost of obtaining price information might affect both the individual and the collective behavior of agent buyers and sellers that will participate in and define these electronic markets. Our work builds on economic models that were originally introduced to explain how search costs can engender price dispersion in markets(Salop and Stiglitz, 1977; Varian, 1980; Burdett and Judd, 1983). In such work, economists typically approximate the relevant aspects of human behavior by mathematical algorithms, and compare the collective behavior that arises in their models with the observed behavior of real markets. In contrast, we regard mathematical algorithms, not as approximate descriptions of human behavior, but as precise prescriptions of software agent behavior.[1] Our goal is not to explain observed market behavior, but to anticipate and influence future market behavior.

The paper is organized as follows. Section 2 presents our model of a simple market populated with electronic sellers, electronic buyers, and a shopbot that readily provides price information. The three main sections analyze the market from the perspective of each type of player in turn: first the sellers, then the buyers, and finally the shopbot. In Section 3, we present a game-theoretic analysis of the sellers' pricing strategies. In Section 4, we explore the dynamic aspects of buyers' search behavior and the co-evolutionary relationship between pricing and search. Specifically, we investigate the effects of linear search costs, nonlinear search costs, and buyers who do not avail themselves of search mechanisms. In Section 5, we extend our model by providing economic incentives to the shopbot itself, investigating how it might strategically price the information service it provides. In Section 6, we discuss related work, and finally in Section 7, we summarize our findings and discuss interesting directions for future work.

2. MODEL

We consider a market in which there is a single commodity that is offered for sale by S sellers and of interest to B buyers (see Figure 7.1). Some or all of the sellers and buyers may be (or be represented by) software agents. Periodically, at the rate ρ_b, buyer b attempts to purchase a unit of the commodity. Each attempted purchase proceeds as follows. Buyer b conducts a search of fixed sample size i, which entails requesting $0 \leq i \leq S$ price quotes.[2] We refer to such a buyer as a *Search-i* buyer, or a buyer of type i. We allow for a distribution of buyer types across the population; this distribution may be fixed exogenously, or it may arise endogenously through the efforts of buyers to adopt optimal search strategies. A search mechanism, which could be manual or shopbot-assisted, instantly provides price quotes for i randomly chosen sellers. Buyer b then selects a seller s whose quoted price p_s is lowest among the

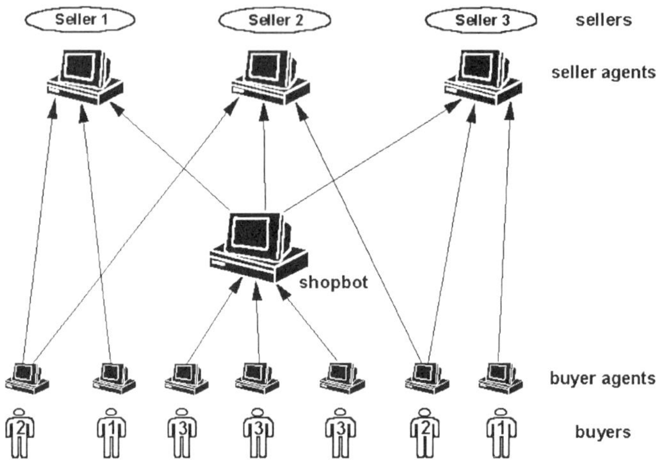

Figure 7.1. Shopbot Model. Buyers' numerical labels indicate the number of price quotes they compare before selecting a seller.

i (ties are broken randomly), and purchases the commodity from seller s if and only if $p_s \leq v_b$, where v_b is buyer b's valuation of the commodity.

In addition to the purchase price, buyers may incur search costs. The cost c_i of using search strategy i does not enter into the buyers' purchasing decisions, however, because buyers must commit to conducting their search before any results become available. In other words, search payments are sunk costs. But when search strategies are determined endogenously, search costs do affect buyers' strategic choices. In this case, buyers are assumed to behave rationally, although they need not be fully informed. In particular, buyer b is assumed to re-evaluate its strategy at random intervals at an average rate $\sigma_b \leq \rho_b$, where typically $\sigma_b \ll \rho_b$. Upon re-evaluation, buyer b estimates a price $\hat{p}_i$ that it would expect to pay for the commodity if it were to abide by strategy i, and then selects the strategy j that minimizes $\hat{p}_j + c_j$, provided that $\hat{p}_j + c_j \leq v_b$. If this condition cannot be satisfied, then the buyer opts out of the market at that time: *i.e.*, $j = 0$. Whenever a rational buyer is fully informed, it makes an optimal decision regarding which search strategy to employ, given the current state of the market.

The buyer population at any given moment is characterized by the strategy vector $\vec{w}$, in which component w_i represents the fraction of buyers employing strategy i and $\sum_{i=0}^{S} w_i = 1$. A seller s's expected profit per unit time π_s depends on the strategy vector $\vec{w}$, the price vector $\vec{p}$ describing all sellers' prices, and the cost of production r_s for seller s. In particular, $\pi_s(\vec{p}, \vec{w}) = D_s(\vec{p}, \vec{w})(p_s - r_s)$, where $D_s(\vec{p}, \vec{w})$ is the rate of demand for the good produced by seller s, in terms of the current price and search strategy vectors. The demand $D_s(\vec{p}, \vec{w})$ is the product of three terms: (i) the overall buyer rate of demand, namely $\rho = \sum_b \rho_b$, (ii) the likelihood that seller s is selected as a potential seller, denoted $h_s(\vec{p}, \vec{w})$, and (iii) the fraction of buyers whose valuations satisfy $v_b \geq p_s$, denoted $g(p_s)$. Specifically, $D_s(\vec{p}, \vec{w}) = \rho B h_s(\vec{p}, \vec{w}) g(p_s)$. Without loss of generality, we define the time scale such that $\rho B = 1$, and we then interpret π_s as seller s's expected profit per unit sold systemwide. Now, seller s's profits are given by:

$$\pi_s(\vec{p}, \vec{w}) = (p_s - r_s) h_s(\vec{p}, \vec{w}) g(p_s). \tag{1}$$

The functions $h_s(\vec{p}, \vec{w})$ and $g(p)$ are analyzed in the following section.

3. SELLERS

In this section, we analyze the shopbot model from the sellers' perspective. We assume that the buyers' strategies $\vec{w}$ are fixed and determined exogenously, and that the sellers seek to maximize profit. In addition, we assume that the sellers have equal production costs, *i.e.*, $r_s = r$ for all s. Under these assumptions, we derive a Nash equilibrium—a vector of prices at which sellers maximize their individual profits, and from which no seller has any incentive to deviate (Nash, 1951).

If $w_1 = 1$, then all buyers choose sellers at random; thus, the unique Nash equilibrium is such that all sellers charge the monopoly price. At the opposite extreme in which $w_1 = 0$, there are multiple Nash equilibria with the property that at least two sellers charge the production cost r and all sellers receive zero profits. It remains to consider the case $0 < w_1 < 1$. If the price quantization is sufficiently fine, then there are no pure strategy Nash equilibria. (A proof of this assertion is given in the appendix.) There does exist a set of Nash equilibria in mixed strategies, the symmetric variant of which can be derived as follows.[3]

Let $f(p)$ denote the probability density function according to which sellers set their equilibrium prices, and let $F(p)$ denote the corresponding cumulative distribution function. Following Varian (Varian, 1980), we note that $F(p)$ has no mass points in the range for which it is defined, since otherwise a seller could decrease its price by an arbitrarily small amount and experience a discontinuous increase in profits. Moreover, there are no gaps in the distribution, since otherwise prices would not be optimal—a seller charging a price at the low end of the gap could increase its price to fill the gap while retaining its market share, thereby increasing its profits. The cumulative distribution function $F(p)$ is computed in terms of the quantity $h_s(\vec{p}, \vec{w})$.

Recall that $h_s(\vec{p}, \vec{w})$ represents the probability that buyers select seller s as their potential seller. This function is expressed in terms of the probabilistic demand for seller s by buyers of type i, namely $h_{s,i}(\vec{p})$, for $0 \leq i \leq S$. The first component $h_{s,0}(\vec{p}) = 0$. Consider the next component, $h_{s,1}(\vec{p})$. Buyers of type 1 select sellers at random; thus, the probability that seller s is selected by such buyers is simply $h_{s,1}(\vec{p}) = 1/S$. Now consider buyers of type 2. In order for seller s to be selected by a buyer of type 2, s must be included within the pair of sellers being sampled—which occurs with probability $(S-1)/\binom{S}{2} = 2/S$—and s must be lower in price than the other seller in the pair. Since, by the assumption of symmetry, the other seller's price is drawn from the same distribution, this occurs with probability $1 - F(p)$. Therefore $h_{s,2}(\vec{p}) = (2/S)\,[1 - F(p)]$. In general, seller s is selected by a buyer of type i with probability $\binom{S-1}{i-1}/\binom{S}{i} = i/S$, and seller s is the lowest-priced among the i sellers selected with probability $[1 - F(p)]^{i-1}$, since these are $i - 1$ independent events. Thus,

$$h_{s,i}(\vec{p}) = \frac{i}{S}[1 - F(p)]^{i-1} \tag{2}$$

and[4]

$$h_s(p, \vec{w}) = \frac{1}{S}\sum_{i=1}^{S} i w_i [1 - F(p)]^{i-1}. \tag{3}$$

Substituting Eq. 3 into Eq. 1, and dropping the subscript s due to symmetry, the expected profit for any given seller is:

$$\pi(p, \vec{w}) = \frac{1}{S}\sum_{i=1}^{S} i w_i [1 - F(p)]^{i-1}(p - r)g(p). \tag{4}$$

A Nash equilibrium in mixed strategies requires that all prices that receive positive probability yield equal profits; otherwise, it would not be optimal to randomize. Thus, the equilibrium profits given by Eq. 4 equal some constant π for all prices p that are included in the support of $f(p)$. The constant π can be derived if we can find a special value of p for which the right-hand side of Eq. 4 can be evaluated. Consider the highest price in the support of $f(p)$: *i.e.*, the smallest value of p for which $F(p) = 1$. Denote this price p_m. Evaluating Eq. 4 at p_m, we obtain $\pi = (w_1/S)(p_m - r)g(p_m)$. This has a simple interpretation. Since $F(p_m) = 1$, a seller that charges p_m is catering solely to buyers of type 1, and $1/S$ of such buyers will consider purchasing from it. For these buyers, the seller is in essence a monopolist. The remaining factor, $(p - r)g(p)$, is the profit function for a monopolist. Since a monopolist sets p so as to maximize this profit function, p_m can be identified as the monopolist price, which yields the monopolist profit $\pi_m \equiv (p_m - r)g(p_m)$. Thus the constant π is simply expressed as

$$\pi = \frac{w_1 \pi_m}{S}. \tag{5}$$

Substituting Eq. 5 into the left hand side of Eq. 4, we find that for any p in the support of $f(p)$,

$$(p - r)g(p) = \frac{w_1 \pi_m}{\sum_{i=0}^{S} i w_i [1 - F(p)]^{i-1}}. \tag{6}$$

Eq. 6 implicitly defines p and $F(p)$ in terms of one another, and in terms of $g(p)$, for all p such that $0 \leq F(p) \leq 1$.

The function $g(p)$ can be expressed as $g(p) = \int_p^\infty \gamma(x)dx$, where $\gamma(x)$ is the probability density function describing the likelihood that a given buyer has valuation x. For example, if the buyers' valuations are uniformly distributed between 0 and v, with $v > 0$, then the integral yields $g(p) = 1 - p/v$, for $p \leq v$. This case was studied in Greenwald, *et al.* (Greenwald et al., 1999).

In this paper, we assume $v_b = v$ for all buyers b, in which case $\gamma(x)$ is the Dirac delta function $\delta(v - x)$, and the integral yields a step function $g(p) = \Theta(v - p)$, defined as follows:

$$\Theta(v - p) = \begin{cases} 1 & \text{if } p \leq v \\ 0 & \text{otherwise.} \end{cases} \tag{7}$$

For this distribution of buyer valuations $g(p)$, the monopolist's profit function is simply $(p - r)$ for $p \leq v$, and 0 for $p > v$. This function attains a maximum of $\pi_m = v - r$ at the price $p_m = v$. For the remainder of this paper, we shall assume (without loss of generality) that $v = 1$ and $r = 0$, and hence $\pi_m = 1$.[5] Inserting these values into Eq. 6 and solving for p in terms of F yields:

$$p(F) = \frac{w_1}{\sum_{i=1}^{S} i w_i [1 - F]^{i-1}}. \tag{8}$$

Eq. 8 has several important implications. In a population in which there are no buyers of type 1 (*i.e.*, $w_1 = 0$) the sellers charge the production cost r and earn zero profits; this is the traditional Bertrand equilibrium (Tirole, 1988). Also of interest is the case in which the the population consists of just two buyer types, 1 and some $i \neq 1$; in this case, it is possible to invert $p(F)$ to obtain:

$$F(p) = 1 - \left[\left(\frac{w_1}{iw_i}\right)\left(\frac{1-p}{p}\right)\right]^{\frac{1}{i-1}}. \tag{9}$$

The case in which $i = S$ was studied previously by Varian (Varian, 1980); in his model, buyers either choose a single seller at random (type 1) or search all sellers and choose the lowest-priced among them (type S).

Since $F(p)$ is a cumulative probability distribution, it is only valid in the domain for which its value lies between 0 and 1. The upper boundary is $p = 1$, since prices above this threshold lead to decreases in market share that exceed the benefits of increased profits per unit. The lower boundary p^* can be computed by setting $F(p^*) = 0$ in Eq. 8, which yields:

$$p^* = \frac{w_1}{\sum_{i=1}^{S} iw_i}. \tag{10}$$

In general, Eq. 8 cannot be inverted to obtain an analytic expression for $F(p)$. It is possible, however, to plot $F(p)$ without resorting to numerical root finding techniques. We can use Eq. 8 to evaluate p at equally spaced intervals in $F \in [0, 1]$; this produces unequally spaced values of p ranging from p^* to 1.

We now consider the probability density function $f(p)$. For the given choice of $g(p)$, the profits for seller s equal $h_s(p, \vec{w})p = w_1/S$, for $p \leq 1$; thus, the probabilistic demand for seller s is given by $h_s(p, \vec{w}) = w_1/pS$. Differentiating both sides of this expression with respect to p and substituting Eq. 3, we obtain an expression for $f(p)$ in terms of $F(p)$ and p that is conducive to numerical evaluation:

$$f(p) = \frac{w_1}{p^2 \sum_{i=2}^{S} i(i-1)w_i[1 - F(p)]^{i-2}}. \tag{11}$$

The values of $f(p)$ at the boundaries p^* and 1 are as follows:

$$f(p^*) = \frac{\left[\sum_{i=1}^{S} iw_i\right]^2}{w_1\left[\sum_{i=2}^{S} i(i-1)w_i\right]} \quad \text{and} \quad f(1) = \frac{w_1}{2w_2}. \tag{12}$$

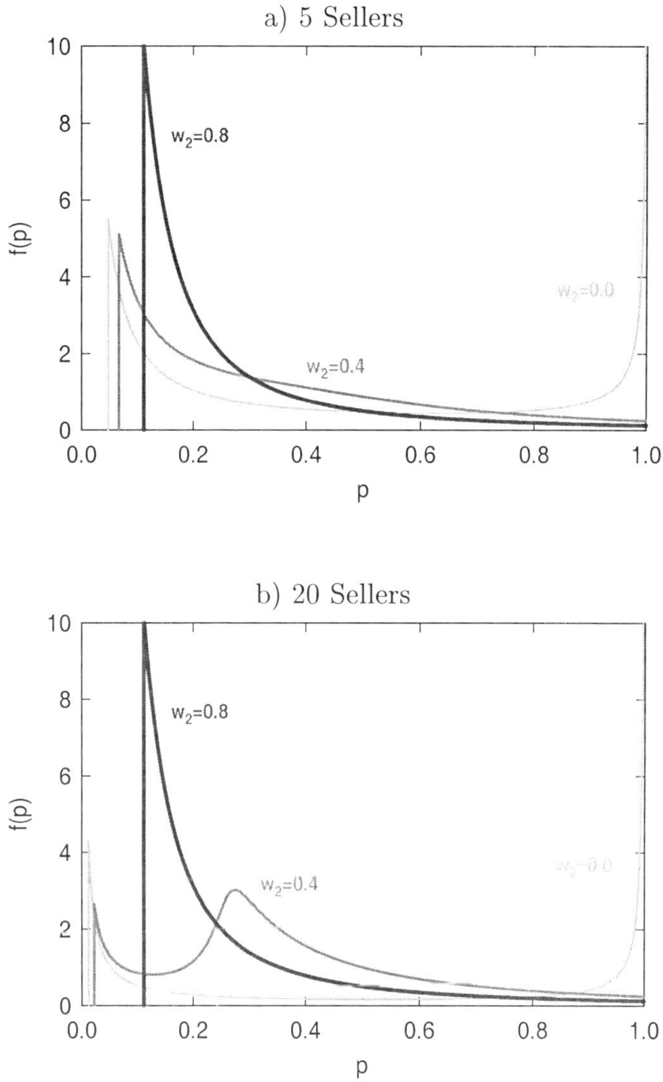

Figure 7.2. Probability density functions $f(p)$ for three buyer strategy vectors: $(w_1, w_2, w_S) = (0.2, 0.0, 0.8)$, $(0.2, 0.4, 0.4)$, and $(0.2, 0.8, 0.4)$. a) $S = 5$. b) $S = 20$.

Fig. 7.2 depicts the probability density function $f(p)$ for three different buyer strategies representing different mixtures of strategies 1, 2, and S, with $S = 5$ in Fig. 7.2(a) and $S = 20$ in Fig. 7.2. When $w_2 = 0$, $f(p)$ is bimodal, with most of the probability density concentrated either just above p^*, where sellers expect low margins but high volume, or just below v, where they expect high margins but low volume. As S increases, the peaks become increasingly accentuated, and $p^* \to 0$. Using Eq. 11, it can be shown that the peak near $p = 1$ is an integrable singularity when $w_2 = 0$, and that the singularity disappears for any finite value of w_2. We have observed that this peak tends to disappear even for fairly small values of w_2. When both w_2 and w_S are of moderate size, one can observe a peak in $f(p)$ at moderate values of p. This peak can be attributed to the viability of a middle-of-the-road strategy in which sellers recognize that they need not undercut all other sellers: if a sizeable proportion of buyers only compare two sellers, then a seller need only undercut the seller with which it is randomly paired.

Since we are assuming that $\pi_m = v - r = 1$, then by Eq. 5 the profit earned by each seller is $\pi = w_1/S$. Note that this quantity is strictly positive so long as $w_1 > 0$. It is as though only buyers of type 1 are contributing to sellers' profits, although the actual distribution of contributions from buyers of type 1 vs. buyers of type $i > 1$ is not as one-sided as it appears. In reality, buyers of type 1 are charged less than $v = 1$ on average, and buyers of type $i > 1$ are charged more than $r = 0$ on average, although total profits are equivalent to what they would be if the sellers practiced perfect price discrimination. In effect, buyers of type 1 exert negative externalities on buyers of type $i > 1$ by creating surplus profits for sellers.

4. BUYERS

Heretofore in our analysis, we have assumed rational decision-making on the part of the sellers, but an exogenous distribution of buyer types. In this section, we allow each buyer to use a rational criterion for choosing its search strategy, and therefore the search strategy vector $\vec{w}$ is determined endogenously.

The rational criterion that buyers employ to select a search strategy is as follows. Recall from Section 2 that search costs are specified by a vector $\vec{c}$, where c_i denotes the cost of comparing the prices of i sellers. A rational buyer estimates the expected price $\hat{p}_i$ that it would pay to the lowest-priced seller among a randomly chosen set of i sellers, given the price distribution $f(p)$, and selects the strategy i^* that minimizes $\hat{p}_i + c_i$. If no strategy i^* satisfies the restriction that $\hat{p}_i + c_i \leq v_b$, then the expected combined cost of discovering prices and purchasing the good exceeds the buyer's valuation, and the buyer opts out of the market, *i.e.*, it selects $i^* = 0$.

How do buyers estimate $\hat{p}_i$? One reasonable method would be to compute estimates using historical price data. In this study, however, we assume that buyers know the price distribution $f(p)$, and therefore compute the expected price $\bar{p}_i$ as the expected value of the lowest of i draws from $f(p)$. Using $\bar{p}_i$ as their estimate of $\hat{p}_i$, buyers can then select an optimal buyer strategy i^* that satisfies

$$i^* \in \arg \min_{0 \leq i \leq S} \bar{p}_i + c_i. \tag{13}$$

The remainder of this section is organized as follows. In Section 4.1, we derive a general expression for the expected price $\bar{p}_i$ in terms of $f(p)$, which allows us to compute the marginal benefit of the ith search, namely $\bar{p}_i - \bar{p}_{i-1}$. Then, in Section 4.2, we assume search costs are linear in the amount of search undertaken. In a market of rational buyers and sellers, $\vec{w}$ and $f(p)$ co-evolve to an equilibrium in which only buyer types 1 and 2 prevail. In the following two sections, we simulate the co-evolutionary dynamics of $\vec{w}$ and $f(p)$ in two more complex scenarios: in Section 4.3 we explore the effect of nonlinear search costs; then, in Section 4.4, we assume that a portion of the buyer population does not have access to search mechanisms.

4.1. EXPECTED BUYER PRICES

Recall that buyers who obtain i price quotes pay the lowest of the i prices observed. (At equilibrium, the sellers' prices never exceed 1 since $F(1) = 1$, so buyers are *always* willing to pay the lowest price.) Thus, the computation of $\bar{p}_i$ depends on the cumulative distribution of the minimal values of i independent samples drawn from the distribution $f(p)$, which is given by $Y_i(p) = 1 - [1 - F(p)]^i$. Differentiating both sides of this equation with respect to p yields the probability distribution: $y_i(p) = if(p)[1 - F(p)]^{i-1}$. The average price for the distribution $y_i(p)$ can be expressed as follows:

$$\bar{p}_i = \int_{p^*}^{v} dp\, p\, y_i(p) = v - \int_{p^*}^{v} dp\, Y_i(p) = p^* + \int_0^1 dF\, \frac{(1-F)^i}{f} \tag{14}$$

where the first equality is obtained via integration by parts, and the second depends on the observation that $dp/dF = [dF/dp]^{-1} = 1/f$. Combining Eqs. 8, 11, and 14 would lead to an integrand expressed purely in terms of F. Integration over the variable F (as opposed to p) is advantageous because F can be chosen to be equispaced, as standard numerical integration techniques require.

Fig. 7.3(a) depicts sample price distributions for buyers of various types: in particular, $y_1(p)$, $y_2(p)$, and $y_{20}(p)$ are plotted for $S = 20$ and $(w_1, w_2, w_{20}) = (0.2, 0.4, 0.4)$. The dashed lines represent the average prices $\bar{p}_i$ for $i \in \{1, 2, 20\}$ as computed by Eq. 14. The line labeled *Search-1*, which depicts the distribution $y_1(p)$, is identical to the line

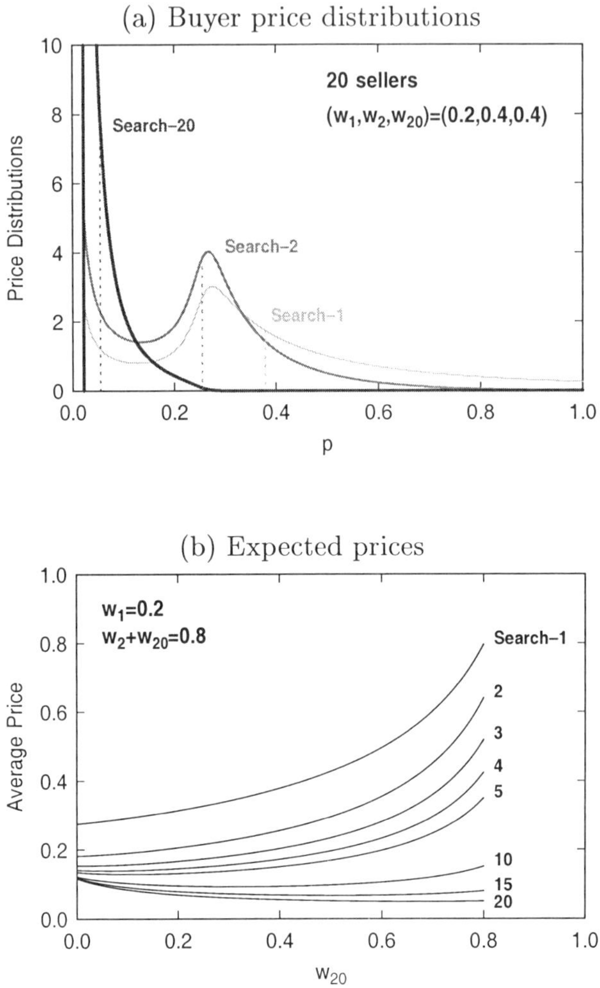

Figure 7.3. (a) Probability distributions of prices paid by buyers performing 1, 2, and 20 searches in a market with 20 sellers. The buyer strategy vector is $w_1 = 0.2, w_2 = 0.4, w_{20} = 0.4$, corresponding to one of the curves in Fig. 7.2(b). (b) Expected prices $\bar{p}_i$ paid by buyers performing the indicated amount of search, as a function of w_{20}. There are 20 sellers. The buyer strategy vector is $w_1 = 0.2$, $w_2 = 0.8 - w_{20}$.

labeled $w_2 = 0.4$ in Fig. 7.2(b), since $y_1(p) = f(p)$. The distributions shift toward lower values of p for those buyers who base their buying decisions on information obtained from more sellers.

Fig. 7.3(b) depicts the expected prices obtained by buyers of various types, assuming w_1 is fixed at 0.2 and $w_2 + w_{20} = 0.8$. The various values of i (*i.e.*, buyer types) are listed to the right of the curves. Notice that as w_{20} increases, the average prices paid by those buyers who perform relatively few searches increases rather dramatically for larger values of w_{20}. This is because w_1 is fixed, which implies that the sellers' profit surplus is similarly fixed; thus, as more and more buyers perform extensive searches, the average prices paid by those buyers decreases, which causes the average prices paid by the less diligent searchers to increase. The situation is slightly different for those buyers who perform extensive but non-maximal searches, *e.g.*, $i = 10$ or $i = 15$. Such buyers initially reap the benefits of increasing the number of buyers of type 20, but when w_{20} becomes sufficiently large their expected prices increase slightly as well. Given a fixed portion of the population designated as buyers of type 1, Fig. 7.3(b) demonstrates that searching S sellers is a superior buyer strategy to searching $1 < i < S$ sellers. Thus, there is value in performing price searches: *shopbots offer added value* in markets in which some buyers shop randomly.

4.2. LINEAR SEARCH COSTS

Initially, we model buyer search costs following Burdett and Judd (Burdett and Judd, 1983), who assume costs are linear in the number of searches; in particular, $c_i = c_1 + \delta(i-1)$, where $c_1, \delta > 0$ are, respectively, fixed and marginal costs of obtaining price quotes. As shown by Burdett and Judd, the buyer cost function $\bar{p}_i + c_i$ is a convex function of i, which is minimized at either a single integer value i^*, or two consecutive integer values i^* and i^*+1. If c_1 is sufficiently large, then $w_0 = 1$ at equilibrium, *i.e.*, all buyers opt out of the market. Otherwise, w_1 is bounded away from 0, since if $w_1 = 0$ then competition drives all sellers' prices to zero (see Eq. 8), making search irrational. But if search is irrational, then $w_1 = 1$, contradicting the original assumption that $w_1 = 0$. Therefore, for sufficiently small values of c_1, equilibrium search strategies are either such that $w_1 = 1$, in which case all sellers charge the monopolistic price, or such that $w_1 + w_2 = 1$, with $w_1 > 0$ and the sellers' prices given by the distribution $f(p)$.

Given that $w_1 + w_2 = 1$, we obtain analytic expressions for the prices expected by buyers of types 1 and 2 by substituting Eq. 9 into Eq. 14:

$$\bar{p}_1(w_2) = p^* + \left(\frac{1-w_2}{2\,w_2}\right)\left[\log\left(\frac{1+w_2}{1-w_2}\right) - \frac{2\,w_2}{1+w_2}\right] \tag{15}$$

$$\bar{p}_2(w_2) = p^* + \left(\frac{(1-w_2)}{2\,{w_2}^2\,(1+w_2)}\right)\left[2\,w_2 + (1-w_2^2)\,\log\left(\frac{1-w_2}{1+w_2}\right)\right]. \tag{16}$$

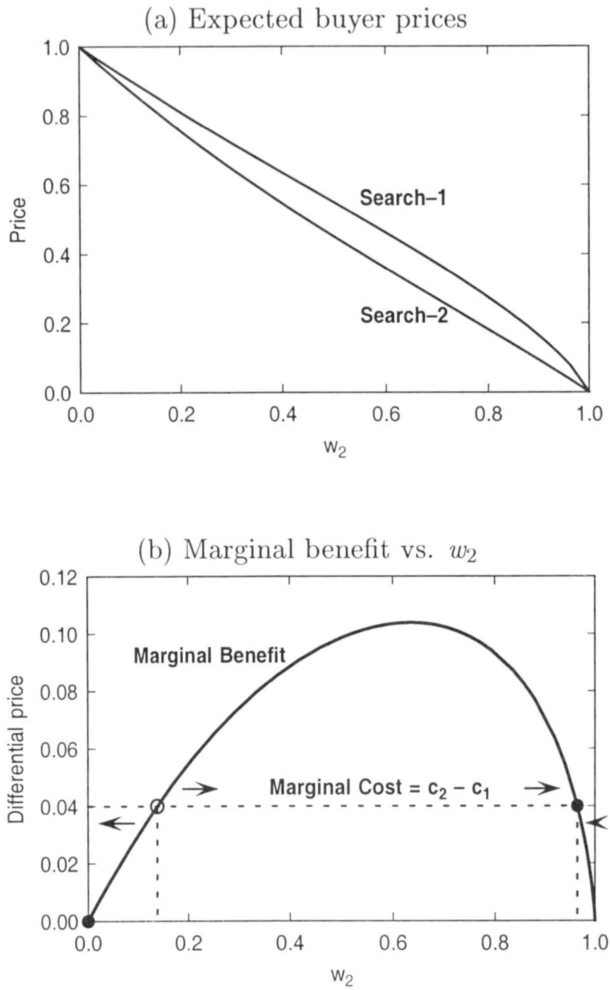

Figure 7.4. Benefits and costs of search when search costs are linear, and hence $w_1 + w_2 = 1$. a) Expected prices for *Search-1* and *Search-2* buyers as a function of w_2. b) Marginal benefit of *Search-2* strategy over *Search-1* as a function of w_2. Equilibria occur at values of w_2 for which marginal benefit equals marginal cost. Stable and unstable equilibria are represented by solid and open circles, respectively. Arrows indicate the direction in which a population of rational buyers would move if they were permitted to change their search strategies.

Fig. 7.4(a) plots $\bar{p}_1$ (*i.e.*, *Search–1*) and $\bar{p}_2$ (*i.e.*, *Search–2*) as a function of w_2. Both curves decrease monotonically with w_2, since sellers' prices are forced downward when more buyers compare prices. Fig. 7.4(b) plots the marginal benefit β of *Search–2* over *Search–1*, which is computed by subtracting Eq. 16 from 15:

$$\beta(w_2) = \left(\frac{1-w_2}{w_2}\right)\left[\frac{1}{2w_2}\log\left(\frac{1+w_2}{1-w_2}\right) - 1\right]. \tag{17}$$

Suppose that all buyers are fully informed and rational, and therefore estimate the expected marginal benefit accurately. Then, if buyers are free to choose between the strategies *Search-1* and *Search-2*, the buyer population reaches equilibrium when the marginal benefit of a second price quote exactly balances its marginal cost. As a specific example, suppose that the marginal cost $\delta = c_2 - c_1 = 0.04$. This is represented by the dotted line in Fig. 7.4(b). The marginal benefit and marginal cost curves intersect twice in this diagram, representing two of the three equilibria that arise in this setting. Above the dotted line, benefit outweighs cost; thus, search is advantageous and there is momentum in the rightward direction. Below the dotted line, cost outweighs benefit; it is therefore not desirable to search, and there is momentum in the leftward direction. Following the direction of the arrows, the filled-in circle that falls on the curve is a stable equilibrium, while the open circle represents an unstable equilibrium. The unstable equilibrium represents a boundary between two basins of attraction: a buyer population in which the initial value of w_2 is greater than this threshold will migrate towards the equilibrium near $w_2 = 1$, while one in which w_2 is initially smaller than this threshold will migrate towards $w_1 = 1$. Finally, there is a second stable equilibrium in the lower left-hand corner of the graph (indicated by a second filled-in circle) where $w_1 = 1$, the equilibrium price is the monopolistic price 1, and $\lim_{w_2 \to 0} \beta(w_2) = \delta = 0$.

4.3. NONLINEAR SEARCH COSTS

A typical shopbot such as the one available at `DealTime.com` permits users to choose the number of sellers among whom to search. Since the service is free for buyers at present, and moreover, since the search is very fast—`DealTime` searches prices at a few dozen book retailers within about 20 seconds—there is only a mild disincentive not to request a large number of price quotes. Thus, the effective search cost is only weakly dependent on the number of searches.

One way to model weak dependence on the number of searches is via a nonlinear search cost schedule of the form

$$c_j = c_1 + \delta(j-1)^\alpha \tag{18}$$

where the exponent α is in the range $0 \leq \alpha \leq 1$. Note that $\alpha = 1$ yields a linear search cost model, while $\alpha = 0$ yields a bundled search cost model, in which the search cost is fixed at $c_1 + \delta$ regardless of the number of searches.

In Section 4.2, we performed an equilibrium analysis for $\alpha = 1$. Here, we examine non-equilibrium behavior for arbitrary α. To do so, we simulate the dynamics of a system in which the strategy vector $\vec{w}$ and the price distribution $f(p)$ co-evolve through the continual re-evaluation of search strategies by buyers and the continual re-setting of the price distribution by sellers. The buyers search strategies evolve via a discrete-time process described below. As the buyers modify their strategies, we assume that the sellers monitor the strategy vector $\vec{w}$, and instantaneously re-compute the symmetric price distribution $f(p)$ according to which they randomly set their prices.

We approximate the co-evolution of $\vec{w}$ and $f(p)$ by a discrete time process in which, at each time step, a fraction η of the buyer population switches to the optimal strategy. The vector $\vec{w}$ evolves as follows:

$$w_i(t+1) = w_i(t) + \eta(\delta_{ij} - w_i(t)), \tag{19}$$

where j is the strategy that minimizes $\bar{p_j} + c_j$ and δ_{ij} represents the Kronecker delta function, equal to 1 when $i = j$ and 0 otherwise. Buyers are assumed to determine j by monitoring $f(p)$ (or by using Eq. 8 to compute $F(p)$ from the current state of the strategy vector $\vec{w}$) and then numerically integrating Eq. 14 to compute $\hat{p_j} = \bar{p_j}$.

Fig. 7.5(a) illustrates the evolution of the components of $\vec{w}$ in a 5-seller system. As in Fig. 7.4(b), the search costs are linear ($\alpha = 1$), with $c_1 = 0.1$ and $\delta = 0.04$. The value of η is chosen to be 0.002. According to Burdett and Judd (Burdett and Judd, 1983), $\vec{w}$ must evolve toward an equilibrium consisting of a finite number of type 1 and type 2 buyers. Indeed, we observe such an equilibrium, but the *trajectory* of the strategy vector $\vec{w}$ *en route* to equilibrium is unexpectedly interesting.

Initially, $\vec{w}(t = 0) = (0.2, 0.3, 0.0, 0.0, 0.5)$. In this situation, the favored strategy is type 3, which begins increasing at the expense of w_1, w_2 and w_5. As w_5 diminishes, however, the total amount of search in the system diminishes, causing $f(p)$ to flatten and shift in such a way that eventually the favored strategy changes from 3 to 2. Thereafter, w_2 grows at the expense of w_3 and the other components until the system nearly stabilizes. In this simulation, near but imperfect equilibrium is achieved: due to the finite size of η, there are small oscillations in w_2 around an average value that is close to the theoretical value of 0.9641721. This value can be derived by identifying the value of w_2 corresponding to $\delta = 0.04$ in Fig. 7.4(b). Although there is a second value of w_2 satisfying $\delta = 0.04$ in this figure, namely $w_2 = 0.1375564$, it is the unstable equilibrium. As discussed in the previous section, it

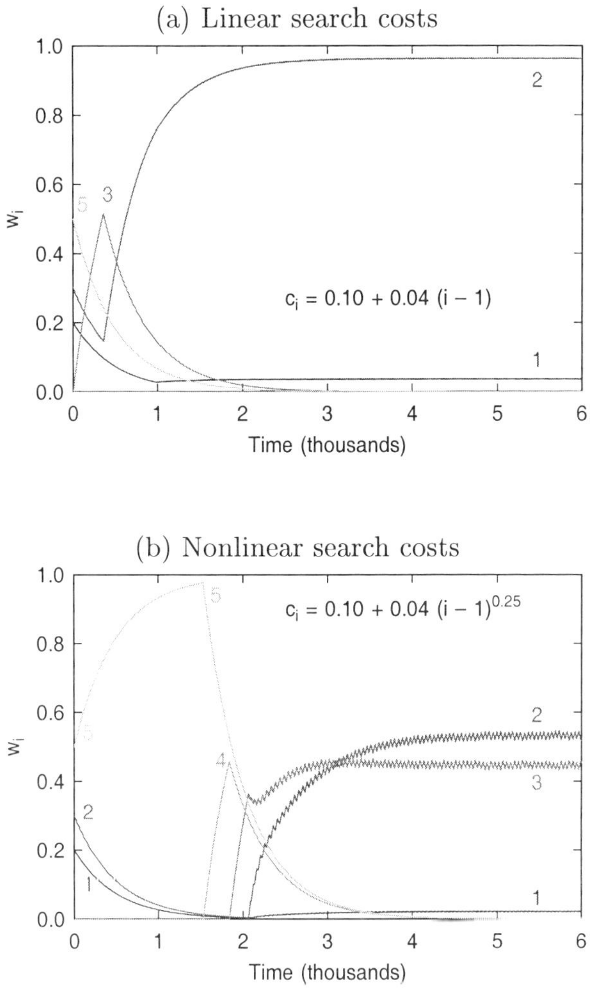

Figure 7.5. (a) Evolution of indicated components of buyer strategy vector $\vec{w}$ for 5 sellers, with linear search costs $c_i = 0.10 + 0.04(i - 1)$. Final equilibrium oscillates with small amplitude around the theoretical solution involving a mixture of strategy types 1 and 2. (b) Evolution of indicated components of buyer strategy vector $\vec{w}$ for 5 sellers, with nonlinear search costs $c_i = 0.10 + 0.04(i - 1)^{0.25}$. Final equilibrium oscillates in a complex small-amplitude limit cycle around a mixture of strategy types 1, 2, and 3.

marks the boundary between two basins of attraction, one in which the final equilibrium is $(w_1, w_2) = (0.0358279, 0.9641721)$, and the other in which $(w_1, w_2) = (1, 0)$.

The derivation of an equilibrium in which only type 1 and type 2 strategies co-exist is based on the assumption that search costs are linear in the amount of search. In order to investigate the effect of nonlinear search costs, we re-ran the previous experiment with identical parameters, except for the exponent α, which was reduced from 1.0 to 0.25. Fig. 7.5(b) depicts the ensuing dynamics. Interestingly, in this case the system evolves to an equilibrium in which *types 1, 2 and 3 co-exist*: $\vec{w}$ oscillates around the value $(0.0217, 0.5357, 0.4426, 0.0000, 0.0000)$ in a complex, small-amplitude limit cycle with period 104. As η is increased, the amplitude of the oscillations grows, but their period diminishes. Although these oscillations are an artifact of the finite size of η, and disappear in the limit $\eta \to 0$, they indicate that the system would undergo large-scale oscillations if the buyers were to revise their strategies synchronously (*i.e.*, $\eta = 1$) rather than asynchronously.

4.4. LOWER LIMIT ON W_1

Today's shopbots are used by only a small fraction of shoppers. This behavior is at least in part due to the fact that many potential users are unaware of the existence of shopbots, and others do not know where to find them or how to use them. One way of modeling buyers who do not use shopbots is to assume that such uninformed buyers are of type 1 and incur a search cost of c_1. This establishes a lower limit on the fraction w_1, which we denote $\lfloor w_1 \rfloor$. In particular, $\lfloor w_1 \rfloor$ represents the fraction of uninformed buyers who guarantee the sellers a strictly positive profit surplus.

Fig. 7.6(a) depicts a simulation in which $\lfloor w_1 \rfloor = 0.04$, with linear search costs $c_i = 0.10 + 0.01(i - 1)$. Starting from an initial strategy vector $\vec{w}(t = 0) = (0.04, 0.20, 0.00, 0.00, 0.76)$, the system evolves to an equilibrium in which only types 1 and 4 co-exist, with $w_1 = 0.04$ and $w_4 = 0.96$. In numerous experiments with w_1 bounded below and linear search costs, we have observed that the system stabilizes with a mixture of buyer types 1 and i, where i is not necessarily 2, as it is in the traditional economic setting that was analyzed in Section 4.2. The particular strategy i that arises depends on the lower bound $\lfloor w_1 \rfloor$ and the incremental search cost δ; this dependence is illustrated in Table 7.1. Higher values of $\lfloor w_1 \rfloor$ lead to higher values of i at equilibrium (more extensive search) while higher values of δ lead to lower values of i at equilibrium (less extensive search).

Multiple equilibria arise for the table entries $(\lfloor w_1 \rfloor, \delta) = (0.04, 0.01)$ and $(\lfloor w_1 \rfloor, \delta) = (0.20, 0.04)$, and precisely which equilibrium is reached is determined by the initial setting of the strategy vector. The effect

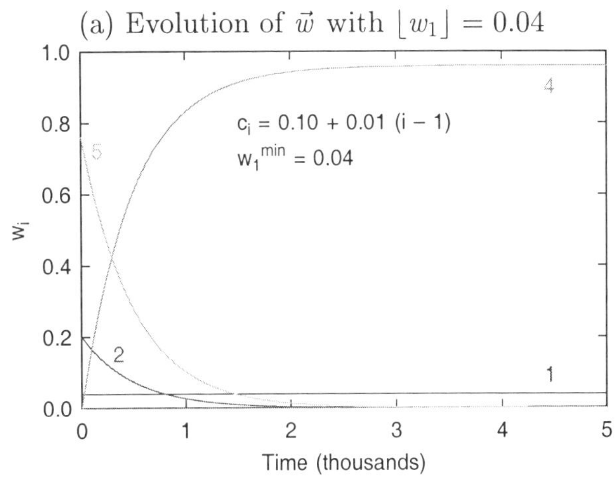

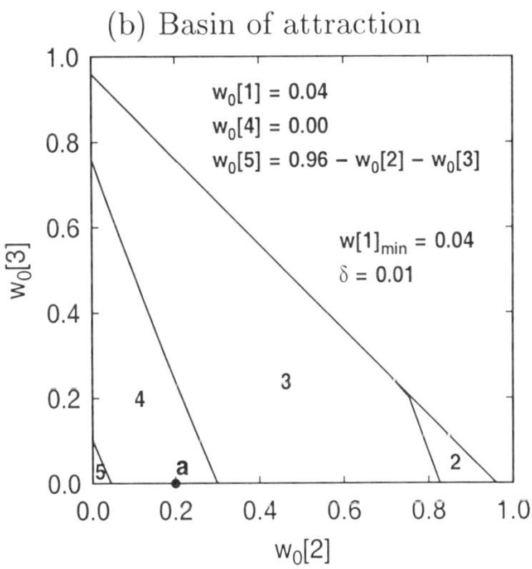

Figure 7.6. (a) Evolution of indicated components of buyer strategy vector $\vec{w}$ for 5 sellers, with linear search costs $c_i = 0.10 + 0.01(i-1)$ and $\lfloor w_1 \rfloor = 0.04$. Starting from an initial state with $w_1 = 0.04$, $w_2 = 0.20$, and $w_5 = 0.76$, $\vec{w}(t)$ evolves towards an equilibrium in which only w_1 and w_4 are nonzero. (b) Two-dimensional cross-section of basin of attraction for $(\lfloor w_1 \rfloor, \delta) = (0.04, 0.01)$, associated with the middle cell of Table 7.1. Numerical labels indicate the strategy i that mixes with strategy 1 in equilibrium, given that $\vec{w}$'s initial state is in the indicated region. The simulation depicted in (a) is associated with the point marked a.

of initial conditions on equilibrium selection in the case $(\lfloor w_1 \rfloor, \delta) = (0.04, 0.01)$ is illustrated in Fig. 7.6(b). Four equilibria are possible, all

Table 7.1 . Search strategies that co-exist with *Search-1*, as a function of the lower limit $\lfloor w_1 \rfloor$ and the incremental cost δ.

$\lfloor w_1 \rfloor$	$\delta = 0.002$	$\delta = 0.010$	$\delta = 0.040$
0.01	5	2	2
0.04	5	2–5	2
0.20	5	5	2–3

of the form $w_1 + w_i = 1$, for $i = 2, 3, 4, 5$. The set of initial conditions leading to equilibrium i—its *basin of attraction*(Schuster, 1988)—forms a contiguous, smoothly bounded region, a two-dimensional cross-section of which is depicted in Fig. 7.6(b).

5. SHOPBOT

Thus far, we have followed the traditional price dispersion literature in regarding the search costs $\vec{c}$ as fixed. Now we shall explore what might happen if a shopbot were to behave as an economically-motivated agent itself—strategically setting the search costs in an effort to maximize its own profit. Although shopbots do not currently use such a business model, it would be legitimate for them to charge buyers some fraction of the savings that they provide.[6] Since we are now viewing the market from the shopbot's perspective, we consider the search cost vector $\vec{c}$ as a *price schedule* that the shopbot sets so as to maximize its profits.

How might a shopbot compute a profit-maximizing price schedule? At first glance, the following method might seem plausible:

1 For each of a large number of candidate price schedules $\vec{c}$, simulate the market dynamics to determine the asymptotic equilibrium $\vec{w}(\vec{c})$.

2 Compute the shopbot's profit, namely[7]

$$\pi_{\text{sh}}(\vec{c}, \vec{w}(\vec{c})) = \sum_{i=1}^{S} c_i w_i(\vec{c}). \tag{20}$$

3 Select the price schedule $\vec{c}$ that maximizes $\pi_{\text{sh}}(\vec{c}, \vec{w}(\vec{c}))$.

However, several complications prevent such an approach from being useful in practice. First, the computational burden associated with this algorithm could be severe, as it entails optimization of a function in $S-1$ dimensions, and each function evaluation requires simulation. Second, recall from Section 4.3 that the evolutionary dynamics of buyer strategies may not lead to an equilibrium. Indeed, as illustrated in Fig. 7.5(b), $\vec{w}(t)$ can take the form of a complex limit cycle when search costs are nonlinear. Third, even when c_i is linear in i, and $\vec{w}$ reaches an equilibrium, there can be multiple equilibria, as illustrated in Figs. 7.4(b) and 7.6(b). In these cases, the chosen equilibrium depends on the initial value of $\vec{w}$ at time $t = 0$, which is beyond the shopbot's control. Finally, the shopbot is unlikely to be the only means by which price information may be obtained. It is more natural to suppose that buyers have alternative mechanisms by which to discover prices, such as manual search, and that buyers weigh the costs and benefits of using the shopbot against those of using any alternative search mechanisms in order to decide which mechanism to use and how much search to undertake.

Taking these factors into account, we allow for a possibly time-dependent expected shopbot profit $\pi_{\text{sh}}(\vec{c}, \vec{w}(t))$, given by:

$$\pi_{\text{sh}}(\vec{c}, \vec{w}(t)) = \sum_{i=1}^{S} c_i\, w_i(t)\, \Theta(c_{\text{alt},i} - c_i)\Theta(1 - (\bar{p}_i + c_i)) \qquad (21)$$

where $\Theta(x)$ represents the step function, equal to 1 for $x \geq 0$ and 0 otherwise[8], and $\vec{c}_{\text{alt}}$ represents the alternative search cost schedule. The step functions in Eq. 21 reflect the fact that, for each search strategy i, the shopbot is motivated to set c_i such that (i) it undercuts $c_{\text{alt},i}$, and (ii) it is low enough to ensure that the sum of the search cost and the expected price of the good does not exceed the buyer valuation.

We explore both linear and nonlinear shopbot price schedules in the remainder of this section. First, in Section 5.1, we assume that the shopbot restricts itself to a linear price schedule of the form $c_i = c_1 + \delta(i - 1)$. This assumption permits an exact analysis that yields the optimal values of c_1 and δ. An interesting complication arises, however, due to the existence of multiple equilibria. It is insufficient for the shopbot to merely fix its prices based on the optimal values of c_1 and δ. It must must also somehow ensure that the market evolves to its preferred equilibrium. We present one technique by which the shopbot can guide buyers and sellers towards this equilibrium in Section 5.2. Finally, in Section 5.3, we lift the restriction of price schedules to a linear form. Rather than undertaking an analysis of time-dependent asymptotic behavior in this case, we introduce an adaptive heuristic for setting the shopbot's price schedule. We observe that the shopbot can earn significantly higher profits by setting its price schedule such that c_i is a highly nonlinear function of i. Under both scenarios, we demonstrate that the shopbot can improve the welfare of buyers and sellers.

5.1. OPTIMAL LINEAR PRICE SCHEDULE

In this section, we derive the optimal linear price schedule, which has the form $c_i = c_1 + \delta(i-1)$. We assume there exists an alternative search mechanism with a cost that is strictly proportional to the number of price quotes: $c_{\text{alt}}(i) = ic'$. This could correspond to the cost of manual search, for example, which is at present the most natural alternative to the shopbot.

The shopbot's goal is to set the values c_1 and δ so as to maximize its expected profits. As discussed in Section 4.2, when costs are linear in the number of comparisons, a population of rational, fully-informed buyers evolves towards an equilibrium in which $w_1 + w_2 = 1$. Suppose that the shopbot takes a long-term view in that it does not factor transients into its calculations, but seeks to set a price schedule $\vec{c}$ that maximizes $\pi_{\text{sh}}(\vec{c}, \vec{w})$, with $\vec{w}$ assumed to be its asymptotic equilibrium value.

Given that $\vec{w}$ stabilizes at an equilibrium in which the marginal benefit β is equal to the marginal cost δ, the shopbot's profit is:

$$\begin{aligned} \pi_{\text{sh}}(\vec{c}) &= w_1 c_1 + w_2 c_2 \\ &= c_1 + w_2\,\delta \\ &= c_1 + w_2 \beta \end{aligned} \tag{22}$$

subject to the following conditions, derived from the step functions in Eq. 21:

$$c_1 \leq c' \tag{23}$$

$$c_2 \leq 2c' \tag{24}$$

$$c_1 \leq 1 - \bar{p}_1 \tag{25}$$

$$c_2 \leq 1 - \bar{p}_2 \tag{26}$$

Note that Cond. 26 can be restated as $c_1 + \delta \leq 1 - (\bar{p_1} - \beta)$, which reduces to Cond. 25, given that $\delta = \beta$ at equilibrium. Thus, only Conds. 23, 24, and 25 need be considered.

First, let us temporarily ignore Conds. 24 and 25. From Cond. 23 and Eq. 22, it is evident that c_1 should be set to c', and that δ should be set so as to maximize $w_2\beta$. We can use Eq. 17, which makes explicit the dependence of β on w_2, to solve numerically for the optimal value of w_2. The result is $w_2^* = 0.781796$, at which point $\beta^* = \beta(w_2^*) = 0.095741$.

It would appear that the shopbot could encourage the buyer population to evolve towards w_2^* by setting its price schedule such that $\delta^* = \beta^* = 0.095741$. However, fixing δ at this value is not guaranteed to have the desired effect. Recall from Fig. 7.4(b) that there are *two* values of w_2 that correspond to $\delta^* = 0.095741$. In addition to the desired value $w_2^* = 0.781796$, there is an unstable solution at $w_2 = 0.465602$. If the initial value of w_2 exceeds 0.465602, the buyer population will evolve towards the desired equilibrium $w_2^* = 0.781796$. However, if the initial value of w_2 is less than 0.465602, the buyer population will evolve towards the undesirable equilibrium solution $w_2 = 0$. Interestingly, as shall be illustrated in Section 5.2, the shopbot can cope with low initial values of w_2 by using a time-dependent pricing strategy to shift the system towards the desired equilibrium. Therefore, we shall proceed with an analysis that assumes that the desired equilibrium can be reached regardless of initial conditions.

According to our analysis thus far, $c_1 = c'$ and $c_2 = c' + \delta^*$. However, this violates Cond. 24 whenever $c' < \delta^*$. In this regime, the shopbot cannot charge as much as δ^* for the second quote because buyers would opt for the cheaper alternative search. Instead, to optimize its profits, the shopbot must match the cost of the alternative search method by setting $\delta = c'$ whenever $c' < \delta^*$. At equilibrium, this implies that $\beta(w_2) = c'$, or inversely, $w_2 = \beta^{-1}(c')$. Again, there are two possible solutions that maximize $w_2\beta(w_2)$. The favored solution is the one for which the value of w_2 is larger. Again, we assume that the shopbot can use time-dependent pricing to manipulate the buyer population toward the favored solution.

Next, consider the effect of Cond. 25. Substitution of Eqs. 10 and 15 into Cond. 25 yields the constraint $c_1 \leq \phi(w_2)$, where

$$\phi(w_2) = 1 - \left[\frac{1 - w_2}{2w_2}\right] \log\left(\frac{1 + w_2}{1 - w_2}\right). \tag{27}$$

This constraint comes into play when $c' > \phi(w_2^*) = 0.706946$.[9] In this regime, c' is so large that, with $c_1 = c'$ and $\delta = \delta^*$, the expected price plus the search cost exceeds every buyer's valuation. Consequently, all buyers opt out of the market, and the sellers and the shopbot earn no profits. To optimize its profit, the shopbot must therefore reduce δ below δ^*, which has the effect of increasing w_2 above w_2^*, thus lowering the expected prices for all buyers. The shopbot's optimal strategy is to set δ to the value at which buyers are marginally interested in shopping, which occurs when Cond. 25 is satisfied with equality, *i.e.*, $c_1 = 1 - \bar{p}_1 = \phi(w_2)$. Thus, the optimal value of w_2 in this regime is determined by inverting $\phi(w_2) = c'$, which yields $w_2 = \phi^{-1}(c')$. To ensure that this value of w_2 is attained, the shopbot must set $\delta = \beta(w_2) = \beta(\phi^{-1}(c'))$.

Finally, note that since $w_2 \leq 1$, the results of the previous paragraphs hold only when $c' \leq 1$. If $c' > 1$, then the alternative search mechanism is not viable because a single price quote costs more than the buyer's valuation of the good. In this case, the shopbot is effectively a monop-

Table 7.2. Optimal linear shopbot price schedule in terms of c_1 and δ, and distribution of the buyer population w_2, as a function of alternative search cost c'. Numerical values of the special constants appearing in the table are: $\delta^* = 0.095741$, $w_2^* = 0.781796$, and $\phi(w_2^*) = 0.706946$.

c'	c_1	δ	w_2
$0 \le c' \le \delta^*$	c'	c'	$\beta^{-1}(c')$
$\delta^* \le c' \le \phi(w_2^*)$	c'	δ^*	w_2^*
$\phi(w_2^*) \le c' \le 1$	c'	$\beta(\phi^{-1}(c'))$	$\phi^{-1}(c')$
$c' > 1$	1	0	1

olist, and its optimal strategy is to set $c_1 = 1$ and $\delta = 0$. This price schedule encourages all buyers to be comparison shoppers ($w_2 = 1$), forcing sellers' prices to zero. Thus, sellers earn no profits, and buyers gain no surplus because the entire amount of their valuation is paid to the shopbot for the first price quote. The shopbot extracts all of the market surplus.

The analytic results for c_1, δ, and w_2 in the four relevant ranges of c' are summarized in Table 7.2 and illustrated in Fig. 7.7. Table 7.3 displays, over the four relevant ranges of c', three quantities of interest that can be computed from the values in Table 7.2. The shopbot's profit π_{sh} is computed using Eq. 22. The total seller profits, $S\pi$, are computed using Eq. 5 with $\pi_m = 1$, which yields $w_1 = 1 - w_2$. Finally, the average buyer surplus σ is computed as follows:

$$\begin{aligned}
\sigma &= w_1(1 - (\bar{p}_1 + c_1)) + w_2(1 - (\bar{p}_2 + c_2)) \\
&= w_1(1 - \bar{p}_1) + w_2(1 - \bar{p}_2) - \pi_{\text{sh}} \\
&= \phi(w_2) + w_2\beta(w_2) - \pi_{\text{sh}} \\
&= w_2 - \pi_{\text{sh}}
\end{aligned}$$

The total social welfare is $\pi_{\text{sh}} + S\pi + \sigma = \pi_{\text{sh}} + 1 - w_2 + w_2 - \pi_{\text{sh}} = 1$.[10] In other words, the market is efficient regardless of the value of c'.

Fig. 7.8(a) graphically displays the information contained in Table 7.2. For comparison, Fig. 7.8(b) depicts the same quantities under a scenario in which no shopbot is present, and price information costs c' per quote. The behavior of the market in the absence of a shopbot has two regimes of interest. If c' is less than the maximal value of $\beta(w_2)$, namely 0.103872, then there are three equilibria. The first one is the trivial stable equilibrium at $w_2 = 0$. The two non-trivial equilibria are the solutions to $\beta(w_2) = c'$. Of these, the one with the larger value of w_2 is stable, while the one with the smaller value of w_2 is unstable. If

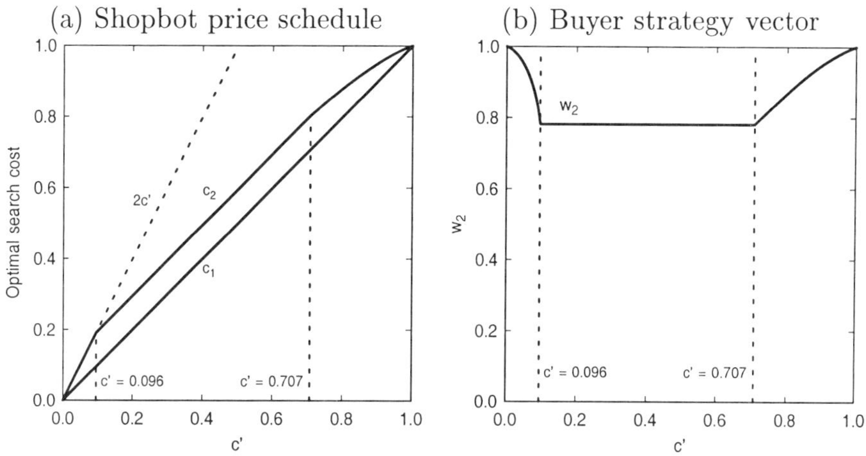

Figure 7.7. (a) Optimal linear shopbot price schedule as a function of alternative search cost c'. (b) Equilibrium values of buyer strategy vector $\vec{w} = (w_1, w_2)$.

Table 7.3 . Shopbot's profit π_{sh}, total seller profits $S\pi$, and average buyer surplus σ.

c'	π_{sh}	$S\pi$	σ
$0 \leq c' \leq \delta^*$	$c'(1+\beta^{-1}(c'))$	$1-\beta^{-1}(c')$	$\beta^{-1}(c')(1-c')-c'$
$\delta^* \leq c' \leq \phi(w_2^*)$	$c' + \delta^* w_2^*$	$1 - w_2^*$	$\phi(w_2^*) - c'$
$\phi(w_2^*) \leq c' \leq 1$	$\phi^{-1}(c')$	$1-\phi^{-1}(c')$	0
$c' > 1$	1	0	0

the initial value of w_2 exceeds that of the unstable non-trivial equilibrium solution, then the buyer population evolves towards the non-trivial stable equilibrium. Otherwise, the system evolves to the trivial stable equilibrium. Assuming that the buyer population stabilizes at the non-trivial equilibrium, the total seller profits and buyer surplus are given by precisely those formulas that appear in the first row of Table 7.3, but the range of validity extends up to $c' = \max \beta(w_2) = 0.103872$ rather than $c' = \delta^* = 0.095741$. This equivalence is apparent for low values of c' in Figs. 7.8(a) and 7.8(b).

If $c' > 0.103872$, then there is no solution such that $\beta(w_2) = c'$. In this case, the system evolves to the trivial stable equilibrium at $w_2 = 0$. Since none of the buyers compare prices, sellers are free to behave as monopolists. If the sellers are unaware of (or do not take into account) the buyers' search cost c'—a tacit assumption in the foregoing analysis[11]—

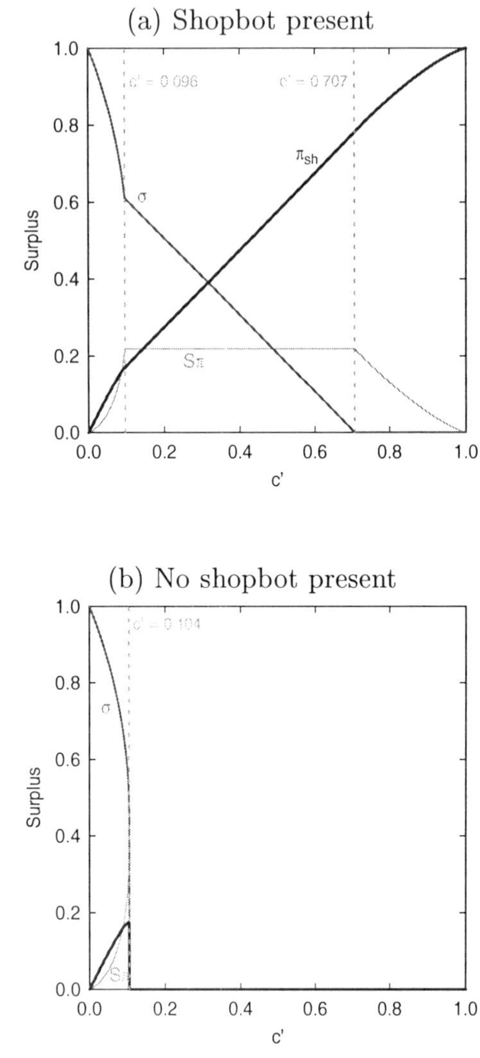

Figure 7.8. Average buyer surplus σ, total seller profits $S\pi$, and total search cost as a function of alternative search cost c'. (a) Shopbot present. Total search cost is equivalent to shopbot profit π_{sh}. (b) No shopbot present. Unlabeled curve that is roughly straight for small c' represents buyers' total expenditure on search.

then they attempt to maximize their profits by charging the buyers exactly their valuations. But since $c' > 0$, the sum of the product and search costs would exceed the buyers' valuations. No goods would be bought or sold, resulting in zero surplus for the buyers, the sellers, and the shopbot.

In the presence of a shopbot (Fig. 7.8(a)), the market behavior is quite different. The shopbot deliberately sets the differential search cost to $\delta < c'$ in order to encourage buyers to search. This lowers prices, making it affordable for buyers to stay in the market. This helps both buyers and sellers. For $c' < 0.706946$, buyers receive a positive surplus, and sellers receive a positive profit. For $c' \geq 0.706946$, buyers receive zero surplus, but the sellers continue to eke out a profit that diminishes monotonically towards zero as $c' \to 1$.

5.2. STRATEGIC MANIPULATION OF EQUILIBRIA

In this section we address the issue that was deferred in Section 5.1: when there are multiple equilibria, how can the shopbot control which equilibrium is reached, regardless of initial conditions? The trick is to use a time-dependent pricing strategy to strategically manipulate the equilibria and their basins of attraction so as to guide the market towards the desired equilibrium.

Figure 7.9 illustrates such a scenario, in which it is assumed that $c' = 0.25$. Initially, the shopbot sets the "optimal" linear price schedule as computed from Table 7.2: $c_1 = 0.25$ and $\delta = \delta^* = 0.095741$. The initial strategy vector is taken to be $\vec{w} = (0.95, 0.05, 0, 0, 0)$, reflecting a situation in which there is very little search on the part of the buyers. This initial state lies in the basin of attraction of the trivial (and unfavorable) equilibrium $w_2 = 0$. Throughout the simulation, the market evolves according to Eq. 19, with $\eta = 0.004$. As the co-evolution of buyers' strategies and sellers' price distributions proceeds, sellers' prices rise towards 1. Any buyer who remains in the market pays nearly its valuation *plus* a search cost, and thus receives a negative surplus. Therefore, the buyers opt out of the market at their first opportunity, as is reflected in the growth of the w_0 component in Fig. 7.9b.

Recognizing that the buyer population is destined for an undesirable equilibrium in which the shopbot earns no profit, the shopbot drops its prices to zero at time $t = 1000$. Now in fact the shopbot earns no profit, but since buyers can obtain an unlimited number of quotes for free, the optimal strategy now shifts to $i = 5$. The buyers (who reevaluate their strategy once every 250 time steps on average) gradually shift to this optimal strategy, and hence w_5 approaches 1. This extensive search pins sellers' prices at low values, resulting in essentially no profits for sellers. On the other hand, the buyers experience both low prices and zero search costs, which causes their surplus to rise towards 1.

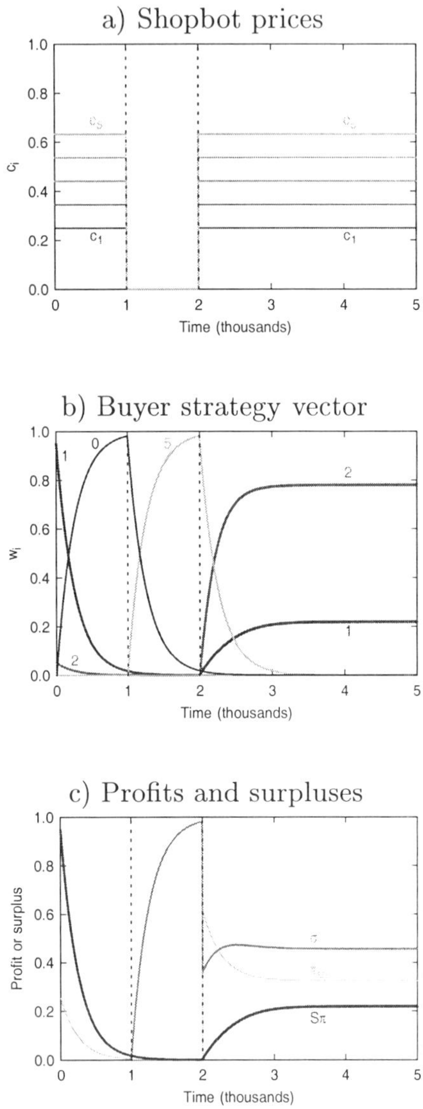

Figure 7.9. Example of how a shopbot can adjust its prices dynamically so as to shift the market from an unfavorable equilibrium to one that maximizes the shopbot's profit. a) Shopbot prices vs. time. b) Evolution of buyer search strategies vs. time, with components $w_i(t)$ labeled according to i. Asymptotic strategy vector is 0.218204 for search strategy 1, 0.781796 for strategy 2, and 0 for all other strategies. c) Shopbot profit, total seller profits, and average buyer surplus vs. time. Asymptotic values are $\pi_{\rm sh} = 0.324850$, $S\pi = 0.218204$, and $\sigma = 0.456946$, respectively.

Finally, at time $t = 2000$, the shopbot reverts back to its original "optimal" linear price schedule. Now that most of the buyers are engaging in extensive search, the buyer population is in a state that lies within the favorable equilibrium's basin of attraction. But since the shopbot is once again charging for search, the buyers search less, ultimately stabilizing at the optimal state in which $w_2 = w_2^* = 0.781796$ and $w_1 = 1 - w_2^*$. In this state, the shopbot realizes its optimal profit of 0.324850, and the total seller profit and the buyer surplus assume the calculated values in Table 7.2 and Fig. 7.8a for $c' = 0.25$: 0.218204 and 0.456946, respectively. In this scenario, the self-interested manipulation of the market by the shopbot benefits the buyers and sellers, who otherwise would not have bought or sold any goods.

5.3. HEURISTIC NONLINEAR PRICE SCHEDULE

In the foregoing analysis, we have assumed that the shopbot adheres to a linear price schedule. In this section, we investigate how a shopbot might attempt to maximize its profits if freed from this constraint. Rather than attempting to compute the optimal nonlinear price schedule (which may be dependent upon initial conditions and time), we introduce an adaptive heuristic by which the shopbot dynamically sets a nonlinear price schedule.

One heuristic for dynamically setting the price schedule is to set $c_1 = c'$, and to set all remaining prices such that $\bar{p}_i + c_i$ is constant for all i. In other words, at time t, set the marginal price of quote i, namely $c_i - c_{i-1}$, equal to the marginal benefit of quote i, given the current vector $\vec{w}(t)$. This heuristic ensures that all search strategies are equally desirable to buyers, and encourages $\vec{c}$ and $\vec{w}$ to reach an equilibrium. It is possible for this heuristic to be overly aggressive in forcing equilibrium, causing the shopbot to settle on a non-optimal price schedule, but in practice it typically yields reasonable solutions.

Figure 7.10 depicts this heuristic in action. Up until time $t = 3000$, the simulation is identical to Fig. 7.9. By this time, the buyer strategy $\vec{w}$ and the shopbot profit have nearly converged to equilibrium. In particular, the shopbot profit has nearly converged to the optimal value that can be attained by a linear price schedule, 0.324850. Then at time $t = 3000$, the heuristic described above is put into effect. On a time scale of a few thousand time steps, the shopbot prices $\vec{c}$ and the buyer strategies $\vec{w}$ co-evolve to an equilibrium in which the c_i are nonlinear in i. The nonlinearity is roughly of the form assumed in Eq. 18, with

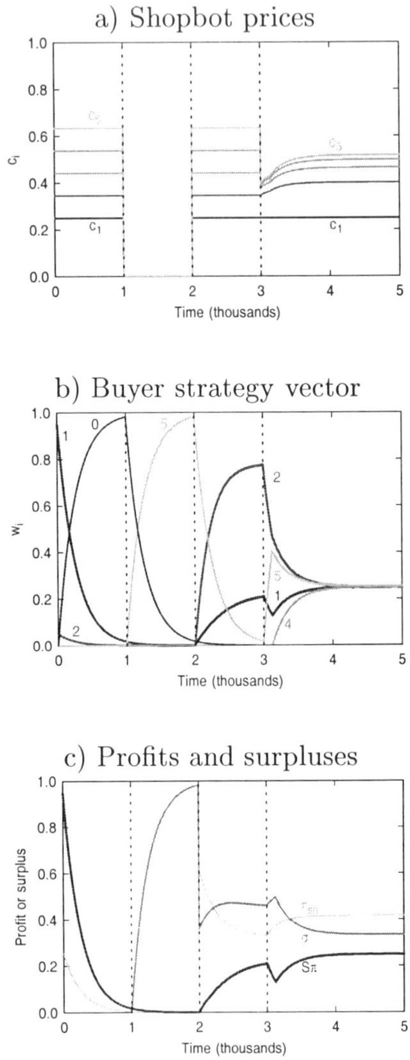

Figure 7.10. Scenario similar to that of Fig. 7.9 except that at times $t \geq 3000$ an adaptive pricing algorithm is employed by the shopbot. The resultant nonlinear price schedule yields higher profits. (a) Shopbot prices vs. time. Asymptotic price schedule is $\vec{c} = (0.250000, 0.401388, 0.465005, 0.498231, 0.517981)$. (b) Evolution of buyer search strategies vs. time, with components $w_i(t)$ labeled according to i. The asymptotic strategy vector is 0.249344 for search strategies 1, 2, and 4, 0.251969 for strategy 5, and 0 for strategies 0 and 3. (c) Shopbot profit, total seller profits, and buyer surplus vs. time. Asymptotic values are $\pi_{\text{sh}} = 0.417165$, $S\pi = 0.249344$, and $\sigma = 0.333491$, respectively.

exponent $\alpha \approx 0.45 < 1$. Thus, the incremental cost of quote i decreases roughly as the square root of i.

The nonlinear price schedule that arises via this heuristic yields the shopbot an equilibrium profit of 0.417165—more than a 28% increase over what is attainable with a linear price schedule. The increase in profits comes about because now about half of the buyers are requesting 4 or 5 price quotes, and paying substantially more for them than they would have paid for 2 price quotes in the linear price schedule. As a side-effect of the shopbot's switch to a nonlinear price schedule, the total seller profit increases by over 14%, from 0.218204 to 0.249344. These increased profits come at the expense of the buyers, whose surplus drops by 27%, from 0.456946 to 0.333491.

In order to understand why the use of a nonlinear price schedule by the shopbot increases the total seller profits, note that, by Eq. 5 and the assumption that $\pi_m = 1$, the total seller profits $S\pi = w_1$. Since the nonlinear pricing heuristic results in a near-even four-way split among search strategies, the asymptotic value of w_1 is very close to 0.25. This is higher than the value $w_1^* = 1 - w_2^* = 0.218204$ that is attained with the optimal linear pricing structure, and the increase in seller profits follows directly from this increase in w_1. The increase in seller profits is not general, however. If the number of sellers increases appreciably beyond 5, then the even splitting of the non-zero components of $\vec{w}$ leads to a *decrease* in w_1, and therefore a decrease in total seller profits as well. We have observed this effect in simulations with 10 and 20 sellers.

The decrease in average buyer surplus after the introduction of the heuristic follows from the fact that the overall surplus in the market is a constant, since the shopbot adjusts its prices to ensure that all buyers purchase the good. Therefore, for $S = 5$, the increase in both the shopbot's and the sellers' profits must be offset by a decrease in the average buyer surplus. The buyer surplus appears to be relatively insensitive to the number of sellers, never varying by more than 5% as S ranges from 5 to 20. As the number of sellers increases, the decrease in total seller profits results in an increase in the shopbot's profit.

The co-evolution of $\vec{c}(t)$ and $\vec{w}(t)$ after $t = 3000$ is somewhat complex. Initially, c_5 drops so low that search strategy 5 is consistently favored. After less than 200 time steps, however, the heuristic begins to equalize the strategies, until ultimately the equilibrium strategy vector is nearly evenly divided among search strategies 1, 2, 4, and 5, with no buyers pursuing search strategies 0 and 3. Since the heuristic is designed to create indifference among the various strategies, the buyers' avoidance of strategy 3 would appear to be a numerical artifact. However, we have consistently obtained the same result under several different experimental conditions, including different initial conditions, different durations of the period of free search, and different values of η. In addition, for related scenarios with more sellers, we typically find that some components of the equilibrium strategy vector tend to zero, and that often the remainder is divided evenly (or nearly so) among the non-zero components. Most likely, there is a subtle effect associated with the finite size of η that prevents exact indifference among the strategies.

In summary, a shopbot can increase its profit by using a well-tuned nonlinear price schedule. This price schedule can influence seller profit and buyer surplus in ways that are not intuitively obvious. Relative to the optimal linear price schedule, the nonlinear price schedule may result in increased or diminished seller profits, depending on the number of sellers, while the average buyer surplus is typically diminished. These results should be regarded as illustrative rather than general. A more general investigation would explore how nonlinear shopbot pricing affects the market under different pricing heuristics assuming different distributions of buyer valuations. Furthermore, as shopbots become more prevalent, one can expect the emergence of competition among shopbots. The ability of shopbots to manipulate the market will be greatly diminished when this occurs, and the resultant shopbot pricing strategies and market dynamics are likely to be radically different from what we have explored here for a monopolist shopbot.

6. RELATED WORK

The study of the economics of information was launched in 1961 with the seminal work of Stigler (Stigler, 1961). Stigler cited several empirical examples of price dispersion, which he attributed to the costly search procedures faced by consumers. He stressed the utility of trade journals and organizations that specialize in the collection and dissemination of product information, such as *Consumer Reports*. Stigler also reminded us that, in medieval times, localized marketplaces thrived in spite of the heavy taxes levied on merchants, demonstrating how worthwhile it was for sellers to participate in such markets rather than undertake individual searches for buyers. Shopbots create analogous localized marketplaces; accordingly, we observe some sellers sponsoring shopbots and paying commissions on sales, essentially paying for the right to participate in the shopbot marketplace.

Since Stigler, many economists have developed and analyzed formal models that attempt to explain the phenomenon of price dispersion.[12] We have already mentioned the work of Varian (Varian, 1980) and Burdett and Judd (Burdett and Judd, 1983). The former study is a special case of our model of shopbot economics in the case of two types of buyers, namely type 1 and type S, existing in fixed proportions. The latter authors, on the other hand, consider independent buyer decisions among a set of search rules of fixed sample size. In this paper, we integrate these two approaches by specifying a fixed minimal proportion of buyers of type 1 while allowing all other buyers to choose their sample sizes. Moreover, we extend these ideas with the notion that shopbots themselves may well price for the information services they provide, acting as economic agents in their own right; this creates an additional strategic variable, namely the cost of search, as determined by shopbots attempting to maximize their profitability.

In the absence of shopbots, a model of sellers' price adjustment was studied in Diamond (Diamond, 1971), in which a somewhat paradoxical outcome arises: for any positive search costs, no consumers search and all sellers charge the monopolistic price v. More recent work on the dynamics of price-setting includes the evolutionary approach of Hopkins and Seymour (Hopkins and Seymour, 1996), where it is argued that the symmetric Nash equilibrium in mixed strategies is dynamically unstable. This raises concerns about the validity of assuming that $f(p)$ describes sellers' behavior. Since we also assume that sellers are represented by software agents, however, it is perhaps plausible to consider pricing in this manner, as $f(p)$ is computed according to a well-specified algorithm. Nonetheless, we are currently studying learning algorithms that evolve stably to *asymmetric* mixed strategy Nash equilibria in this model (Greenwald and Kephart, 2001); these may be more likely to arise as equilibrium outcomes in a dynamic economy of software agents.

Bakos (Bakos, 1997) has recently considered the potential impact of reduced buyer search costs on the electronic marketplace. His model allows for product differentiation, but it does not allow for a variety of buyer types. It remains to incorporate product differentiation into our model of buyer search. Baye and Morgan (Baye and Morgan, 2001) consider the consequences of introducing into a commodity market a profit-seeking "gatekeeper", an intermediary that charges both buyers and sellers for price information. Although their assumptions about the market and the nature of the intermediary differ in detail from ours, they observe similar effects. In particular, a gatekeeper can wield enormous power over market dynamics, encouraging buyer search and improving social welfare. But to effectively maximize profits, gatekeepers have to contend with the existence of multiple equilibria.

7. CONCLUSIONS AND FUTURE WORK

In our explorations of the potential economic impact of shopbots on electronic markets, we have proposed a model that is similar to several that have been investigated by economists interested in understanding the phenomenon of price dispersion. However, in contrast to these previous studies, our goals are *prescriptive* rather than *descriptive*. We are interested in designing economically-motivated software agents, and in understanding the market behavior that is likely to emerge from their collective interactions. Our different orientation is reflected in many ways, including our emphasis on the constructive computation of price distributions (as opposed to theorems and lemmas regarding their basic properties), and our extensive study of non-equilibrium behavior (as opposed to proofs of the existence of equilibria).

Arguing that nonlinear search cost schedules are likely to arise naturally, or could even be adopted intentionally by shopbots, we studied their effect within the context of our model. Our findings reveal that nonlinear search costs can lead to more complicated mixtures of buyer strategies and more extensive search than occur with linear costs. Another practical assumption, namely the existence of a positive number of uninformed buyers who do not take advantage of search mechanisms, can lead to similar outcomes. Finally, the evolutionary dynamics of buyer strategies can be complicated, possibly settling into a complex limit cycle. Even when buyer strategies stabilize, there may be multiple equilibria, with the choice of equilibrium governed by the initial state of the buyer strategies, among other factors.

Placing ourselves in the role of shopbot designers, we explored the strategic pricing of price information. Through modeling, analysis, and simulation, we validated one of our earlier assumptions by showing that nonlinear cost schedules come about as the natural consequence of endowing shopbots with economic incentives. Even in the face of competition from an alternative search mechanism, shopbots can wield a good deal of control over markets; specifically, they can manipulate search costs so as to extract an appreciable fraction of the market surplus. In addition, we showed how buyers can benefit from the shopbot's self-interested price manipulation. The power of an individual shopbot would of course be diminished considerably if there were competition among two or more shopbots. A study of coupled markets in which shopbots compete as providers of price and product information services to buyers or buyer agents would be fascinating.

In closing, we mention two important future directions of our work. First, it is necessary to develop a more faithful representation of buyer behavior that explicitly takes both product and vendor attributes into account. By conducting an empirical study of the behavior of buyers who use the shopbot provided at `DealTime.com` (known as `evenbetter` at the time of their study), Brynjolfsson and Smith (Bynjolfsson and Smith, 2000) have found that brand and consumer loyalty can lead buyers to select higher-priced vendors. Shopbots like `mysimon.com` are already starting to provide additional information about product attributes, and moreover, they have experimented with eliciting utility functions from buyers that can be used to compute personalized rankings of product offerings. It would be of both theoretical and practical interest to analyze and simulate a model that accounts for these various aspects of buyer behavior, perhaps along the lines of Bakos' analysis of horizontal differentiation in electronic markets (Bakos, 1997) or the analysis of vertical differentiation of Sairamesh *et al.* (Sairamesh and Kephart, 2000). In addition to studying the effect of more realistic buyer behavior, it would be desirable to introduce more realistic models of seller pricing behavior. Previous studies of asynchronous (Greenwald et al., 1999) and adaptive (Greenwald and Kephart, 2001) pricing in markets with fixed-strategy buyers have revealed interesting nonequilibrium behavior, which may well become even more complex in nature when the buyers are adaptive.

Acknowledgments

The authors gratefully acknowledge the support of IBM's Institute for Advanced Commerce.

Appendix: No Pure Strategy Nash Equilibrium

In this appendix, it is shown that no pure strategy Nash equilibria exist in our model of shopbot economics whenever (i) prices are chosen from a sufficiently large (or continuous) set, and (ii) $0 < w_1 < 1$.[13] If $w_1 = 1$, then the unique Nash equilibrium is such that all sellers charge the monopoly price p_m. If $w_1 = 0$, then at equilibrium at least two sellers charge the competitive price r and all sellers earn zero profits (see (Greenwald, 1999; Greenwald and Kephart, 1999)). Our proof proceeds by contradiction: we assume the existence of a pure strategy Nash equilibrium, derive the unique form of such an equilibrium, and then argue that a strategy profile of this form is not individually optimizing for all sellers.

Assuming the existence of a pure strategy Nash equilibrium, we rederive Eq. 3 of Section 3, which describes the probability $h_s(\vec{p}, \vec{w})$ that buyers choose seller s. As in Section 3,

$$h_s(\vec{p}, \vec{w}) = \sum_{i=0}^{S} w_i h_{s,i}(\vec{p}) \tag{7.A.1}$$

where $h_{s,i}(\vec{p})$ denotes the demand for seller s by buyers of type i. Since we are now assuming pure rather than mixed strategies, it is convenient to define the following functions in deriving $h_{s,i}(\vec{p})$, a quantity whose probabilistic analog was previously expressed in terms of cumulative distribution functions:

- $\mu_s(\vec{p})$ is the number of sellers charging a higher price than s;
- $\tau_s(\vec{p})$ is the number of sellers charging the same price as s, excluding s itself;
- $\lambda_s(\vec{p})$ is the number of sellers charging a lower price than s.

The quantity $h_{s,i}(\vec{p})$ is is the product of the probability x_s that s is among i potential sellers selected at random, and the conditional probability y_s that s is either the lowest-priced seller or the lucky one in a set of lowest-priced sellers, given a set of i potential sellers that includes s. The probability x_s that s is one of a set of i potential sellers selected at random is simply $x_s = \binom{S-1}{i-1}/\binom{S}{i}$.

The conditional probability y_s is a function of $\lambda_s(\vec{p}), \tau_s(\vec{p})$, and $\mu_s(\vec{p})$. For convenience, fix seller s, and abbreviate $\nu \equiv \nu_s(\vec{p})$, for $\nu \in \{\lambda, \mu, \tau\}$. Given τ sellers charging the same price as seller s, the probability y_s that s is one of the lowest-priced sellers among i potential sellers, and moreover, that s is randomly selected from among $0 \leq t \leq \tau$ such lowest-priced sellers, is given by:

$$y_s(\lambda, \tau, \mu) = \sum_{t=0}^{\tau} \frac{1}{t+1} x_s(\lambda, \tau, \mu, t) \tag{7.A.2}$$

where $x_s(\cdot, t)$ denotes the probability that s is one of the t lowest-priced sellers of i potential sellers. The quantity $x_s(\cdot, t)$ is computed by determining the number of ways in which to arrange $i-1$ of the remaining $S-1$ sellers, exclusive of seller s, such that t sellers charge the same price as s, but no seller charges a price less than s, divided by the total number of ways in which to choose $i-1$ other potential sellers from the remaining $S-1$: *i.e.*,

$$x_s(\lambda, \tau, \mu, t) = \frac{\binom{\lambda}{0}\binom{\tau}{t}\binom{\mu}{i-1-t}}{\binom{S-1}{i-1}}. \tag{7.A.3}$$

Combining the expressions for x_s, y_s, and z_s and simplifying, we arrive at the following:

$$h_{s,i}(\vec{p}) = \binom{S}{i}^{-1} \frac{1}{\tau+1} \sum_{t=0}^{\tau} \binom{\tau+1}{t+1} \binom{\mu}{i-(t+1)}. \tag{7.A.4}$$

Finally, as in Section 3, let $\rho B = 1$ and $g(p) = \Theta(v - p)$, which yields the sellers' profit function $\pi = (p - r)h_s(\vec{p}, \vec{w})$ with $h_s(\vec{p}, \vec{w})$ defined by Eqs. 7.A.1 and 7.A.4.

We now proceed to derive the unique form of a possible pure strategy Nash equilibrium. Suppose that the sellers are ordered $s_1, \ldots, s_j, \ldots, s_S$ such that the indices $i < j$ whenever equilibrium prices $p_i^* \leq p_j^*$. First, note that equilibrium prices $p_j^* \in (r, v]$, since $p_j < r$ yields strictly negative profits, but $p_j = r$ and $p_j > v$ yield zero profits, and $p_j \in (r, v]$ yields strictly positive profits. The following observation describes the form of a pure strategy Nash equilibrium whenever $0 < w_1 < 1$: *at equilibrium, no two sellers charge identical prices.*

Suppose that, on the contrary, two distinct sellers offer equivalent prices: *i.e.*, $r < p_1^* < \ldots < p_j^* = p_{j+1}^* < \ldots < p_S^* \leq v$. In this case, seller j stands to gain by undercutting seller $j+1$ by ϵ, implying that p_j^* is not in fact an equilibrium price. In particular,

$$\begin{aligned}\pi_j(p_j^* - \epsilon, p_{-j}^*) &= \textstyle\sum_{i=0}^{S} w_i \binom{S}{i}^{-1} \binom{\mu+1}{i-1} (p_j - \epsilon - r) \\ &> \textstyle\sum_{i=0}^{S} w_i \binom{S}{i}^{-1} \left[\binom{\mu}{i-1} + \frac{1}{2}\binom{\mu}{i-2}\right] (p_j - r) \\ &= \pi_j(p_j^*, p_{-j}^*)\end{aligned}$$

for sufficiently small values of ϵ. The price quantum ϵ is sufficiently small precisely when the set of possible prices is sufficiently large: the set of prices can be continuous or it can be discretized such that $\epsilon = (v - r)/(|P| - 1)$, where $|P|$ is the cardinality of the price set.

Further, we observe that seller S charges price v at equilibrium, since $\pi_S(v, p_{-S}^*) = \frac{1}{S} w_A(v - r) > \frac{1}{S} w_A(p_S - r) = \pi_S(p_S, p_{-S}^*)$, for all $p_{S-1}^* < p_S < v$. Therefore, the relevant price vector consists of S distinct prices ordered such that $r < p_1^* < \ldots < p_{S-1}^* < p_S^* = v$.

The price vector $(p_1^*, \ldots, p_j^*, \ldots, p_S^*)$, however, is not in fact a Nash equilibrium. It is the case that $p_S^* = v$ is an optimal response to p_{-S}^*, since the profits of seller S are maximized at v given that there exists lower priced sellers $1, \ldots, S-1$. However, price $p_j^* < p_S = v$ is *not* an optimal response to p_{-j}^*. On the contrary, for all sellers $1 \leq j \leq S-1$, seller j has an incentive to deviate, since whenever $\epsilon < p_{j+1} - p_j$,

$$\begin{aligned}\pi_j(p_{j+1}^* - \epsilon, p_{-j}^*) &= \textstyle\sum_{i=0}^{S} w_i \binom{S}{i}^{-1} \binom{\mu}{i-1} (p_{j+1} - \epsilon - r) \\ &> \textstyle\sum_{i=0}^{S} w_i \binom{S}{i}^{-1} \binom{\mu}{i-1} (p_j - r) \\ &= \pi_j(p_j^*, p_{-j}^*).\end{aligned}$$

It follows that there is no pure strategy Nash equilibrium in our model of shopbot economics, assuming sellers choose prices from a large (or continuous) set and $0 < w_1 < 1$.

Notes

1. Typically, these prescriptions are based on an effort to maximize each individual agent's economic welfare.

2. This is referred to as a *nonsequential* (Burdett and Judd, 1983) or *fixed sample size* (Hopkins and Seymour, 1996) search strategy in the economics literature. We permit a search strategy of 0 to allow buyers to opt out of the market entirely, which may be desirable if search costs are prohibitive.

3. There also exist *asymmetric* Nash equilibria that can arise dynamically in this shopbot model (Greenwald and Kephart, 2001).

4. In Eq. 3, $h_s(p, \vec{w})$ is expressed as a function of seller s's scalar price p, given that probability distribution $F(p)$ describes the other sellers' expected prices.

5. Results for general v and r are easily derived by renormalizing any derived prices appropriately. This price normalization, however, is specific to this particular form of $g(p)$, and is not necessarily applicable to more general distributions of valuations.

6. Baye and Morgan (Baye and Morgan, 2001) propose an interesting alternative in which a "gatekeeper" intermediary charges fees to sellers who advertise their prices via the gatekeeper, and charges buyers to access the advertised prices. Markopoulos and Ungar (Markopoulos and Ungar, 2001) study a different scenario in which the sellers themselves charge for their own price information.

7. Eq. 20 depends on the assumption that the shopbot has no production costs.

8. By letting $\Theta(x)$ be 1 for $x \geq 0$, rather than for $x > 0$, we are in effect assuming that buyers will prefer the shopbot to the alternative mechanism if the prices are equal. One could offer plausible rationales supporting preference for either the shopbot or the alternative search mechanism when prices are equal; this choice is more cosmetic than anything else as it allows us to avoid carrying infinitesimals along through the analysis.

9. Thus Conds. 24 and 25 are never violated simultaneously, and can be treated separately, as we have done.

10. For non-normalized values of the buyer valuation v and the production cost r, these quantities can be rescaled to yield total social welfare of $v - r$. This simple rescaling assumes that all buyers have equal valuations v, which motivates the shopbot to manipulate the market to ensure that all buyers buy the good.

11. If the sellers were to take into account the effect of search costs upon buyers' search behavior, they could not legitimately regard $\vec{w}$ as fixed and independent of their own pricing decisions. This additional consideration would require an extensive overhaul of the game-theoretic analysis of Section 3.

12. We mention only a handful of papers that make up this large body of literature, but refer the reader to the bibliography included in Hopkins and Seymour (Hopkins and Seymour, 1996) for additional sources.

13. The case of small, discrete price sets is considered for a related shopbot model in Appendix A of Greenwald and Kephart (Greenwald and Kephart, 2000). In some cases, it is possible to obtain pure strategy Nash equilibria.

References

Bakos, J. Y. (1997). Reducing buyer search costs: Implications for electronic marketplaces. *Management Science*, 43(12).

Baye, M. R. and Morgan, J. (2001). Information gatekeepers on the internet and the competitiveness of homogeneous product markets. *American Economic Review*, 91:454–474.

Burdett, K. and Judd, K. L. (1983). Equilibrium price dispersion. *Econometrica*, 51(4):955–969.

Bynjolfsson, E. and Smith, M. D. (2000). The great equalizer? consumer choice behavior at internet shopbots. Working paper. Draft available on World Wide Web at http://ecommerce.mit.edu/erik.

Chavez, A. and Maes, P. (1996). Kasbah: an agent marketplace for buying and selling goods. In *Proceedings of the First International Conference on the Practical Application of Intelligent Agents and Multi-Agent Technology*, London, U.K.

DeLong, J. B. and Froomkin, A. M. (1998). The next economy? In Hurley, D., Kahin, B., and Varian, H., editors, *Internet Publishing and Beyond: The Economics of Digital Information and Intellecutal Property*. MIT Press, Cambridge, Massachusetts.

Diamond, P. (1971). A model of price adjustment. *Economic Theory*, 3:156–168.

Eriksson, J., Finne, N., and Janson, S. (1996). Information and interaction in MarketSpace — towards an open agent-based market infrastructure. In *Proceedings of the Second USENIX Workshop on Electronic Commerce*.

Green, H. (1998). Good-bye to fixed pricing? *Business Week*, pages 71–84.

Greenwald, A. (1999). *Learning to Play Network Games.* PhD thesis, Courant Institute of Mathematical Sciences, New York University, New York.

Greenwald, A. and Kephart, J. (1999). Shopbots and pricebots. In *Proceedings of Sixteenth International Joint Conference on Artificial Intelligence*, volume 1, pages 506–511.

Greenwald, A. and Kephart, J. (2000). Shopbots and pricebots. In Moukas, A., Sierra, C., and Ygge, F., editors, *Lecture Notes on Artificial Intelligence: Proceedings of the IJCAI Workshop on Agent-mediated Electronic Commerce.* Springer-Verlag.

Greenwald, A. and Kephart, J. (2001). Probabilistic pricebots. In *Proceedings of the Fifth International Conference on Autonomous Agents*, pages 560–567.

Greenwald, A., Kephart, J., and Tesauro, G. (1999). Strategic pricebot dynamics. In *Proceedings of First ACM Conference on Electronic Commerce*, pages 58–67.

Hopkins, E. and Seymour, R. M. (1996). Price dispersion: An evolutionary approach. Unpublished Manuscript.

Kephart, J. O., Hanson, J. E., Levine, D. W., Grosof, B. N., Sairamesh, J., Segal, R. B., and White, S. R. (1998). Dynamics of an information filtering economy. In *Proceedings of the Second International Workshop on Cooperative Information Agents.*

Krulwich, B. (1996). The BargainFinder agent: Comparison price shopping on the Internet. In Williams, J., editor, *Agents, Bots and Other Internet Beasties*, pages 257–263. SAMS.NET publishing (MacMillan). URLs: http://bf.cstar.ac.com/bf, http://www.geocities.com/ResearchTriangle/9430.

Lewis, T. G. (1997). *The Friction-Free Economy: Marketing Strategies for a Wired World.* HarperCollins, New York.

Markopoulos, P. and Ungar, L. (2001). Pricing price information in e-commerce. In *Proceedings of Third ACM Conference on Electronic Commerce.*

Nash, J. (1951). Non-cooperative games. *Annals of Mathematics*, 54:286–295.

Sairamesh, J. and Kephart, J. O. (2000). Price dynamics of vertically differentiated information markets. *Decision Support Systems*, 28(1,2).

Salop, S. and Stiglitz, J. (1977). Bargains and rip-offs: a model of monopolistically competitive price dispersion. *Review of Economic Studies*, 44:493–510.

Schuster, H. G. (1988). *Deterministic Chaos: An Introduction.* VCH, New York.

Stigler, G. J. (1961). The economics of information. *Journal of Political Economy*, 69(3):213–225.

Tirole, J. (1988). *The Theory of Industrial Organization.* The MIT Press, Cambridge, MA.

Tsvetovatyy, M., Gini, M., Mobasher, B., and Wieckowski, Z. (1997). MAGMA: an agent-based virtual market for electronic commerce. *Applied Artificial Intelligence.*

Varian, H. (1980). A model of sales. *American Economic Review, Papers and Proceedings*, 70(4):651–659.

Chapter 8

FINDING THE BEST WAY TO JOIN IN: A DYNAMIC ACCESSION GAME

László Á. Kóczy

Abstract In this paper we define and solve the accession game, a dynamic game with a per-member partition function that satisfies a set of conditions as used by (Yi 1997) containing a union and a set of applicants. The solution gives an equilibrium partition of the players as well as, after (Morelli and Penelle 1997), the optimal path to it, a subgame-perfect sequence of partitions, where each player maximises the present value of its payoffs subject to others' moves.

While our findings are largely theoretical, our motivation and main example throughout the paper is the ongoing extension of the European Union.

Keywords: Per-member partition function game, integration, European Union.

1. INTRODUCTION

Economic and political integration has gained much attention recently. Mergers and acquisitions exceed all previous volumes, bourses coordinate their trading hours, bigger and stronger trading blocks develop just to mention a few. This paper takes the European integration as the main example, mainly focusing on the current, Eastern extension.

With its rapidly developing literature coalitional game theory contributes a lot to the better understanding of integration. Coalition formation games are usually defined in the characteristic function form, that besides its merits, has some shortcomings: As in this form the payoff of a coalition depends only on its members the concept ignores the side effects of coalition formation. (Thrall and Lucas 1963) defined the *partition function*, a generalisation of the characteristic function that assigns a payoff to a coalition given the entire coalition structure or *partition* and hence accounts for externalities too (See (Lucas 1971, pp509-511) for an introduction.). The per-member partition function is its special case although in a different sense than the characteristic function: players are assumed to be symmetric. This simplification enables us to observe the outside effects of coalition formation at a computationally lower cost. We follow (Yi 1997) in assuming a set of conditions.

After the introduction of the notation, terminology and some examples we will discuss the *accession game*, where a number of applicants try to get membership in a dominant coalition larger than the others, called the Union following an optimal path (Morelli and Penelle 1997), or sequence of coalition structures that maximises the present value of future payoffs not independent of the behaviour of other players. We give the solution algorithm of the accession game with an arbitrary number of applicants.

2. GAME THEORETIC FOUNDATIONS

2.1. THE EXTENSION OF YI'S MODEL

(Yi 1997) has introduced a two-stage non-cooperative game using a per-member partition-function with three conditions. For more details on the game and for further references in this subsection see (Yi 1997). We use a cooperative dynamic extension of this game:

- *Stage 1* The initial partition is given exogenously.
- *Stage 2a* Players form their cooperative (binding) strategies.
- *Stage 2b* Players execute the move they have "agreed" upon in stage 2a.

The game is repeated with Stage 2 of the kth game determining the initial partition for Stage 1 of the $k+1$st game. It is important to note that this is a coalition formation game, and hence a strategy really means a partition.

Definition 1 *A* coalition structure $\mathcal{P} = \{P_1, ..., P_k\}$ *is a partition of the player set* $N = \{1, ..., n\}$*;* Π *is the set of coalition structures,* $|\Pi| < \infty$*. We define* payoff *for individual players and coalitions too.*

Assumption 2 (Assumption 2.1, (Yi 1997)) *Players are* ex-ante *identical, and share the coalition payoff equally [p.205].*

Yi [p.209] also shows the following:

Proposition 3 *The ordered pair* $(m, \mathcal{P})$ *fully describes the player belonging to a coalition of size* m *in the partition* $\mathcal{P}$ *including its possible* *strategies. Furthermore, if only pure strategies are allowed, a partition will always be followed by the same partition.*

As players of all coalitions of the same size are identical, only the size of the coalitions they belong to matters. So from now on coalitions are identified by a positive integer, their size, and a partition is a set of positive integers that sum up to n.

Now we can explain the actual play of the game in more detail. By Proposition 3 all players belonging to coalitions of equal size are identical, but we allow identical players to make a decision in which they have different future positions and hence payoffs. The players are assigned to these different roles randomly, by fair draws, and hence this strategy is preferred if its expected "worth" exceeds that of other strategies.

This argument might seem unrealistic in the case of the EU extension as in the case of countries we, in general, attach great importance to individual features. However there are examples when, for instance the first "former Eastern Bloc country" joins a respected international organisation. A company may look for an Asian partner or large companies are often looking for possible acquisitions.

In one *move* we allow one "action" per player: an agreement is settled, the necessary draws are made, the proposed coalitions are formed and payoffs are paid out. Note that the agreement is signed *before* the draws are made. Less relevant approaches would allow interaction between two players at a time only or allow only self-enforcing strategies (too slow/too fast communication). Players who enforce the move are called *perpetrators*, while the rest are the *residuals* (after (Ray and Vohra 1997)). As the definition of the residual coalition is often arbitrary, we use the term informally. We assume that the perpetrators' set is always minimal, meaning that if the set of coalitions of size k is not smaller in partition $\mathcal{P}$ than in $\mathcal{P}'$ and the two are separated by one move, all players in a coalition k in $\mathcal{P}'$ are in k in $\mathcal{P}$, too.

Definition 4 *The* per-member partition function *is the function*

$$v_i : \Pi \longrightarrow \mathbf{R}^N, \quad v = (v_1, \dots, v_n)$$

The pair (N, v) *is a* per-member partition game.

We use Yi's conditions [Section 4.1.] on the per-member partition function.

Condition 1 $v(m, \mathcal{P}) > v(m, \mathcal{P}')$*, where* $m \in \mathcal{P} \cap \mathcal{P}'$ *and* $\mathcal{P}$ *is a refinement of* $\mathcal{P}'$*. This expresses that mergers hurt residual players.*

Condition 2 $v(m_j, \mathcal{P}) < v(k, \mathcal{P}')$, *where* $k = \sum_{i=1}^{j} m_i$, *for all* $i = 1, \ldots, j$, $m_i \geq m_j$ *and* $\mathcal{P} = \mathcal{P}' \cup \{m_1, m_2, \ldots, m_j\} \setminus \{k\}$, *that is, a merger with coalitions that are not smaller is beneficial to the members of the coalition.*

Condition 3 $v(m, \mathcal{P}) < v(m'+1, \mathcal{P}')$, *where* $\mathcal{P}' = \mathcal{P} \cup \{m-1, m'+1\} \setminus \{m, m'\}$, $m \leq m'$, *that is, a member of a coalition is strictly better off by leaving the coalition and joining another that is not smaller.*

2.2. PATH DEPENDENCE

The concept of examining path alternatives leading to the same outcome is due to (Morelli and Penelle 1997). We give the basic definitions, introduce a more general notation and proceed to the set-up of our own model.

Let $\Pi^f(\mathcal{P})$ be the exogenously given set of *feasible partitions* after a move starting from $\mathcal{P}$. (We will allow $\Pi^f(\mathcal{P}) \neq \Pi$ for some $\mathcal{P} \in \Pi$.)

Definition 5 (Path) *A* path π *is a sequence* $\{\mathcal{P}_i\}_{i>0}$ *of partitions such that* $\mathcal{P}_{i+1} \in \Pi^f(\mathcal{P}_i)$. *By Proposition 3 it can be written in the general form:*

$$\pi = \left\{\mathcal{P}_1, \mathcal{P}_2, \ldots, \overline{\mathcal{P}_{\lambda-\kappa+1}, \ldots, \mathcal{P}_\lambda}\right\},$$

where the length *of the path* $\lambda \in \mathbf{N}$ *is maximal, such that for all* $i, j \leq \lambda$, $\mathcal{P}_i = \mathcal{P}_j$ *implies* $i = j$.

The last κ partitions in a path form the equilibrium outcome that is repeated forever. Unless $\kappa = 1$ this is a non-trivial *loop*. The partition after playing the game t times along path π is denoted by $\mathcal{P}_t(\pi)$.

The present value for the player $(m, \mathcal{P})$ along path π from $\mathcal{P}$ is the discounted average of the payoffs along path π that players of type $(m, \mathcal{P})$ obtain, and is denoted by $w^m(\pi)$. Let Φ be the set of paths that can be enforced by the same player, assuming rational behaviour from the others. This means that if this player proposes the next partition along any path $\pi \in \Phi$ others are either willing to cooperate or cannot block the formation of the new partition.

Definition 6 (Value) *The* expected present value *or shortly* value *is defined by*

$$w^m = \max_{\pi \in \Phi} w^m(\pi), \tag{1}$$

and the optimal path

$$\pi^m \in \arg\max_{\pi \in \Phi} w^m(\pi), \tag{2}$$

A path is a *solution* if the outcome cannot be improved and the corresponding game is coalition-proof and subgame perfect ((Bernheim 1987; Morelli and Penelle 1997)).

If the payoff of a single player is considered as its annual revenue the value gives the total revenue potential that can be achieved as a minimal figure. The optimal path, in turn, gives the path that should be followed to obtain this value. Note that not all moves depend on the player in question, sometimes this player is in the residual set, but we may count on other players being also rational and thus that they follow their corresponding "instructions", namely this equilibrium path.

3. THE ACCESSION GAME

3.1. INTRODUCTION

In this section we define the accession game: a game of the extension of the union, a dominant coalition denoted by S_t, with $|S_t| = s_t$ at time t. It will lead to no confusion if we write simply S with s referring to the actual size of S.

Assumption 7 (Monotonicity) *We will restrict our attention to paths where the union does not secede, that is, $S_t \subseteq S_{t+1}$ for all t, the size of the union is monotone increasing.*

As a result for all nontrivial games there exists $\mathcal{P} \in \Pi$ such that $\Pi^f(\mathcal{P}) \neq \Pi$. Assumption 7 enables us to solve the games in a simple way. While this is a strong assumption, in the case of the EU we certainly feel justified to make it.

We can have two definitions of the applicants' sets, A: the set of initial applicants, $N \setminus S_0$ for all t is the natural definition in the sense that it does not change in the course of the game. The alternative we use, the set of current applicants or outsiders, $N \setminus S_t$, on the other hand does not preserve history; as soon as some of the applicants join S, a new game is considered with fewer applicants, giving rise to an inductive solution. Similarly we could make a distinction between the original and current members of the union, but their interests coincide. We define values for the sets A and S, as the mean of their respective members' (elements') values.

The focus of the solution is on the conflict between the interests of the applicants' set A and the union S. Although we do not aim to model the formation of the cooperative agreement, each step of the game can be pictured as a bargaining process: If no offers are made by the union or the offers are not accepted, the applicants play a *disagreement strategy* by repartitioning themselves to obtain the highest possible value without acceptance. The union makes its most preferred offer if this does not give it highest value. The offer is either accepted if a subset of applicants is willing to take it, and is able to enforce it, or it is not, in which case the union makes its next most preferred offer as long as this gives an improvement over the disagreement strategy.

For an illustration we turn to the relation of the EU (or its predecessors) and the European Free Trade Association (EFTA). When in 1958 the six countries signed the Treaty of Rome, some left-out countries found that forming a counter-alliance –the EFTA– was the optimal reaction: the disagreement strategy. When later some EFTA members got a membership offer in 1970 most of them joined disregarding the probably negative effects of this move on the remaining EFTA members as they were willing to join, and could enforce this. Norway had, since then several times the possibility of joining, but was unwilling to take the offer.

The first of the two approaches we consider is the pure non-transferable utility (NTU) game, that is, where individual payoffs are defined by the payoff function only. In the second we allow payoff transfers among applicants, so there is a way to reward players for their cooperation. In this case the applicants' aggregated preference is expressed by their total expected present value in the proposed partition. This approach maximizes the applicants' value along the optimal path.

Here we must note that the second approach is not the transferable utility or TU case, as here we permit transfers across coalitions, while in TU games the transfers are allowed among members within, coalitions, an option that is not possible in a per-member partition function game.

3.2. AN EXAMPLE

The table below shows the payoffs of game G. The headings refer to the size of the coalition, so $v(s+1, \{s+1,1,1\})$ must be read from the column with $\geq s$ at the top, and along line $\mathcal{P}_3 = \{s+1,1,1\}$.

v_G		1	2,3	$\geq s$
$\mathcal{P}_0=$	$\{s+3\}$			4
$\mathcal{P}_1=$	$\{s+2,1\}$	0		5
$\mathcal{P}_2=$	$\{s+1,2\}$		2	6
$\mathcal{P}_3=$	$\{s+1,1,1\}$	1		7
$\mathcal{P}_4=$	$\{s,3\}$		3	3
$\mathcal{P}_5=$	$\{s,2,1\}$	1	4	4
$\mathcal{P}_6=$	$\{s,1,1,1\}$	2		5

As the union cannot secede, increasing Π by looking further back into the past of S to partitions, where some of its currents members were "still" applicants does not increase the set of feasible outcomes Π^f. Thus we can have an inductive argument: first solve for the trivial case when we have 0 applicants (partition $\mathcal{P}_0$), and then given the solution for i-applicants, we can solve for $i+1$ applicants. The number of applicants is finite, so in a finite number of steps we arrive to the case we aim to solve. For 0 applicants the solution is trivial.

1 applicant By monotonicity, the set of strategies is $\{\mathcal{P}_1, \mathcal{P}_0\}$, and the only possible disagreement strategy is $\mathcal{P}_1$, with $w^S = \frac{5}{1-\delta}$. At partition $\mathcal{P}_0$ (the grand coalition) the dominant coalition would get $\frac{4}{1-\delta}$ as calculated for the game with zero applicants in the previous step. As $\frac{5}{1-\delta} > \frac{4}{1-\delta}$ it prefers the disagreement strategy, that is, not to make an offer of acceptance to the applicant.

2-applicants The disagreement strategy of the applicants is $\mathcal{P}_2$ giving $\frac{6}{1-\delta}$ to the union. As this value is not exceeded by any partitions including an offer, this strategy is played. Note that the values at $\mathcal{P}_2$ have not been calculated yet, but once players agreed to disagree, no further move is possible, so with the assumption that this will be the disagreement strategy it can be calculated from the payoff ($w = \frac{v}{1-\delta}$).

3-applicants We deal with partitions $\mathcal{P}_6$ and $\mathcal{P}_4$ first. If the union makes no offers, $\mathcal{P}_4$ is preferred to $\mathcal{P}_6$. Furthermore, although the applicants are indifferent between the values of $\mathcal{P}_4$ and $\mathcal{P}_5$, $\mathcal{P}_4$ gives the union a lower payoff while it is still credible. Observe that this is not a Pareto efficient outcome, but at the same time the union has no means enforce $\mathcal{P}_5$. The disagreement payoffs are therefore $w^A = 3\frac{1}{1-\delta}$ and $w^S = 3\frac{1}{1-\delta}$ which is the least favourable of all strategies for S. The applicants prefer $\mathcal{P}_0$ most and hence this is played.

Partition $\mathcal{P}_5$ is the only case in this simple game, where the applicants have inhomogeneous preferences. Since the 2-coalition does not benefit from a merger with the singleton ($v(3, \mathcal{P}_4) < v(2, \mathcal{P}_5)$), the disagreement outcome is $\mathcal{P}_5$, giving $w^1 = \frac{1}{1-\delta}$, $w^2 = \frac{4}{1-\delta}$, $w^s = \frac{4}{1-\delta}$.

The table below summarises the steps as the different preference orderings are evaluated with the relation signs expressing preferences. In Stage I we can remove the strategies that are dominated by the disagreement strategy for the union (only $\mathcal{P}_0$).

From here the strategies are eliminated from backwards. The last chance to improve payoffs before disagreement is $\mathcal{P}_1$. In Stage II the doubleton benefits from accepting the last offer $\mathcal{P}_1$, so the disagreement strategy $\mathcal{P}_5$ is never played.

Stage I	1	$\mathcal{P}_2 > \mathcal{P}_0 > \mathcal{P}_3 > \mathcal{P}_5 > \mathcal{P}_1$
	2	$\mathcal{P}_1 > \mathcal{P}_0 = \mathcal{P}_3 = \mathcal{P}_5 > \mathcal{P}_2$
	s	$\mathcal{P}_3 > \mathcal{P}_2 > \mathcal{P}_1 > \mathcal{P}_5 = \mathcal{P}_0$
Stage II	1	$\mathcal{P}_2 > \mathcal{P}_3 > \mathcal{P}_5 > \mathcal{P}_1$
	2	$\mathcal{P}_1 > \mathcal{P}_3 > \mathcal{P}_5 = \mathcal{P}_2$
	s	$\mathcal{P}_3 > \mathcal{P}_2 > \mathcal{P}_1 > \mathcal{P}_5$
Stage III	1	$\mathcal{P}_2 > \mathcal{P}_3 > \mathcal{P}_1$
	2	$\mathcal{P}_2 > \mathcal{P}_3 > \mathcal{P}_1$
	s	$\mathcal{P}_3 > \mathcal{P}_2 > \mathcal{P}_1$
Stage IV	1	$\mathcal{P}_2 > \mathcal{P}_3$
	2	$\mathcal{P}_2 > \mathcal{P}_3$
	s	$\mathcal{P}_3 > \mathcal{P}_2$
Stage V	1	$\mathcal{P}_3$
	2	$\mathcal{P}_3$
	s	$\mathcal{P}_3$

Foreseeing these actions, the singleton will accept the previous offer in Stage III, $\mathcal{P}_2$, as it improves its payoff, and it can enforce it. This is the worst possible outcome for the doubleton, it prefers the preceding offer $\mathcal{P}_3$, and can enforce it, and hence this is the outcome for the game. The union exploited the tension among applicants very well: its first offer is accepted, while for the applicants $\mathcal{P}_0$ Pareto-dominates this outcome. The table below summarises the results for all i-s in game G with nontransferable utilities.

Game G	$\mathcal{P}$	m	π^*	w
0-applicants	$\mathcal{P}_0$	$s+3$	$\{\overline{\mathcal{P}_0}\}$	$4\frac{1}{1-\delta}$
1-applicant	$\mathcal{P}_1$	1	$\{\overline{\mathcal{P}_1}\}$	0
		$s+2$		$5\frac{1}{1-\delta}$
2-applicants	$\mathcal{P}_2$	2	$\{\overline{\mathcal{P}_2}\}$	$2\frac{1}{1-\delta}$
		$s+1$		$6\frac{1}{1-\delta}$
	$\mathcal{P}_3$	1	$\{\mathcal{P}_3, \overline{\mathcal{P}_2}\}$	$1+2\frac{\delta}{1-\delta}$
		$s+1$		$7+6\frac{\delta}{1-\delta}$
3-applicants	$\mathcal{P}_4$	3	$\{\mathcal{P}_4, \overline{\mathcal{P}_0}\}$	$3+4\frac{\delta}{1-\delta}$
		s		$3+4\frac{\delta}{1-\delta}$
	$\mathcal{P}_5$	1	$\{\mathcal{P}_5, \mathcal{P}_3, \overline{\mathcal{P}_2}\}$	$1+\delta+2\frac{\delta^2}{1-\delta}$
		2		$4\frac{1}{1-\delta}$
		s		$4+7\delta+6\frac{\delta^2}{1-\delta}$
	$\mathcal{P}_6$	1	$\{\mathcal{P}_6, \overline{\mathcal{P}_0}\}$	$2+4\frac{\delta}{1-\delta}$
		s		$5+4\frac{\delta}{1-\delta}$

If we allow transfers among applicants, $\mathcal{P}_4$ becomes the disagreement strategy as the singleton can compensate the 2-coalition for its loss when moving to $\mathcal{P}_4$, and thus the strategy offering the lowest value to the union S becomes a credible threat, and offer $\mathcal{P}_0$ is made. Remarkably, transfers never take place, as the threat is never executed; if the union believes that transfers *could* take place, a better outcome is achieved.

In the real life, however, it is hard to make a clear distinction between the two games. When we talk about countries the payoffs are in many different shapes and forms: the GDP, trade volume, budget deficit or surplus are just as well part of it as military or political benefits even in the case of a primarily economic union as the EU. Transfers can then be loans, preferential treatment or support of political interests.

While the above was only an example in more general games with more applicants there are more partitions where the lack of transfers is disadvantageous. Therefore the unfriendly, competitive behaviour among the EU applicants can turn to their disadvantage by ensuring the EU that transfers are unlikely to take place among them.

3.3. THE GENERAL FORM

In the general accession game we allow transfers among the applicants. Such a game, even with an arbitrary number of applicants simplifies to a two-player game between the union S and the applicants A. At each partition $\mathcal{P}$, given the corresponding $\Pi^f(\mathcal{P})$ the next move is determined the same way: Given $\mathcal{P}$, both S and A can assign a value to any partition in $\Pi^f(\mathcal{P})$. Given these, the union proposes its favoured partition. The applicants can accept or reject this in which case the union makes further proposals as long as these are better than the disagreement strategy:

$$\mathcal{P}_D \in \arg \max_{\substack{\mathcal{P}' \in \Pi^f \\ |S| \in \mathcal{P}'}} \left\{ w^A \left(\mathcal{P}' \right) \right\} \tag{3}$$

By perfect knowledge, applicants may choose their most preferred offer. This is an equilibrium by construction and formally we have the solution:

$$\mathcal{P}^* = \arg \max_{\mathcal{P} \in \Pi^f(\mathcal{P})} \left\{ w^A \left(\mathcal{P} \right) \middle| w^S \left(\mathcal{P} \right) > w^S \left(\mathcal{P}_D \right) \right\} \tag{4}$$

3.4. SOLVING THE GENERAL ACCESSION GAME

Our solution will be inductive. Let $\Pi^a = \left\{ \mathcal{P} \in \Pi \;\middle|\; n - s = a \right\}$ denote the set of partitions with exactly a applicants. For $a = 0$, $\Pi = \{\{n\}\}$ and the solution is trivial. When solving for $a + 1$ we assume that for all $\mathcal{P} \in \Pi^k$ with $0 \leq k < a$ the solution is known. Now let the *disagreement set*, the set of disagreement strategies –that do not have to be necessarily unique– be defined as

$$\Pi_D^a = \arg\max_{\mathcal{P}\in\Pi^a} v^A(\mathcal{P})$$

$$\Pi_+ = \left\{\mathcal{P} \in \bigcup_{a=0}^{n-s_0} \Pi^a \;\middle|\; w^S(\mathcal{P}) \geq \min_{\mathcal{P}_D\in\Pi_D^a} w^S(\mathcal{P}_D)\right\} \text{ the \textit{set of offers},}$$

$$\mathcal{P}^a \in \arg\max_{\mathcal{P}\in\Pi_+} w^A(\mathcal{P}) \text{ the \textit{accepted offer} in the case of } a \text{ applicants.}$$

Here, just as in the selection of the disagreement strategy we cannot assume uniqueness. Often it is assumed that the moves are Pareto-optimal, but this does not have to be the case, applicants might willingly hurt the union. Let $b \in \mathbf{N}$ such that $\mathcal{P}^a \in \Pi^b$. Starting from the initial partition $\mathcal{P} \in \Pi^a$ we have the following results:

$$\pi^*(\mathcal{P}) = \begin{cases} \{\mathcal{P}, \mathcal{P}^a\} & \text{if } b = a \\ \{\mathcal{P}\} \cup \pi^*(\mathcal{P}^a) & \text{otherwise,} \end{cases} \tag{5}$$

$$w^A(\mathcal{P}) = \begin{cases} v^A(\mathcal{P}) + \delta\frac{v^A(\mathcal{P}^a)}{1-\delta} & \text{if } b = a \\ v^A(\mathcal{P}) + \delta w^A(\mathcal{P}^a) & \text{otherwise,} \end{cases} \tag{6}$$

$$w^S(\mathcal{P}) = \begin{cases} v^S(\mathcal{P}) + \delta\frac{v^S(\mathcal{P}^a)}{1-\delta} & \text{if } b = a \\ v^S(\mathcal{P}) + \delta w^S(\mathcal{P}^a) & \text{otherwise,} \end{cases} \tag{7}$$

So if no applicants get membership, they move into “defense” and stay there forever. Otherwise the solution is determined by the solution of the game thus created, having a smaller number of applicants.

4. SOLVING MORE GENERAL GAMES

4.1. MOTIVATION

Like (Morelli and Penelle 1997), having discussed the extensions of the EU, we must raise the question: how is the union, the large coalition whose existence we assumed in the first part of the paper formed? So far we have very strongly used its assumed dominance, but now we want to know what happens if it is smaller, or does not exist.

In this section we consider games without a dominant coalition. By removing it, we also remove the preordering of the strategies that was so far determined by the union's preferences. In the lack of a dominant player it would be very arbitrary to assign this role to any other player, and so a similar solution cannot be established.

4.2. SOLVING GAMES WITHOUT A DOMINANT COALITION

In a game with a monotonically increasing union the grand coalition was a fixed point where we could "grab" the path and find the place of other partitions along the equilibrium path or paths by an inductive argument, only to find sometimes that the game does not reach the grand coalition, we get "stuck" somewhere. In the present games this does not work, there is no natural choice for a natural partition that we could build our paths from.

If we assume that the solution contains no non-trivial loops the final partition is a coalition-proof, subgame-perfect equilibrium. While such outcomes cannot, in general, be pointed out at once, by Condition 2 coalitions of the same size benefit from merging so partitions with coalitions of the same size can be excluded; the set of remaining partitions will be denoted by Π^∞. For a large n this leaves relatively few partitions to test. (If $n = 15$ we have $|\Pi^\infty| = 27$ such partitions out of a total of 176.)[1] Partitions in Π^∞ are interesting even if they fail as equilibrium outcomes, as a further state can only be achieved by cooperation between coalitions of different sizes in which case –by rationality– all benefit, or by secession, where, by Condition 1 all outsiders get to a state with a higher payoff that often means a higher expected present value, too.

From the partitions in Π^∞ we will build paths in a similar recursive way as before, but not knowing whether we are on the right track until the end.

We describe the algorithm intuitively: Assume that there is a solution and this path has a trivial loop. If $\mathcal{P}_\lambda \subset \Pi^\infty$ and λ is the length of the path as defined in definition 5 then from $\mathcal{P}_{\lambda-1}$ partition $\mathcal{P}_\lambda$ is the optimal move and so on. Therefore when players at $\mathcal{P}_{\lambda-1}$ have to make a choice between paths in Π^∞ they *should* choose $\mathcal{P}_\lambda$. We can argue similarly at $\mathcal{P}_{\lambda-2}$, etc.

Unfortunately, by assuming that $\mathcal{P} \in \Pi^\infty$ are equilibrium outcomes, we have assumed that they last for ever, and this determines some values, that can differ from the true values for them if they turn out not to be an equilibrium. Because of this we take all partitions and look for the best path from all of them.

Let

$$\Omega_1 = \left\{\{\mathcal{P}\} \;\middle|\; \mathcal{P} \in \Pi^\infty\right\} \tag{8}$$

Now we create paths $\pi' = \{\mathcal{P}\} \cup \pi$, where $\mathcal{P} \in \Pi$ and $\pi \in \Omega_1$. If $\mathcal{P}$ is indeed an equilibrium outcome the path selected from $\mathcal{P}$ will be $\{\overline{\mathcal{P}}\}$, and hence the new path will be the same. Therefore

$$\Omega_2 = \left\{\pi' = \{\mathcal{P}\} \cup \pi \;\middle|\; \mathcal{P} \in \Pi^\infty, \pi \in \Omega_1\right\} \tag{9}$$

contains paths of length at most 2. In a similar way we can define Ω_k from Ω_{k-1} for all k.

Note that as the path is already infinite at one end, we want it to be finite at the other. Indeed, in the definition of the path we allow repetition in the loop part only.

On one hand we can say the by assumption we exclude nontrivial loops, and so from $\mathcal{P}$ we do not consider paths $\pi \in \Omega_k$ such that $\mathcal{P} \in \pi$ and then only paths of finite length can be built from a finite number of partitions.

On the other hand we can show that this algorithm works for paths with nontrivial loops, too. In this case there are only two ways that the optimal path $\pi^* \in \Omega_{k-1}$ from $\mathcal{P}$ already contains $\mathcal{P}$: either the entire π^* is a loop with $\mathcal{P}_\lambda(\pi) = \mathcal{P}$ or we have just broken up a loop and created another consisting of $\mathcal{P}_1(\pi), \ldots, \mathcal{P}_k(\pi) = \mathcal{P}$. Values are a little harder to compute in the case of loops, there is often a need for separate calculations. Otherwise loops should be treated as normal, so as a new path from a member in the middle of the path changes the path, the same, happening in a loop, breaks it and creates a new path, possibly another loop. For example, if $\pi = \left\{\mathcal{P}_5, \overline{\mathcal{P}_4, \mathcal{P}_3}\right\}$, and players prefer $\left\{\mathcal{P}_2, \mathcal{P}_1, \overline{\mathcal{P}_0}\right\}$ to $\left\{\overline{\mathcal{P}_3, \mathcal{P}_4}\right\}$ at $\mathcal{P}_4$ then the loop $\overline{\mathcal{P}_4, \mathcal{P}_3}$ breaks up, and we have a new path from $\mathcal{P}_5$, too, $\pi' = \left\{\mathcal{P}_5, \mathcal{P}_4, \mathcal{P}_2, \mathcal{P}_1, \overline{\mathcal{P}_0}\right\}$.

The algorithm terminates in both cases if for some k, $\Omega_k = \Omega_{k-1}$, that is, no new paths are created. As the number of paths and loops is finite, this is reached in finite time. By construction the paths are the ones we have been looking for.

It is not the aim of this paper to define a concept to aggregate the players' interests. We can discuss some simple examples, where the solution is evident, but the general problem is open for further research. Essence of the problem is to have a unique aggregated preference ordering. If uniqueness fails the second part of Proposition 3 does not hold and then the number of possible paths and loops is not finite any more and hence the algorithm above does not terminate.

4.3. EXAMPLES

We solve a game to illustrate the introduced concepts and to draw attention to some difficulties encountered when solving games. The table below shows the payoff values for game A. Games with 5 players have 6 possible partitions. We aim to find the optimal path from $(1, \mathcal{P}_6)$.

	v_A	1	2	≥ 3
$\mathcal{P}_0 =$	$\{5\}$			6
$\mathcal{P}_1 =$	$\{4, 1\}$	-3		7
$\mathcal{P}_2 =$	$\{3, 2\}$		2	5
$\mathcal{P}_3 =$	$\{3, 1, 1\}$	1		6
$\mathcal{P}_4 =$	$\{2, 2, 1\}$	1	4	
$\mathcal{P}_5 =$	$\{2, 1, 1, 1\}$	2	5	
$\mathcal{P}_6 =$	$\{1, 1, 1, 1, 1\}$	3		

Looking at game A, beyond the standard properties, we notice that in partition $\mathcal{P}_2$ all players benefit from cooperating and playing $\mathcal{P}_0$. Thus Π^∞ reduces to $\{\mathcal{P}_1, \mathcal{P}_0\}$. For each player we compute the value along all 1-path alternatives. The table below summarises our calculations: The sign # marks enforceable values, that is, values that can be achieved without cooperation with coalitions of different sizes.

		Ω_1			$\Omega_2 = \{\{\mathcal{P}_i\} \cup \pi^*\}$	
$\mathcal{P}_i$	m	$\{\mathcal{P}_2\}$	$\{\mathcal{P}_1\}$	$\{\mathcal{P}_0\}$	π^*	$w(\pi^*)$
$\mathcal{P}_0$	5	$\frac{19}{5}\frac{1}{1-\delta}$ #	$5\frac{1}{1-\delta}$ #	$6\frac{1}{1-\delta}$ #	$\pi_0 = \{\overline{\mathcal{P}_0}\}$	$6\frac{\delta}{1-\delta}$
$\mathcal{P}_1$	1	$2\frac{1}{1-\delta}$	$-3\frac{1}{1-\delta}$ #	$6\frac{1}{1-\delta}$	$\pi_1 = \{\overline{\mathcal{P}_1}\}$	$-3\frac{1}{1-\delta}$
	4	$\frac{17}{4}\frac{1}{1-\delta}$	$7\frac{1}{1-\delta}$ #	$6\frac{1}{1-\delta}$		$7\frac{\delta}{1-\delta}$
$\mathcal{P}_2$	2	$2\frac{1}{1-\delta}$ #	$2\frac{1}{1-\delta}$	$6\frac{1}{1-\delta}$	$\pi_2 = \{\mathcal{P}_2, \mathcal{P}_0\}$	$2 + 6\frac{\delta}{1-\delta}$
	3	$5\frac{1}{1-\delta}$ #	$7\frac{1}{1-\delta}$	$6\frac{1}{1-\delta}$		$5 + 6\frac{\delta}{1-\delta}$
$\mathcal{P}_3$	1	$2\frac{1}{1-\delta}$ #	$2\frac{1}{1-\delta}$	$6\frac{1}{1-\delta}$	$\pi_3 = \{\mathcal{P}_3, \mathcal{P}_0\}$	$1 + 6\frac{\delta}{1-\delta}$
	3	$5\frac{1}{1-\delta}$	$7\frac{1}{1-\delta}$	$6\frac{1}{1-\delta}$		$6\frac{1}{1-\delta}$
$\mathcal{P}_4$	1	$5\frac{1}{1-\delta}$	$-3\frac{1}{1-\delta}$	$6\frac{1}{1-\delta}$	$\pi_4 = \{\mathcal{P}_4, \mathcal{P}_1\}$	$1 - 3\frac{\delta}{1-\delta}$
	2	$\frac{7}{2}\frac{1}{1-\delta}$	$7\frac{1}{1-\delta}$ #	$6\frac{1}{1-\delta}$		$4 + 7\frac{\delta}{1-\delta}$
$\mathcal{P}_5$	1	$5\frac{1}{1-\delta}$ #	$\frac{11}{3}\frac{1}{1-\delta}$	$6\frac{1}{1-\delta}$	$\pi_5 = \{\mathcal{P}_5, \mathcal{P}_0\}$	$2 + 6\frac{\delta}{1-\delta}$
	2	$2\frac{1}{1-\delta}$	$7\frac{1}{1-\delta}$	$6\frac{1}{1-\delta}$		$5 + 6\frac{\delta}{1-\delta}$
$\mathcal{P}_6$	1	$\frac{19}{5}\frac{1}{1-\delta}$ #	$5\frac{1}{1-\delta}$ #	$6\frac{1}{1-\delta}$ #	$\pi_6 = \{\mathcal{P}_6, \mathcal{P}_0\}$	$3 + 6\frac{\delta}{1-\delta}$

$\mathcal{P}_6$ **and** $\mathcal{P}_0$ Players are in a symmetric situation so they chose the path with the highest value.

$\mathcal{P}_5$ The singletons can get to $\mathcal{P}_2$ and obtain $5\frac{1}{1-\delta}$ without cooperating with the 2-coalition. This is better than $\mathcal{P}_1$, so the 2-coalition will never obtain the value $7\frac{1}{1-\delta}$. Given these, $\{\mathcal{P}_0\}$ is an improvement for all players.

$\mathcal{P}_4$ The 2-coalition wants and can enforce strategy $\{\mathcal{P}_1\}$.

$\mathcal{P}_3$ Similarly to $\mathcal{P}_5$ the singletons can get to $\mathcal{P}_2$, and so, although this is not strictly better than playing $\mathcal{P}_1$ it is a credible threat forcing the 3-coalition to play $\mathcal{P}_0$.

$\mathcal{P}_2$ The same argument applies here.

		Ω_2					$\lambda = 1$	Ω_3	
$\mathcal{P}_i$	m	π_6	π_5	π_4	π_3	π_2		π^*	$w(\pi^*)$
$\mathcal{P}_0$	5	$< 0\#$	$< 0\#$	$< 0\#$	$< 0\#$	$< 0\#$	$6\frac{1}{1-\delta}\#$	π_0	6γ
$\mathcal{P}_1$	1	$6 + 9\gamma$	$5 + 9\gamma$	4	$4 + 9\gamma$	$5 + 9\gamma$	$-3\frac{1}{1-\delta}\#$	π_1	$-3\frac{1}{1-\delta}$
	4	$< 0\#$	$< 0\#$	$< 0\#$	< 0	< 0	$7\frac{1}{1-\delta}$		7γ
$\mathcal{P}_2$	2	< 0	< 0	$-2 + \gamma$	< 0	-	$6\frac{1}{1-\delta}$	π_2	$2 + 6\gamma$
	3	< 0	$< 0\#$	$< 0\#$	0		$6\frac{1}{1-\delta}$		$5 + 6\gamma$
$\mathcal{P}_3$	1	< 0	< 0	< 0	-	$< 0\#$	$6\frac{1}{1-\delta}$	π_3	$1 + 6\gamma$
	3	< 0	< 0	$-2 + \gamma$		< 0	$6\frac{1}{1-\delta}$		$6\frac{1}{1-\delta}$
$\mathcal{P}_4$	1	$6 + 9\gamma$	$5 + 9\gamma$	-	$4 + 9\gamma$	$8 + 9\gamma$	$-3\frac{1}{1-\delta}$	π_4	$1 - 3\gamma$
	2	$< 0\#$	$< 0\#$		< 0	< 0	$7\frac{1}{1-\delta}\#$		$4 + 7\gamma$
$\mathcal{P}_5$	1	< 0	-	$< 0\#$	< 0	$< 0\#$	$6\frac{1}{1-\delta}$	π_5	$2 + 6\gamma$
	2	< 0		$-2 + \gamma$	0	< 0	$6\frac{1}{1-\delta}$		$5 + 6\gamma$
$\mathcal{P}_6$	1	-	$< 0\#$	$< 0\#$	$< 0\#$	$< 0\#$	$6\frac{1}{1-\delta}\#$	π_6	$3 + 6\gamma$

where $\Omega_3 = \{\{\mathcal{P}_i\} \cup \pi^*\}$

In the table we compare the values along the new paths with each other and with the values of the old ones. Only the improvements are shown, and where this is less than 0 we only indicate this fact. $\gamma = \frac{\delta}{1-\delta}$. Arguing as before we find that no possible extensions are better than paths in Ω_2, and column π^* contains the optimal paths. In particular the solution of the game, and the corresponding value are:

$$\pi = \{\mathcal{P}_6, \overline{\mathcal{P}_0}\} \tag{10}$$

$$w_6^1 = 3 + 6\frac{\delta}{1-\delta} \tag{11}$$

4.4. LOOPS

To illustrate the significance of loops, before moving to more complex issues, first we will solve the simplest game L (its payoff function is below) that contains a loop. This game has four players. Notice the relatively high value of v_1^2. We can guess that the players want to exploit this.

v_L		1	≥ 2
$\mathcal{P}_0 =$	$\{4\}$		9
$\mathcal{P}_1 =$	$\{3, 1\}$	0	13
$\mathcal{P}_2 =$	$\{2, 2\}$		8
$\mathcal{P}_3 =$	$\{2, 1, 1\}$	4	15
$\mathcal{P}_4 =$	$\{1, 1, 1, 1\}$	7	

In our solution we use the game's simplicity, the fact that at each stage there are coalitions of at most 2 different sizes and there is always a unique disagreement strategy where the two sets of players form two coalitions, alternatively they cooperate in some way. The game lacks partitions with more than one cooperative strategies that dominate disagreement. The table below shows the first three steps of our algorithm in the case of loops. Further steps do not change Ω_3, all paths end in stable loops, hence we have the optimal paths.

	Ω_1		Ω_2		Ω_3	
	w	π^*	w	π^*	w	π^*
$(1, \mathcal{P}_4)$	$\frac{7}{1-\delta}$	$\{\mathcal{P}_4\}$	$7 + \frac{19}{2}\frac{\delta}{1-\delta}$	$\{\mathcal{P}_4, \mathcal{P}_3\}$	$7 + \frac{\frac{19}{2}\delta + 9\delta^2}{1-\delta}$	$\{\mathcal{P}_4, \overline{\mathcal{P}_3, \mathcal{P}_0}\}$
$(1, \mathcal{P}_3)$	$\frac{4}{1-\delta}$	$\{\mathcal{P}_3\}$	$4 + 9\frac{\delta}{1-\delta}$	$\{\mathcal{P}_3, \mathcal{P}_0\}$	$4 + \frac{9\delta + \frac{19}{2}\delta^2}{1-\delta}$	$\{\overline{\mathcal{P}_3, \mathcal{P}_0}\}$
$(2, \mathcal{P}_3)$	$\frac{13}{1-\delta}$		$13 + 9\frac{\delta}{1-\delta}$		$13 + \frac{9\delta + \frac{19}{2}\delta^2}{1-\delta}$	
$(2, \mathcal{P}_2)$	$\frac{8}{1-\delta}$	$\{\mathcal{P}_2\}$	$8 + \frac{19}{2}\frac{\delta}{1-\delta}$	$\{\mathcal{P}_2, \mathcal{P}_3\}$	$8 + \frac{\frac{19}{2}\delta + 9\delta^2}{1-\delta}$	$\{\mathcal{P}_2, \overline{\mathcal{P}_3, \mathcal{P}_0}\}$
$(1, \mathcal{P}_1)$	0	$\{\mathcal{P}_1\}$	0	$\{\mathcal{P}_1\}$	0	$\{\overline{\mathcal{P}_1}\}$
$(3, \mathcal{P}_1)$	$\frac{13}{1-\delta}$		$\frac{13}{1-\delta}$		$\frac{13}{1-\delta}$	
$(4, \mathcal{P}_0)$	$\frac{9}{1-\delta}$	$\{\mathcal{P}_0\}$	$7 + \frac{19}{2}\frac{\delta}{1-\delta}$	$\{\mathcal{P}_0, \mathcal{P}_3\}$	$\frac{9\delta + \frac{19}{2}\delta^2}{1-\delta}$	$\{\overline{\mathcal{P}_0, \mathcal{P}_3}\}$

Game M has 5 players. Its outcome, a loop does not include partitions where all players become identical, so differences once occurring never vanish. As a result values have to be calculated separately.

v_M		1	2	≥ 3
$\mathcal{P}_0 =$	$\{5\}$			4
$\mathcal{P}_1 =$	$\{4, 1\}$	0		6
$\mathcal{P}_2 =$	$\{3, 2\}$		2	5
$\mathcal{P}_3 =$	$\{3, 1, 1\}$	1		8
$\mathcal{P}_4 =$	$\{2, 2, 1\}$	1	5	
$\mathcal{P}_5 =$	$\{2, 1, 1, 1\}$	2	11	
$\mathcal{P}_6 =$	$\{1, 1, 1, 1, 1\}$	3		

The table below shows the process of solving game M. By the time the set Ω_3 is defined, it becomes clear that all paths lead to the loop $\overline{\mathcal{P}_5, \mathcal{P}_1}$, but further stages give different values for the expected present values.

	Ω_1		Ω_2		Ω_3	
	w	π^*	w	π^*	w	π^*
$(1, \mathcal{P}_6)$	$\frac{3}{1-\delta}$	$\{\mathcal{P}_6\}$	$3 + \frac{28}{5}\frac{\delta}{1-\delta}$	$\{\mathcal{P}_6, \mathcal{P}_5\}$	$3 + \frac{28}{5}\delta + \frac{24}{5}\frac{\delta^2}{1-\delta}$	$\{\mathcal{P}_6, \mathcal{P}_5, \mathcal{P}_1\}$
$(1, \mathcal{P}_5)$	$\frac{2}{1-\delta}$	$\{\mathcal{P}_5\}$	$2 + 4\frac{\delta}{1-\delta}$	$\{\mathcal{P}_5, \mathcal{P}_1\}$	$2 + 4\delta + 5\frac{\delta^2}{1-\delta}$	$\{\overline{\mathcal{P}_5, \mathcal{P}_1}\}$
$(2, \mathcal{P}_5)$	$\frac{11}{1-\delta}$		$11 + 6\frac{\delta}{1-\delta}$		$11 + 6\delta + \frac{13}{2}\frac{\delta^2}{1-\delta}$	
$(1, \mathcal{P}_4)$	$\frac{1}{1-\delta}$	$\{\mathcal{P}_4\}$	$1 + 2\frac{\delta}{1-\delta}$	$\{\mathcal{P}_4, \mathcal{P}_5\}$	$1 + 2\delta + 4\frac{\delta^2}{1-\delta}$	$\{\mathcal{P}_4, \mathcal{P}_5, \mathcal{P}_1\}$
$(2, \mathcal{P}_4)$	$\frac{5}{1-\delta}$		$5 + \frac{13}{2}\frac{\delta}{1-\delta}$		$5 + \frac{13}{2}\delta + 5\frac{\delta^2}{1-\delta}$	
$(1, \mathcal{P}_3)$	$\frac{1}{1-\delta}$	$\{\mathcal{P}_3\}$	$1 + 3\frac{\delta}{1-\delta}$	$\{\mathcal{P}_3, \mathcal{P}_1\}$	$1 + 3\delta + \frac{17}{4}\frac{\delta^2}{1-\delta}$	$\{\mathcal{P}_3, \mathcal{P}_1, \mathcal{P}_5\}$
$(3, \mathcal{P}_3)$	$\frac{8}{1-\delta}$		$8 + 6\frac{\delta}{1-\delta}$		$8 + 6\delta + \frac{13}{2}\frac{\delta^2}{1-\delta}$	
$(2, \mathcal{P}_2)$	$\frac{2}{1-\delta}$	$\{\mathcal{P}_2\}$	$2 + 3\frac{\delta}{1-\delta}$	$\{\mathcal{P}_2, \mathcal{P}_1\}$	$2 + 3\delta + \frac{17}{4}\frac{\delta^2}{1-\delta}$	$\{\mathcal{P}_2, \mathcal{P}_1, \mathcal{P}_5\}$

$(3, \mathcal{P}_2)$	$\frac{3}{1-\delta}$		$5 + 6\frac{\delta}{1-\delta}$		$5 + 6\delta + \frac{13}{2}\frac{\delta^2}{1-\delta}$	
$(1, \mathcal{P}_1)$	0	$\{\mathcal{P}_1\}$	$2\frac{\delta}{1-\delta}$	$\{\mathcal{P}_1, \mathcal{P}_5\}$	$2\delta + 4\frac{\delta^2}{1-\delta}$	$\{\overline{\mathcal{P}_1, \mathcal{P}_5}\}$
$(4, \mathcal{P}_1)$	$\frac{6}{1-\delta}$		$6 + \frac{13}{2}\frac{\delta}{1-\delta}$		$6 + \frac{13}{2}\delta + 5\frac{\delta^2}{1-\delta}$	
$(5, \mathcal{P}_0)$	$\frac{4}{1-\delta}$	$\{\mathcal{P}_0\}$	$4 + \frac{28}{5}\frac{\delta}{1-\delta}$	$\{\mathcal{P}_0, \mathcal{P}_5\}$	$4 + \frac{28}{5}\delta + \frac{24}{5}\frac{\delta^2}{1-\delta}$	$\{\mathcal{P}_0, \mathcal{P}_5, \mathcal{P}_1\}$

While repeating this recursive algorithm gives better and better approximations; we have to solve the following simultaneous equations in order to obtain the exact values of the players in the loop.

$$\begin{aligned} w_1^1 &= v_1^1 + \delta\frac{2w_5^4 + w_5^1}{3} \\ w_1^2 &= v_1^2 + \delta w_5^4 \\ w_5^1 &= v_5^1 + \delta w_1^1 \\ w_5^4 &= v_5^4 + \delta\frac{w_1^1 + w_1^2}{2} \end{aligned}$$

Solving these equations we get the final results:

$(m, \mathcal{P})$	w	π^*
$(5, \mathcal{P}_0)$	$\frac{4}{5}\frac{5+7\delta+\delta^2}{1-\delta^2}$	$\{\mathcal{P}_0, \overline{\mathcal{P}_5, \mathcal{P}_1}\}$
$(1, \mathcal{P}_1)$	$4\frac{\delta(3+6\delta+4\delta^2)}{6-7\delta^2+\delta^4}$	$\{\overline{\mathcal{P}_1, \mathcal{P}_5}\}$
$(4, \mathcal{P}_1)$	$\frac{36+39\delta-12\delta^2+11\delta^3}{6-7\delta^2+\delta^4}$	
$(2, \mathcal{P}_2)$	$\frac{1}{2}\frac{24+36\delta+23\delta^2+12\delta^3+9\delta^4}{6-7\delta^2+\delta^4}$	$\{\mathcal{P}_2, \overline{\mathcal{P}_1, \mathcal{P}_5}\}$
$(3, \mathcal{P}_2)$	$2\frac{9+18\delta+9\delta^2-6\delta^3+-4\delta^4}{6-7\delta^2+\delta^4}$	
$(1, \mathcal{P}_3)$	$\frac{1}{2}\frac{12+36\delta+37\delta^2+12\delta^3+\delta^4}{6-7\delta^2+\delta^4}$	$\{\mathcal{P}_3, \overline{\mathcal{P}_1, \mathcal{P}_5}\}$
$(3, \mathcal{P}_3)$	$\frac{48+36\delta-17\delta^2-12\delta^3-3\delta^4}{6-7\delta^2+\delta^4}$	
$(1, \mathcal{P}_4)$	$\frac{6+12\delta+17\delta^2+16\delta^3+\delta^4}{6-7\delta^2+\delta^4}$	$\{\mathcal{P}_4, \overline{\mathcal{P}_5, \mathcal{P}_1}\}$
$(2, \mathcal{P}_4)$	$\frac{30+39\delta-5\delta^2-11\delta^3-\delta^4}{6-7\delta^2+\delta^4}$	
$(1, \mathcal{P}_5)$	$4\frac{3+6\delta+4\delta^2}{6-7\delta^2+\delta^4}$	$\{\overline{\mathcal{P}_5, \mathcal{P}_1}\}$
$(2, \mathcal{P}_5)$	$2\frac{33+18\delta-19\delta^2-6\delta^3}{6-7\delta^2+\delta^4}$	
$(1, \mathcal{P}_6)$	$\frac{1}{5}\frac{15+28\delta+9\delta^2}{1-\delta^2}$	$\{\mathcal{P}_6, \overline{\mathcal{P}_5, \mathcal{P}_1}\}$

5. CONCLUSIONS

This paper demonstrates a number of problems one has to face when trying to solve the general accession game. If we exclude loops and assume a monotone increasing union, the game can be solved. We are yet to find solution concept that allows us to give up restrictions such as monotonicity or the identity of the players. However, despite its limitations the model matches the selected application rather well, and at the same time demonstrates a clear and sensible mechanism how small players can (sometimes) enforce their interest even against the will of the much larger union purely by cooperation, while the lack of cooperation may allow the union to exploit the division of the applicants.

The difficulties encountered give rise to some questions that we leave open. Loops are very uncommon in games and they might offer a new way to explain cyclic behaviour by firms or countries in forming alliances. (Morelli and Penelle 1997) discuss how different utility transfers affect the accession path; our example shows that even the uncertainty about the transfers can influence the outcome.

There are a number of ways to extend this model. Many of the restrictions introduced are to reduce the level of complexity and to match our limited tools. Multi-level integration as of (Baldwin 1994) is another interesting problem possibly modelled by games, with gradual membership in coalitions such as games with fuzzy coalitions (Butnariu and Klement 1993).

Acknowledgements

I thank Luc Lauwers for his help and guidance in preparing this work, an anonymous referee for the valuable suggestions and Massimo Morelli for turning my attention to Yi's paper. The joint support (SOR-H/99/23 and SOR-H/00/01) of the Katholieke Universiteit Leuven and the Soros Foundation is gratefully acknowledged. This paper is partly based on an earlier version (Kóczy 2000).

Notes

1. Generated using `http://sue.csc.uvic.ca/~cos/gen/nump.html`.

References

Robert J. Aumann and Jacques H. Drèze. Cooperative games with coalition structures. *International Journal of Game Theory*, 3:217–237, 1975. Also, as UCL-CORE Reprint 217.

Richard E. Baldwin. *Towards an Integrated Europe.* CEPR, London, 1994.

Douglas B. Bernheim, Bezabel Peleg, and Michael D. Whinston. Coalition-proof Nash equilibria: I. Concepts. *Journal of Economic Theory*, 42:1–12, 1987.

Dan Butnariu and Erich Peter Klement. *Triangular Norm-Based Measures and Games with Fuzzy Coalitions.* Kluwer Academic Publishers, Dordrecht – Boston – London, 1993.

Maarten F. Cornet. *Game Theoretic Models of Bargaining and Externalities.* Thesis Publishers, Amsterdam, 1998.

Joseph Greenberg. Coalition structures. In Robert J. Aumann and Sergiu Hart, editors, *Handbook of Game Theory*, volume 2, pages 1305–1337. Elsevier Science Publications B.V., 1994.

László Á. Kóczy. Accession games: A dynamic per-member partition function approach. Discussion Paper 90/2000, LICOS - Center for Transition Economies, Leuven, April 2000.

William F. Lucas. Some recent developments in n-person game theory. *SIAM Review*, 13(4):491–523, October 1971.

Massimo Morelli and Philippe Penelle. Economic integration as a partition function game. Discussion Paper 9785, CORE, Louvain-la-Neuve, 1997.

Debraj Ray and Rajiv Vohra. Equilibrium binding agreements. *Journal of Economic Theory*, 73:30–78, 1997.

Robert M. Thrall. Generalised characteristic functions for n-person games. In *Proceedings of the Princeton University Conference of October 1961*, pages 157–160, 1962. Privately printed for members of the conference.

Robert M. Thrall and William F. Lucas. n-person games in partition function form. *Naval Research Logistic Quarterly*, 10:281–298, 1963.

Robert J. Weber. Games in coalitional form. In Robert J. Aumann and Sergiu Hart, editors, *Handbook of Game Theory*, volume 2, pages 1285–1303. Elsevier Science Publications B.V., 1994.

Sang-Seung Yi. Stable coalition structures with externalties. *Games and Economic Behavior*, 20:201–237, 1997.

Chapter 9

SHOPBOTS AND PRICEBOTS IN ELECTRONIC SERVICE MARKETS

Panos M. Markopoulos
Computer and Information Science Dept.
University of Pennsylvania
markopou@unagi.cis.upenn.edu

Lyle H. Ungar
Computer and Information Science Dept.
University of Pennsylvania
ungar@cis.upenn.edu

Abstract This paper explores the role of shopbot and pricebot software agents in electronic service markets, where users have to wait to obtain the service or product because sellers can only process requests at a finite rate. In our model, we consider a stream of customers that arrive in a market and, based on their expected utility costs, choose a seller from which they receive service, and we explore the effects of changing the information and capabilities available to buyers and sellers. We show that widely held beliefs such as "sellers are better off colluding by not accepting shopbots" and "removing information asymmetries and lowering search costs results to higher overall market performance", are in reality dependent on the model structure and efficiency measures used.

Keywords: Shopbots , pricebots, e-commerce, information asymmetry, multiagent systems.

1. INTRODUCTION AND DEFINITIONS

The importance of service markets has grown substantially in the past few years with the development of internet commerce. New electronic marketplaces are rapidly being introduced with new means of exchanging information. In such markets, costs related to information acquisition (search costs, processing delays) can effect the overall efficiency of the market and sellers' and buyers' realized utility.

E-commerce sites that sell products on the internet can be described by the service model. We use the term "service" in a classical Operations Research context, where customers wait in a queue for their turn to be processed. This has proven a powerful modeling tool, as many different sources of delay can be aggregated and their cumulative effects studied by accepting the presence of a queue and uncertainty in a customer's processing time. Potential sources of delay in e-commerce transactions are among others, bandwidth limitations, database searches, and the credit card approval process.

In our model, customers arrive in the market with a rate that depends on their expected realized utility[1], and choose a seller from which they receive service. The way they choose the seller depends on whether they deploy shopbots or not. When they visit the seller they consider their total expected costs (price plus waiting cost) and they have the option of accepting the server's offer for service or leave the market. Should they decide to accept service, they join the seller's queue and wait their turn to receive service.

This model is inspired by observations of shoppers' behavior in an article in *Fortune* magazine that has identified technical problems, including unacceptably slow response times as the number one reason that customers take their business elsewhere (Mardesich, 1999). The importance of buyers' beliefs about sellers in e-commerce is underlined in (Green, 1999) where it is claimed that 28 percent of Netizens that formed a bad opinion about an e-commerce site, left the site, never to return.

Shopbots and pricebots change the information and capabilities available to buyers and sellers, and thus alter the market structure. We use the following definitions of these terms:

A *shopbot* is a software agent attached to a single user and has the ability to query multiple servers on the network (on behalf of the user) to gather information about prices and other service characteristics, like service quality or expected waiting time (Greenwald and Kephart, 1999). We assume that shopbot and user interests are identical and the shopbot's sole purpose is to serve the user's needs.

A *pricebot* is a software agent attached to a single service provider and has the ability to dynamically change the price of the product/service to maximize the provider's expected profit. The pricebot's interests are identical to the service provider's (Greenwald et al., 1999).

It is reasonable to expect that when technology allows it, users will employ shopbots to search for the best deal. Sellers have the option of

Market configurations	Buyers use shopbots?	Sellers use pricebots?
XX Simple market that serves as a basis model.	No	No
SX Sellers can only set their prices relatively slowly. All buyers are bargain hunters.	Yes	No
XP Sellers offer each buyer a customized price. All buyers visit sellers at random.	No	Yes
SP Perfect price discrimination. All buyers are bargain hunters.	Yes	Yes

Table 9.1 . The different market configurations that we explore.

responding with the employment of pricebots that price their services dynamically to maximize expected total profit.

The key attributes of the markets described in the paper are the following:

- The customers incur waiting costs while they wait to receive service.
- There is uncertainty regarding the processing time the user's service request will require. This is almost always the case with any real market and especially the internet.
- Transportation and/or communication costs are small enough to justify customer choice between different service providers, based on price and expected waiting time. In the physical world a customer may have no choice but to visit the closest service provider, but on the internet the costs of visiting multiple vendors are negligible.

Table 1 describes the four different market configurations that we explore. We name them XX, SX, XP and SP, where S stands for Shopbots, P for Pricebots and X denotes the absence of a class of agents from the market. For example SX stands for: Shopbots but no Pricebots. The terms seller, server and service provider are used interchangeably, as are the terms buyer, customer, shopper and client.

We proceed by formalizing our model and describing the different market configurations in section 2. In section 3 we compare the different market configurations based on buyer, seller and total market utility. In section 4 we present related literature and conclude in section 5.

2. OUR MODEL OF AN ELECTRONIC SERVICE MARKET

We consider a market with N identical service sellers with zero production costs and M identical risk neutral service buyers, with $N << M$. Buyers incur waiting costs while waiting for service, which include the time during which they actually receive service. The total cost c the buyers incur is the sum of the price they pay for service p plus the waiting cost G, $c = p + G$. We assume a linear waiting cost of the form $G = v \cdot t$, where t is time and v a constant. Buyers have a reservation price u for total cost. They will not accept incurring an expected cost higher than u and they will rather leave the market. Customers' service time is exponentially distributed with mean $1/\mu$ (all customers have the same expected service time) and service follows a FCFS (First Come First Serve) discipline. The buyers' interarrival time (the time between two successive arrivals) is exponentially distributed with mean λ, where λ is taken to be linearly decreasing in the expected value of c, $E(c)$, of the form

$$\lambda = \alpha - (\alpha/u) \cdot E(c) \tag{1}$$

In equation 1, α is a constant that represents the arrival rate of customers in a market where expected total cost is zero. The formula for the arrival rate is chosen so that the arrival rate is α for expected total cost equal to zero and zero for expected total cost equal to u, the buyers' reservation price.

This model reflects an assumption that the buyers have a "feeling" for what the market is like, based on previous experiences or reputation. Thus, the probability that they will try the market (the rate of arrivals) decreases with the expected total cost.

When buyers do not use shopbots, they do not know the prices that sellers charge or each seller's expected waiting time before they actually visit a seller. They choose a seller at random and only then are price and expected waiting time for this particular seller revealed to them. They choose to accept service or leave the market depending on whether expected total cost exceeds u. They know, however, what the expected total cost $E(c)$ is, given that they pick a server at random. This knowledge is used by the customers indirectly, as it only effects the rate with which they arrive in the market. In other words, sellers choose to enter the market based on their knowledge of $E(c)$, but they choose to accept service based on their updated knowledge for the expected total cost for a specific seller which they have chosen at random.

When buyers use shopbots, they have complete information on prices and expected waiting times on all sellers. In this case, they will choose to visit the most promising seller, since they try to minimize their cost. However, it is not guaranteed that they make the correct decision because waiting time is stochastic and they can only choose the minimum expected total cost.

When sellers do not use pricebots, they pick a price p for the service they provide and charge p for a "long" period of time. By this we mean that sellers do not assume that the market conditions will remain the same after they can readjust their price, and we do not consider multi-period games. When sellers use pricebots, prices are instantly updated and each potential buyer can be offered a different price.

All these markets assume the existence of intermediary agents that support the flow of information and help customers locate the service providers. Service providers can advertise their services through these intermediaries and thus their location becomes known to interested customers. This is the matchmaker/yellow page model of intermediation described in (Sycara et al., 1997). These intermediaries focus on the flow of information and are essential for the market's operation.

2.1. XX MARKET CONFIGURATION — NO SHOPBOTS, NO PRICEBOTS

The market where neither the buyers nor the sellers use agents to search for best opportunities or dynamically adjust prices, serves as a basis model against which the others are compared. Here, customers arrive in the market and choose a seller at random. The mean arrival rate depends linearly on the average total cost of receiving service. When the customer visits the server, she is informed about the price and total expected waiting time. If this sum is less than the customer's reservation price u, the customer enters the server's system for service. Otherwise she leaves the market.

In the appendix we show that in equilibrium sellers charge the monopolistic prices.

2.2. SX MARKET CONFIGURATION — SHOPBOTS ONLY

In the second case, clients use shopbots, while servers do not use pricebots. Customers want to visit the service provider that gives them the lowest expected cost, maximizing their utility. All other things being equal (prices, quality, job sizes etc) that would mean picking the service provider with the smallest job queue. A shopbot is an excellent helper, since it can ask all service providers for their queue sizes, monitor servers broadcasts on their current queue status or communicate with another information intermediary that gathers this information for all service providers. Then taking into account prices, expected waiting time and quality of service, the shopbots can channel their users' request to the most promising service provider. In our model only prices and waiting time are considered, since quality is assumed to be the same across sellers (sellers are identical).

However, nothing can guarantee that service providers will reveal truthfully their queue size when asked. Actually they have a strong incentive to announce a smaller queue size, convincing more customers to visit them. A simple solution is to have an intermediary agent in the market that will monitor the difference between the expected service time that the seller reports to the actual service time that the user experienced. The intermediary can then inform the buyers of any sellers that consistently mislead their customers. [2]

Assuming that service providers have incentives to reveal their queue sizes truthfully, shopbots can help the user make a better choice, by suggesting the most promising server. This puts the customer in a more advantageous position than before.

It can be shown that there exists no symmetric pure Nash equilibrium in the one shot game that the sellers face. We will provide an intuitive proof, ignoring the case of asymmetric and mixed strategy equilibria, since the sellers are identical and will only set their price once, making such equilibria unrealistic.

If there existed a pure Nash equilibrium with more than one seller charging the same price a seller could always increase profits with one of two actions: for a price other than zero the seller can undercut his competitors and for a price of zero the seller can always increase profits by increasing his price. Assume that some sellers charge $p > 0$. One of them can offer an ϵ discount and capture all buyers that would be indifferent when the expected service time is the same among the sellers. This is a situation that occurs with some probability higher than zero since the expected service time of all shoppers is the same. The

ϵ discount can be arbitrarily small so that the revenue from capturing the buyers that would be indifferent when the expected service time is the same among the sellers, is higher than the lost revenue due to the price decrease. However pricing at zero is also not an equilibrium since instead of making zero profits a seller would raise his price and increase revenue from buyers that occasionally come and find all other sellers with non zero expected queue waiting time. Consequently, the market cannot be in equilibrium.

The collusive price in this market configuration is more difficult to compute than the equilibrium in the XX case, as the number of equations is larger. All sellers will charge the same price and end up having the same expected unconditional waiting time[3]. Here we have a finite Markov chain of N dimensions, where N is the number of sellers in the market. Each dimension can take K possible values, where K is the maximum number of users each service provider can simultaneously support and is defined in the Appendix for the XX case. Intuitively, K is the number of customers that makes waiting cost prohibitive high for the next customer that will consider visiting the seller. With collusion, sellers will charge the same price and when a customer finds two or more sellers with the same expected service time (which occurs quite often, since all sellers have the same unconditional expected waiting time), the buyer will pick a server at random. The resulting system will have NK equations (Gross and Harris, 1985) which we can solve numerically. We will not present the equations in this paper.

In the simulations to be presented below for the SX case, it is assumed that sellers charge the collusive prices, to demonstrate a basic result of this paper, that sellers are always better off when they collude with shopbots in the market than colluding by not accepting shopbots. Furthermore, the price instability and steep competition hinted by the absence of a pure Nash equilibrium is qualitatively captured by a strategy that we propose below for the SP market.

2.3. XP MARKET CONFIGURATION — PRICEBOTS ONLY

When sellers use pricebots, while clients do not use shopbots, sellers can guarantee a total fixed cost including the cost of the waiting time. This is done by pricing the service in a way that total expected cost remains constant. They would offer a discount when the expected service time is above the average and require a premium otherwise. In our model the clients are risk neutral and thus not attracted by this extra service. Simulations show that sellers' profit remain the same since the arrival rate does not increase. This application for pricebots has not been sufficiently explored but it appears that it can increase the satisfaction of the customers that are risk averse in a given market and thus lead to higher expected future profits. Another possible use for the pricebots in a no-shopbots environment is to implement a profit maximizing mixed strategy equilibrium, which we will not pursue here.

2.4. SP MARKET CONFIGURATION — SHOPBOTS AND PRICEBOTS

For our final case we study a perfectly competitive market, where shoppers use shopbots to search for the lowest expected cost and sellers are allowed to dynamically adjust their prices by using pricebots. Sellers can now easily undercut each other and immediately attract buyers, unlike the markets where shoppers do not use shopbots and cannot be informed about lowest priced sellers.

We assume that the sellers follow a greedy strategy: they always want the next customer[4]. Pricebots will take advantage of demand fluctuations in the market to maximize their owners' expected profits. For example, given that the customers are using shopbots, if a service provider is idle while all other servers have large queues, he would have a strong incentive to raise his price, knowing that users will still view him as their best choice, given the high expected waiting cost on all other servers. In fact, this server can safely increase his price up to the point where he will still be preferred by utility maximizing customers, even when the server that has the second smallest queue does not charge anything for service. This service provider will be the one that the next customer is going to visit and we will call his price *prevailing price.* If the two least busy servers have the same number of customers in their queues the price will drop to zero. The information of what queue sizes the other servers have, can be obtained by the interested service provider, much in the same way that customers find out about it, by the use of a shopbot from the service provider.

This, however, will lead to sharp price competition that will eventually lower profits. This effect is even more apparent when there are more servers in the market. Even if we observe systems with the same load (the ratio of the processing power used to the total processing power available), we see that the average price that the clients face drops with the number of servers. This is plotted for two different system loads in figure 1.

The prevailing price is determined at any particular time period (defined as the time between two successive events such as an arrival of a

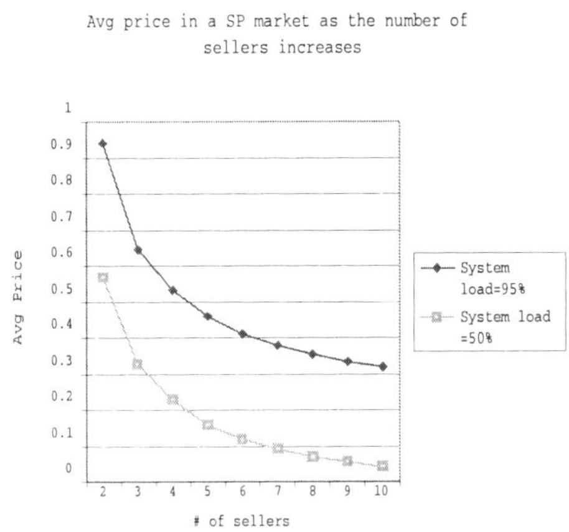

Figure 9.1. Average prevailing price in a market with shopbots and pricebots as the number of sellers increases. For the simulations u=10, v=1, μ=1 and α is such that system load is either 50% or 95%

new customer or the departure of one that just finished service) by the difference in the queues of the two least busy servers.

The SP market with a greedy seller strategy exhibits price fluctuations, unlike the markets cases discussed previously. If we denote the probability distribution of the discrete random variable that represents the difference in the queue sizes of the two least busy servers to be $Q(d)$, where d is the difference, then the distribution of prices over time in the market is $Q(vd/\mu)$, where $1/\mu$ is the average processing time, v is the waiting cost per unit of time and $vd/\mu < u$, since the least busy server can safely charge $vd/\mu - \epsilon$, knowing that he will still get the next customer even if the second least busy server charges nothing. A typical prevailing price distribution can be seen in figure 2.

The greedy strategy described above is not an optimal one for the sellers to follow. In an optimal strategy the seller would never price at zero, since he would rather wait for the chance to attract a customer when all other servers are busy. In fact there does not exist a pure optimal strategy since two or more servers with the same number of customers in their queues would face the pricing problem described for the SX market configuration. The greedy strategy captures qualitatively the price competition and instability that the introduction of pricebots ignites.

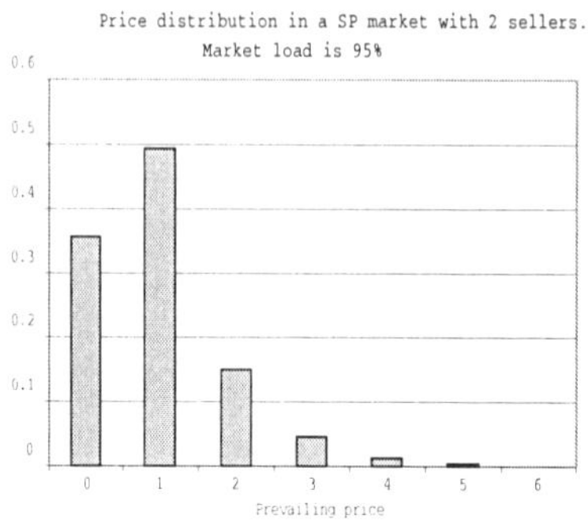

Figure 9.2. Price distribution in a market with 2 servers with shopbots and pricebots. For the simulations u=10, v=1, μ=1 and α is such that system load is 95%

3. COMPARISONS

Different market configurations may exist for many reasons. For example pricebots might not be allowed in a certain market because of legal restrictions or technological considerations might impede the use of shopbots. Overall market efficiency can also be important, especially in systems where both buyers and sellers agents belong to the same organization and the market configuration decision is taken in a centralized manner. However, the buyers' and sellers' incentives play a critical role in all decentralized environments in determining whether shopbots or pricebots are employed. It is thus important to understand the incentives that buyers and sellers have to introduce agents and the effect that their actions will have in overall market performance.

3.1. BUYERS AND SELLERS UTILITY

In figure 3 we can see both buyers' and sellers' utility and overall market utility. The results used are derived by simulations. They match almost perfectly the theoretical results that we were expecting. For the XX case the market is in equilibrium, for the SX case we have assumed collusion[5] and finally, for the SP case, we used the greedy pricing strategy that we described.

We see that total client utility increases as we move from the XX to the SX configuration. Naturally, since clients can avoid unnecessary long waiting times when they use shopbots, their expected utility will increase. Thus their arrival rate will also increase. The increase in clients' utility is even bigger when the number of sellers increases. This

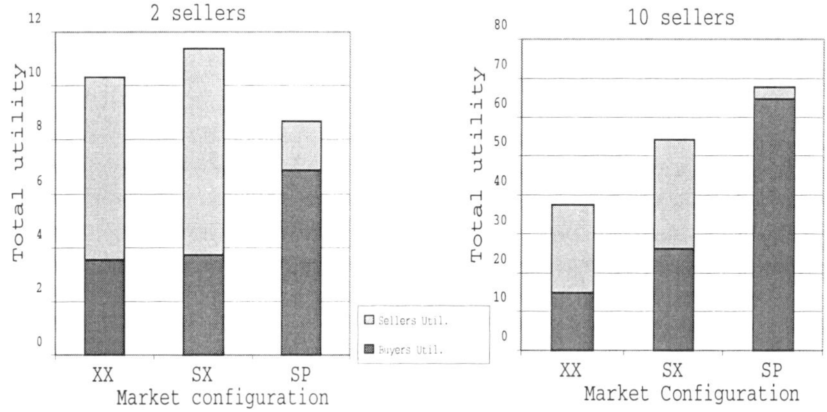

Figure 9.3. Servers and clients utility. For the simulations u=10, v=1, μ=1 and α is such that both series have 95% load when there are shopbots and pricebots

is intuitively due to the fact that clients have now more choices and are more likely to find a good deal among the servers.

The sellers also benefit when the buyers introduce shopbots and the configuration changes from XX to SX. The increase in the arrival rate contributes directly to their profits. Thus, sellers have the incentive to facilitate buyers using shopbots and provide all the software and hardware needed for the shopbots operation.

In fact the sellers are strictly better off when they collude in this new market structure: for a given service price p, the buyers have a lower expected waiting cost when they choose the least busy seller, instead of choosing a seller at random. Therefore the buyers "effective" arrival rate $\lambda_{effSX}(p)$ (defined as the arrival rate of buyers that choose to receive service) as a function of price is strictly above the "effective" arrival rate as a function of price for the XX case $\lambda_{effXX}(p)$ (where buyers visit servers at random). So the profit rate under collusion $max_p\{p \cdot \lambda_{eff}(p)\}$ is strictly higher for the SX case where buyers use shopbots.

Sellers are thus always better off colluding in a shopbot environment than colluding by not accepting shopbots. Generalizing, we can say that the sellers always want buyers to be able to search for service or product dimensions that are important to buyers but do not effect sellers' profits directly. For example, travel agencies want their customers to be able to search for the availability of different seat or meal preferences as this increases customer satisfaction and leads to more sales. Hotels should want customers to search for different room preferences and real-estate agents should want prospective buyers to be able to search for different house preferences, such as school proximity.

We saw that in the SX case, collusion is not the Nash equilibrium of the market. Since the market is not in equilibrium it is likely to become an SP market. Sellers are tempted to undercut each other and start competing by dynamically adjusting their price. They would thus want to introduce pricebots. This leads, however, to the SP market configuration where sellers see their profits significantly reduced. In other words, sellers want buyers to use shopbots as long as they can successfully collude, but collusion might be difficult. In this sense shopbots can be a mixed blessing for sellers, while they are always prefered by buyers.

3.2. TOTAL MARKET UTILITY

It might seem surprising that the overall system utility drops for the two server case when servers introduce pricebots. Total system utility is less for the SP market configuration than it is for the XX market. This is because waiting time is an exponentially increasing function of the effective customer arrival rate. Total utility rate is:

$$U/sec = p \cdot \lambda_{eff} + (u - p - G) \cdot \lambda_{eff} = (u - G) \cdot \lambda_{eff} \qquad (2)$$

where u is the buyers reservation price (how much they value service), G is the average waiting cost that customers that receive service incur, $p \cdot \lambda_{eff}$ is the sellers profit rate, and $(u - p - G) \cdot \lambda_{eff}$ is the buyers utility rate. If the effective arrival rate is already high enough, the introduction of shopbots and pricebots can cause, along with the increase of the effective arrival rate, a much larger increase in waiting costs, driving social welfare down. For small effective arrival rate values the expected waiting time (and consequently the waiting cost) increases slower than the effective arrival rate, but the situation is reversed for higher effective arrival rate values. Thus introducing shopbots and pricebots in a market where the effective arrival rate is already high enough will drive social welfare, as defined in this paper, down. The intuition that shopbots and pricebots are needed most when systems are overloaded is not verified.

Average waiting cost G, as a function of λ_{eff} will change in different market configurations. We cannot compare total utility in two different market configurations unless we know the market parameters. In figure 3 we can see that introducing pricebots in one market increases total utility while it decreases it in another.

The result is surprising because by moving from an XX market configuration to an SP, we are making more information available to the agents and reduce search costs: buyers in the SP case have complete information about prices and waiting times as opposed to the XX case where no such information is available. Furthermore, moving to the SP market configuration we make more information available to sellers: one can view the XX case as a market where a seller does not know what the other sellers are charging and the SP case as a market where sellers have complete information about their competitors. In the SP market agents' capabilities are increased and agents have more options (now sellers can price dynamically) and the assumption that sellers collude is dropped. It is a widely shared belief that more information, reduced search costs and competition always leads to higher total market utility. In our model by removing all "exogenous" friction from our market we expose the "endogenous" friction caused by waiting costs.

As a final remark, it is important how one chooses to measure market efficiency. It is not always the case that overall market utility captures the social welfare aspect of our model. It is implicitly assumed that lost clients can find the service somewhere else. This is not always the case, especially when the resources and clients all belong to the same organization as is the case with agent based manufacturing and highly integrated supply chains. In those cases, the system load, defined as the service rate used divided by the total service rate available in the market, could be a better indicator of social welfare. System load for the different market configurations is plotted in figure 4, where it can be seen that it strictly increases as we move from the simple market, to the market with just shopbots and finally to the market with both shopbots and pricebots.

4. RELATED LITERATURE

The work presented in this paper is related to three active research areas: service and service competition, multiagent systems and information asymmetry.

Markets of service providers have been explored extensively in the operations research literature. Most of the research had the direction of dealing with social optimizations or optimizations within the organization - owner of the servers (Dewan and Mendelson, 1990). A model where information about expected service time is provided to the customers is found in (Whitt, 1999). The author models a telephone call center and has the intuition that informing customers about delays (for example by allowing shopbots) can lead to increased business. However he does not model the increase in the arrival rate to obtain this result and competition with other firms is not studied.

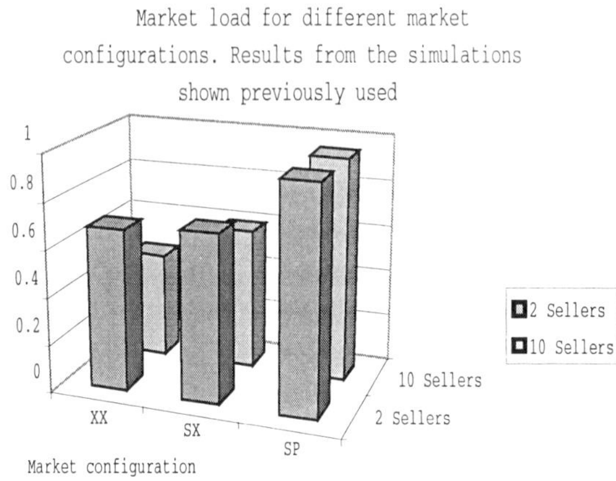

Figure 9.4. System load for different market configurations. For the simulations u=10, v=1, μ=1 and α is such that both series have 95% load when there are shopbots and pricebot

The literature on service competition is also relevant to our work. In (Daniel and Mulligan, 1995) the authors discuss the existence of competitive equlibria in a service market. In (Davidson, 1988) service competition is studied and two models — one where customers have complete price information and one where only the price distribution is known — are compared. However the results do not apply easily to the markets considered in this paper because customers can only visit sellers sequentially and do not have the ability to query multiple sellers in parallel.

Two main results can be identified in this body of literature: First, in the presence of search costs the equilibrium is characterized by all firms charging the monopoly price. This is also true in this paper, as it was shown for the XX market. The second result is that sellers would rather collude by not accepting shopbots but have incentives to deviate. This result has led to the false belief that sellers are always better off by colluding and not accepting shopbots. This paper argues that sellers' collusion while buyers use shopbots, is the best the sellers can hope for.

The literature on shopbots and pricebots has grown substantially recently, but the waiting cost dimension has been neglected. In (Kephart and Greenwald, 1999) the authors address the economic implications of the widespread utilization of shopbots, and in (Kephart et al., 1998) the dynamics of the price-wars caused by the shopbot - pricebot interaction are studied. Two results of the existing shopbots - pricebots literature are related to our work. First, a Nash equilibrium in the one-stage game is unlikely to exist in a market with shopbots and pricebots, a result replicated in this paper. Second, myopic seller strategies lead to price wars. This effect is also present in this paper as the greedy (myopic) strategy studied for the SP market drives prices much lower than in other market configurations, substantially reducing sellers' revenues.

Finally the literature on customer search and information asymmetries is also extensive. A prevailing result in this body of literature is that increased competition and lower search costs (more information) lead to higher overall market utility, for example see (Baron, 1981). In (Nelson, 1993) the author discovers that it is possible that increased competition can lead to reduced total market surplus, when uninformed sellers face informed buyers. In this paper we do not hold the assumption that the sellers are uninformed and they are allowed to have complete information in the SP market configuration. The fact that the total market surplus might fall as we introduce shopbots and pricebots is surprising when contrasted with the existing literature.

5. CONCLUSIONS

As we believe that waiting time is an important dimension of buyers' total cost in e-commerce markets, we have presented a simplified model where shopbots and pricebots software agents operate in an environment where shopper's time is important. We studied the effect that the employment of these agents has on the market's performance and their owners' utility.

We saw that if sellers collude, they are always better off with shopbots in the market than they are without. Colluding in a shopbot environment means that sellers agree to fix prices while allowing buyers to search their system for other product or service dimensions that are important to buyers but do not effect sellers' profits. In other words, sellers provide a service that does not effect their profits directly, but does increase customer satisfaction. They should expect, as a result, a long term increase in sales. So it is arguably better for sellers and service providers such as hotels, or travel and real-estate agents to allow for shopbots to search for different customer preferences.

The widely shared belief that "sellers are better off by colluding and not accepting shopbots" should be rephrased as: "sellers are better off by colluding in a shopbot environment". Letting buyers use shopbots, however, can be potentially dangerous for sellers. The presence of shopbots in the market may induce sellers to start using pricebots and ignite price-wars that will eventually cause a substantial profit decline. The sellers are thus better off by accepting shopbots, as long as they can trust each other not to engage in price-wars, by utilizing pricebots.

As we moved to increasingly complex market organizations, reducing market friction on the way, we saw that total market utility can either increase or decrease. There is no simple threshold of the effective arrival rate that determines if introducing shopbots and pricebots will actually decrease total market utility. Rather we have to look at the market equilibrium before and after the shopbots' and pricebots' introduction to observe the change in market utility. As a rule of thumb, shopbots and pricebots increase total market utility when the system load is relatively small and decrease it otherwise, because total waiting costs increase slowly when market load is low but increase rapidly when market load is high.

Acknowledgements

The authors would like to thank Professor Ravi Aron and one anonymous reviewer for their valuable comments and suggestions on an initial draft of the paper. This research was funded in part by National Science Foundation Grant SBR 96-02053.

Appendix: The Derivation of Equilibrium for the XX Market Configuration

The sellers' problem is how to price service and the buyers will respond by settling to a specific arrival rate as described by equation 3 below. The sellers, in the absence of pricebots, have to play a single period game[6]. Assume without loss of generality that server N is trying to maximize profit given that all other sellers are charging prices $p_1, p_2, \ldots, p_{N-1}$ and have average waiting costs of $G_1, G_2, \ldots, G_{N-1}$. Seller N will chose a price $p \equiv p_N$ and end up having a queue with average waiting cost of $G \equiv G_N$. If we denote average price with $\bar{p} = \frac{\sum_{i=0}^{N} p_i}{N}$ and average waiting cost with $\bar{G} = \frac{\sum_{i=0}^{N} G_i}{N}$, the arrival rate is from equation 1:

$$\lambda = \alpha - (\alpha/u) \cdot (\bar{p} + \bar{G}) \tag{9.A.1}$$

Each service provider can be modeled as an $M/M/1/K$ queue which stands for: exponentially distributed arrivals, exponentially distributed service time and one server with a capacity of K customers. The server can be viewed as having a finite queue capacity because no buyer will visit a seller where the queue size is such that total cost exceeds u. So given that the seller has chosen price p the capacity K, i.e. the maximum number of customers in the service provider's system is:

$$K = \lfloor \frac{\mu \cdot (u - p)}{v} \rfloor \tag{9.A.2}$$

where, again, v represents the cost to the shopper of waiting one time unit, $1/\mu$ is the shopper's expected processing time, so Kv/μ is the expected waiting cost for a customer who has K customers ahead in the queue, u is the customer's reservation price (how much she values the service) and p is the price the server charges.

Each server experiences an exponentially distributed arrival rate with mean $\lambda_S = \lambda/N$. The equations that describe the market are given below (Gross and Harris, 1985):

The expected number of customers in a server is:

$$L = \frac{\rho}{1-\rho} - \frac{(K+1)\rho^{K+1}}{1-\rho^{K+1}} \tag{9.A.3}$$

where ρ is λ_S/μ. The expected waiting time is given by:

$$W = \frac{L}{\lambda'_S}, \qquad \lambda'_S = \lambda_S(1-P_K) \tag{9.A.4}$$

where λ'_S can be interpreted as the arrival rate of customers that choose to receive service — the effective arrival rate as was defined. P_K is simply the probability that exactly K customers are in the server's system, and thus $1 - P_K$ is the probability that the server can accept at least one more customer. P_K is given by:

$$P_K = \frac{(1-\rho)\rho^K}{1-\rho^{K+1}} \tag{9.A.5}$$

with $\rho \neq 1$ as is always the case in our model, since waiting time would be infinite and thus λ would be zero, a contradiction.

To compute the equilibrium we should set $p_1 = p_2 = \ldots = p_N = p$ and $G_1 = G_2 = \ldots = G_N = G$ and solve the system of equations 3 through 7 to acquire λ'_S as a function of price p, and then maximize profit rate $p \cdot \lambda'_S(p)$. However only numerical solutions are possible since some equations are transcendental.

The servers end up charging the monopolistic prices (the prices that maximize each seller's revenues). To see this, imagine that the servers are willingly colluding. The problem that they would face is exactly the same as the one described above and they would derive the same solution. It is a well known result in economics, that in the presence of search costs (shoppers do not use shopbots) sellers charge the monopolistic prices.

If we assume that the percentage of customers that choose not to accept service is negligible the approximation model is easily solved to give:

$$\lambda_S(p) = \frac{\alpha(1-\frac{p}{u}) + \mu - \sqrt{[\alpha(1-\frac{p}{u})+\mu]^2 + 4\alpha \cdot [\frac{v}{u} - \mu(1-\frac{p}{u})]}}{2} \tag{9.A.6}$$

with $\alpha < \mu$, $v < u \cdot \mu$ and $p < u \cdot \mu$. It turns out to be a very good approximation in many cases that we studied, including the simulations that we have presented.

Notes

1. "Utility" is defined as the difference between how much customers value the service and how much they pay for it.

2. The downside of this approach is that it requires the buyers' cooperation and might result in some buyers providing false information on their service time to deliberately damage a seller's reputation. In (Zacharia et al., 2000) the authors propose a solution where the buyers are themselves rated according to the accuracy of the information they provide.

3. We repeat the distinction between the unconditional expected waiting time, i.e. the expected waiting time of a service provider when no other information is available and the expected waiting time of a service provider with a specific queue length.

4. We can also use the term "myopic" for this strategy since sellers only consider the next customer.

5. Again, this is to show one of the main findings of this paper, that sellers are always better off when they successfully collude while shoppers use shopbots, instead of colluding by not accepting shopbots.

6. We repeat our assumption that the sellers must choose a price and charge it for a "long" period of time. They cannot assume that market conditions will remain the same after they can readjust their price, so we do not consider multiperiod games.

References

Baron, D. P. (1981). Price regulation, product quality, and asymmetric information. *The American Economic Review*, 71(1):212–220.

Daniel, J. I. and Mulligan, J. G. (1995). Price and waiting time dispersion among service firms. Technical report, University of Delaware.

Davidson, C. (1988). Equilibrium in servicing industries: An economic application of queueing theory. *Journal of Business*, 61(3):347–367.

Dewan, S. and Mendelson, H. (1990). User delay costs and internal pricing for a service facility. *Management Science*, 36(12):1502–1517.

Green, H. (1999). The great yuletide shakeout. *Business Week*, Nov 1st:18–28.

Greenwald, A. and Kephart, J. O. (1999). Shopbots and pricebots. IJCAI 1999.

Greenwald, A., Kephart, J. O., and Tesauro, G. J. (1999). Strategic pricebot dynamics. ACM Conference on Electronic Commerce.

Gross, D. and Harris, C. M. (1985). *Fundamentals of Queueing Theory.* John Wiley and Sons, 2 edition.

Kephart, J. O. and Greenwald, A. (1999). Shopbot economics. Autonomous Agents 1999.

Kephart, J. O., Hanson, J. E., and Sairamesh, J. (1998). Price-war dynamics in a free-market economy of software agents. ALIFE VI.

Mardesich, J. (1999). The web is no shopper's paradise. *Fortune*, November 8th:188–198.

Nelson, J. (1993). Market power, information and efficiency: The link between solutions and equilibrium distortions. *Journal of Industrial Economics*, 41(3):277–293.

Sycara, K., Decker, K., and Williamson, M. (1997). Middle-agents for the internet. IJCAI 1997.

Whitt, W. (1999). Improving service by informing customers about anticipated delays. *Management Science*, 45(2):192–207.

Zacharia, G., Moukas, A., Boufounos, P., and Maes, P. (2000). Dynamic pricing in a reputation brokered agent mediated knowledge marketplace. Hawaii International Conference on System Sciences, Wailea Maui, Hawaii.

Chapter 10

POLYNOMIAL TIME MECHANISMS FOR COLLECTIVE DECISION MAKING

Thomas C. O'Connell*

Department of Mathematics and Computer Science, Skidmore College, 815 North Broadway, Saratoga Springs, NY 12866

oconnellT@acm.org

Richard E. Stearns

Department of Computer Science, State University of New York at Albany, Albany, NY 12222, USA.

res@cs.albany.edu

Abstract We investigate the problem of designing mechanisms to control collective decisions made by self-interested autonomous agents. In particular, we examine how results in the economics literature on mechanism design apply to collective decisions involving NP-hard optimization problems. We formalize the idea of polynomial time mechanism design and investigate mechanism design for a multiagent version of MAXSAT. We prove that there exists a polynomial time mechanism for multiagent MAXSAT that guarantees the outcome to be within a factor of 1/2 of the optimal outcome. We also show that, in general, a 1/2 approximation is the best approximation possible for dominant strategy, Nash, undominated Nash and subgame perfect implementation. Our analysis highlights some of the difficulties that arise in applying results from mechanism design to computational problems. Our results suggest that we may be much less successful using approximation to overcome hardness results in multiagent settings than in traditional computational settings especially when we require game theoretic guarantees on the outcomes.

*Supported in part by National Science Foundation Grant CCR-97-34936.

Keywords: Game Theory, Mechanism Design, Computational Complexity, Approximation Algorithms.

1. INTRODUCTION

Consider a warehouse inhabited by several robots that have different and possibly conflicting goals.[1] Each robot is concerned only with satisfying its own goals and does not care whether any of the other robots satisfy their goals. Rather than spending time negotiating with one another when a conflict arises, the robots rely on an outside arbitrator to resolve the conflict quickly and equitably. The arbitrator's only goal is that its decisions satisfy some measure of social desirability called a *social choice rule.* For example, the decision might be required to be *Pareto optimal* – no other possible decision makes one of the robots better off without making another robot worse off.

In the economics literature the arbitrator in the example above is called a *mechanism.* When a mechanism guarantees that a social choice rule is satisfied, the mechanism is said to *implement* the social choice rule. The problem of defining a mechanism to implement a particular social choice rule is known as the *mechanism design* or *implementation problem.* The field of mechanism design has recently attracted the interest of researchers in multiagent systems, particularly those studying automated negotiation (Rosenschein and Zlotkin, 1994; Sandholm, 1999). Before mechanism design can make practical contributions to multiagent system design, however, the computational issues involved must be clearly understood.

The literature on mechanism design is vast (see (Moore, 1992) and Chapter 23 of (Mas-Colell et al., 1995) for surveys). However, in (Moore, 1992), Moore points out that much of this research ignores issues of practicality. He cites two areas for potential research along these lines:

1. modeling bounded rationality on the part of the agents

2. designing mechanisms that are robust to errors in the specification of the problem.

In this paper, we address the idea of practicality in mechanism design from the point of view of theoretical computer science. We model bounded rationality by limiting the agents to polynomial time computation. Since we are interested in decisions that can be computed quickly, we also limit the mechanism to polynomial time computation. Our goal is to investigate the difficulties that arise when trying to implement NP-hard social choice rules and to determine what can be done to overcome these difficulties. Section 2 provides a short introduction to mechanism design. Section 3 formalizes what we mean by a polynomial time mechanism. Sections 4 and 5 look at designing polynomial time mechanisms using dominant strategy and Nash equilibrium respectively for a multiagent version of MAXSAT. In multiagent MAXSAT, each agent's preferences over the set of possible outcomes can be described by a disjunction over a set of Boolean variables. Since MAXSAT is NP-hard, the mechanism does not have time to find an outcome that satisfies the maximum number of simultaneously satisfiable agents so it must settle for an outcome that is approximately optimal. Section 6 provides a proof that the best a mechanism can guarantee is that half of the maximum number of simultaneously satisfiable agents will be satisfied. This is true for implementation in the four main equilibrium concepts used in the study of mechanism design in complete information environments.

We are not the first to study approximation algorithms in mechanism design. Nisan and Ronen (Nisan and Ronen, 1999) recently studied mechanism design for the task allocation problem under dominant strategy equilibrium. In (Nisan and Ronen, 2000), Nisan and Ronen show that, under certain conditions, dominant strategy implementation of social choice rules that approximately maximize the sum of the agents' utilities is impossible. Our work extends the idea of computational mechanism design to Nash implementation. The negative results of Section 6 combined with Nisan and Ronen's impossibility result indicate that our ability to find good approximation mechanisms for a problem may be limited even when there exist good approximation algorithms for the non-multiagent version of the same problem.

It certainly could be argued that the multiagent MAXSAT problem as we define it is somewhat restrictive since each agent is limited to preferences defined by simple disjunctions rather than more general Boolean formulae. Our goal, however, is not to develop mechanisms for a particular application. Rather, our interest is in determining how restricting the agents and the mechanism to polynomial time computation affects the existing results in the economics literature on mechanism design. In particular, we would like to determine which mechanisms in the literature are polynomial time when the social choice rule consists of approximately optimal solutions to an NP-hard optimization problem. With these goals in mind and in order to focus our analysis, we need a conceptually simple NP-hard problem with many known approximation algorithms. Multiagent MAXSAT satisfies these requirements. On the other hand, we know of no approximation algorithms for the version of this problem in which the agents' preferences are defined by more general Boolean formulae.

2. MECHANISM DESIGN

We begin with a formal description of the mechanism design problem. This section is based on material from (Mas-Colell et al., 1995), (Maskin, 1985), and (Moore, 1992).

Definition 4 *A* **mechanism design problem** *consists of the following:*

- *a set of I* **agents** *that must make a collective choice over some finite set X of possible outcomes.*
- *for each agent i, a set Θ_i of possible* **types**. *An agent's type determines its preferences over the outcomes. It can be thought of as the characteristics of the agent. A vector indicating the type of each agent is called a* **type profile**.
- *for each agent i, a* **utility function** *$u_i : X \times \Theta_i \rightarrow \Re$ that represents the agent's preferences over the outcomes given the agent's type. Each agent is assumed to be trying to maximize its utility.*
- *a* **social choice rule** *$F : \Theta_1 \times \cdots \times \Theta_I \rightarrow 2^X \setminus \{\emptyset\}$. If $F(\cdot)$ is single valued it is called a* **social choice function**.[2] *A social choice rule maps a type profile to a nonempty set of outcomes that are considered socially desirable given that type profile for the agents.*

Definition 5 *A* **mechanism** *$\Gamma = (A_1, \ldots, A_I, g(\cdot))$ consists of a collection of action sets $A_1, \ldots, A_I$ and an outcome function $g : A_1 \times \cdots \times A_I \rightarrow X$. Each agent i chooses an action from set A_i. The mechanism then sets the outcome to $g(a_1, \ldots, a_I)$.*

The goal of the mechanism designer is to ensure that the outcomes of the mechanism are socially desirable as defined by the social choice rule $F(\cdot)$. For example, consider the following multiagent version of the MAXSAT problem:

Definition 6 *The* **Multiagent** MAXSAT *problem is defined as follows:*

- *The parameters of the problem consist of a set of Boolean variables.*
- *The set of outcomes consists of all truth assignments to those variables.*
- *An agent's type consists of preferences over the truth assignments. The preferences are restricted to those that can be described by a simple disjunction over a subset of the variables and their negations. We will refer to such a disjunction as a clause. Each agent*

prefers that its clause be satisfied but does not differentiate between satisfying truth assignments. A type profile, therefore, is a vector of clauses where each clause represents a single agent's type.

- *The goal of the mechanism is to maximize the total number of satisfied clauses. The social choice rule for this problem is defined by:*

$$\text{MAXSAT}(\theta) = \{t : t \textit{ is an outcome that satisfies } m(\theta) \textit{ agents}\}$$

where $m(\theta)$ is the maximum number of simultaneously satisfiable clauses in the type profile θ.

Consider the robot example again and assume that the states of the world can be represented by truth assignments over a set of Boolean variables. For example, suppose there are two blocks B_1 and B_2 and one table. Let x_i = True represent B_i being on the table and let x_i = False represent B_i being on the floor for $i = 1, 2$. A robot's type corresponds to its goal since the robot's goal determines its preferences over the outcomes. Suppose each robot's goal can be represented by a disjunction over the two variables and their negations. Let the mechanism's goal be to satisfy as many of the robots as possible. This is an example of a multiagent MAXSAT problem. Suppose the mechanism requires the robots to declare their goals. In other words, the action sets are just the set of clauses over the Boolean variables. Nothing prevents the robots from lying to the mechanism since the mechanism has no way to verify what the robots' goals really are. Notice, however, that the social choice rule which the mechanism is trying to satisfy is a mapping of the robots' true types not the declared types. The mechanism wants to choose a truth assignment that maximizes the total number of true clauses that are satisfied regardless of the clauses the robots declared. To achieve its goal, the mechanism must be designed in such a way that the robots are enticed to reveal enough information about their true types for the mechanism to make an appropriate decision.

The situation that we study in this paper is one in which the mechanism does not have time to choose an optimal outcome so it must settle for an approximately optimal outcome. For example, suppose the mechanism uses Johnson's first approximation algorithm[3] to determine the outcome. This is a greedy algorithm that takes the literal that appears in the most clauses and sets it to True. It then repeatedly chooses the unassigned literal that appears most in the remaining unsatisfied clauses and sets that to True. Let ties be broken by choosing the least numbered variable first and assigning True before False. If there are five robots with goals defined by the vector $\theta = (x_1, \bar{x}_1 \vee \bar{x}_2, \bar{x}_1 \vee \bar{x}_2, x_2, x_2)$

then the mechanism chooses $t = \bar{x}_1 x_2$. Robot 1 is not satisfied by this outcome. However, if Robot 1 declared its type to be $x_1 \vee \bar{x}_2$, the outcome would be $t = x_1 \bar{x}_2$ which does satisfy Robot 1. Therefore, in this instance, it is better for Robot 1 to lie about his goal.

The literature considers two types of environments with regard to the information that the agents possess about each other. In the *complete information* case, which is the case we consider in this paper, each agent knows his own type as well as the types of all the other agents. In the *incomplete information* case, each agent has no information or only partial information about the other agents' types. In either case, the mechanism does not know the agents' types. A mechanism $\Gamma = (A_1, \ldots, A_I, g(\cdot))$ combined with a set of possible types Θ_i and a utility function u_i for each agent i induces a game of complete or incomplete information depending on the information available to the agents. At the beginning of the game, each agent somehow comes to know its type. Formally, we say each agent receives a signal that indicates its type. Each agent then chooses an action based on its type. In the complete information case, a strategy for agent i is a mapping $s_i : \Theta_1 \times \cdots \times \Theta_I \rightarrow A_i$ while in the incomplete information case, a strategy for agent i is a mapping $s_i : \Theta_i \rightarrow A_i$.

The mechanism designer's goal is to design the mechanism in such a way that, for every possible type profile, some equilibrium outcome of the induced game is socially desirable. There are many competing equilibrium concepts. In this paper, we will be discussing dominant strategy and Nash equilibrium which are defined below. In these definitions and in the remainder of this paper, we use the notation a_{-i} to indicate the vector $a = (a_1, \ldots, a_I)$ with the i-th component removed. We use (a_i', a_{-i}) to represent the vector a with the i-th component replaced by a_i'.

Definition 7 *A strategy profile $s^*(\cdot) = (s_1^*(\cdot), \ldots, s_I^*(\cdot))$ is a* **dominant strategy equilibrium** *of a mechanism $\Gamma = (A_1, \ldots, A_I, g(\cdot))$ if, for each agent i and all type profiles $\theta = (\theta_1, \ldots, \theta_I)$,*

$$u_i(g(s_i^*(\theta), a_{-i}), \theta_i) \geq u_i(g(a_i', a_{-i}), \theta_i)$$

for all $a_i' \in A_i$ and all $a_{-i} \in A_{-i}$.

In other words, a strategy profile is a dominant strategy equilibrium if no matter what the other agents do, no agent has another action that can increase its utility.

Definition 8 *A strategy profile $s^*(\cdot) = (s_1^*(\cdot), \ldots, s_I^*(\cdot))$ is a* **Nash equilibrium** *of a mechanism $\Gamma = (A_1, \ldots, A_I, g(\cdot))$ if, for each agent i*

and all type profiles $\theta = (\theta_1, \ldots, \theta_I)$,

$$u_i(g(s_i^*(\theta), s_{-i}^*(\theta)), \theta_i) \geq u_i(g(a_i', s_{-i}^*(\theta)), \theta_i)$$

for all $a_i' \in A_i$.

In other words, a strategy profile is a Nash equilibrium if no agent can unilaterally deviate from the prescribed action and increase its utility.

Definition 9 *An* **equilibrium outcome** *of a mechanism* Γ *for some type profile* θ *is the outcome produced by an equilibrium strategy profile* $s^*(\theta)$ *of* Γ.

There are different degrees to which a mechanism can satisfy a social choice rule. These are defined as follows:

Definition 10 *Given a mechanism* Γ, *let* $E(\theta)$ *be the set of equilibrium outcomes for a type profile* θ. *We say*

(i) Γ **implements** $F(\cdot)$ *if* $E(\theta) \cap F(\theta) \neq \emptyset$ *for all* θ.

(ii) Γ **strongly implements** $F(\cdot)$ *if* $\emptyset \neq E(\theta) \subset F(\theta)$ *for all* θ.

(iii) Γ **fully implements** $F(\cdot)$ *if* $E(\theta) = F(\theta)$ *for all* θ.

If multiple equilibria exist, it might be difficult to argue that one equilibrium will be played while another will not since all equilibria are equally rational according to any particular equilibrium criterion. The definition of implementation does not exclude the possibility that undesirable equilibrium outcomes are possible. Strong implementation eliminates mechanisms that do not guarantee that every equilibrium outcome is socially desirable. Full implementation receives a great deal of attention in the literature on Nash implementation. However, in line with the philosophy of approximation algorithms, we want to guarantee that all outcomes reach some threshold of acceptability. We are not necessarily concerned with making all acceptable outcomes possible. In designing mechanisms to control the collective decision making process of a group of autonomous agents, strong implementation should suffice.

3. POLYNOMIAL TIME MECHANISMS

From a computational perspective, the mechanism design problem consists of two parts: the computation performed by the mechanism and the computation performed by the agents. From the mechanism's point of view, an instance of the problem consists of the description of the problem's parameters plus a description of the actions chosen by the agents. From the agent's point of view, an instance of the problem consists of the description of the problem's parameters and a description of the agent's type.[4] Hence, we have the following definitions:

Definition 11 $\Gamma = (A_1, \ldots, A_I, g(\cdot))$ *is a* **polynomial time mechanism** *if there is some polynomial $p(\cdot)$ such that $g(a_1, \ldots, a_I)$ is computable in time which is $O(p(|a_1| + \cdots + |a_I| + l))$ where $|a_i|$ is the size of the binary description of action a_i and l is the size of the binary description of the problem's parameters.*

Definition 12 *A strategy profile $s(\cdot) = (s_1(\cdot), \ldots, s_I(\cdot))$ is a* **polynomial time strategy profile** *if there is a polynomial $p(\cdot)$ such that for all i, $s_i(\theta)$ is computable in time which is $O(p(|\theta| + l))$ where $|\theta|$ is the size of the binary description of the agents' type profile and l is the size of the binary description of the problem's parameters.*

We are interested in situations in which the computation performed by the mechanism and the agents is restricted to polynomial time. This leads to the following variations on Definition 10.

Definition 13 *Given a polynomial time mechanism Γ, let $\mathrm{PE}(\theta)$ be the set of polynomial time equilibrium outcomes for a type profile θ. If Γ is a polynomial time mechanism then we say that,* **in polynomial time for polynomial time bounded agents**,

(i) Γ **implements** $F(\cdot)$ *if* $\mathrm{PE}(\theta) \cap F(\theta) \neq \emptyset$ *for all* θ.

(ii) Γ **strongly implements** $F(\cdot)$ *if* $\emptyset \neq \mathrm{PE}(\theta) \subset F(\theta)$ *for all* θ.

(iii) Γ **fully implements** $F(\cdot)$ *if* $\mathrm{PE}(\theta) = F(\theta)$ *for all* θ.

3.1. REVELATION MECHANISMS

Mechanisms in which $A_i = \Theta_i$ for all i form an important class of mechanisms known as *revelation mechanisms*. In other words, revelation mechanisms require the agents to declare their types. We say that a social choice rule $F(\cdot)$ is truthfully implementable if there is a revelation mechanism for which truthful type declarations by all the agents constitutes an equilibrium with an outcome in $F(\cdot)$. More formally:

Definition 14 *A social choice rule $F(\cdot)$ is* **truthfully implementable** *if there is a revelation mechanism $\Gamma = (\Theta_1, \ldots, \Theta_I, g(\cdot))$ that has an equilibrium strategy profile $s^*(\cdot)$ such that $s_i^*(\theta) = \theta_i$ and $g(s^*(\theta)) \in F(\theta)$ for all type profiles θ and all agents i.*

Definition 15 *A revelation mechanism Γ is said to be* **truthful** *if truth telling is a dominant strategy for Γ.*

The following result known as the *Revelation Principle* states that truthfully implementable social choice rules are the only social choice rules that can be implemented. This principle applies to many equilibrium concepts including dominant strategy and Nash equilibrium so we state it without specifying a particular equilibrium concept (see the discussion in (Maskin, 1985) on p. 182-183):

Proposition 1 (The Revelation Principle) *Suppose there is a mechanism Γ that implements social choice rule $F(\cdot)$. Then $F(\cdot)$ is truthfully implementable.*

Proof. See for example (Mas-Colell et al., 1995). ∎

The Revelation Principle has a polynomial time analog if the agents are restricted to polynomial time strategies. This result holds for any equilibrium concept for which the standard Revelation Principle applies.

Proposition 2 (The Polynomial Time Revelation Principle) *If a social choice rule $F(\cdot)$ is implementable in polynomial time for polynomial time bounded agents then $F(\cdot)$ is truthfully implementable in polynomial time for polynomial time bounded agents.*

Proof. See (O'Connell and Stearns, 2000). ∎

Using the Polynomial Time Revelation Principle, we can show that MAXSAT$(\cdot)$, defined in Definition 6, is not polynomial time implementable.

Proposition 3 MAXSAT$(\cdot)$ *is not polynomial time implementable for polynomial time bounded agents assuming* P $\neq$ NP.

Proof. See (O'Connell and Stearns, 2000). ∎

Since MAXSAT$(\cdot)$ is not implementable in polynomial time for polynomial time bounded agents, the best one can hope for is to find some approximation to MAXSAT$(\cdot)$ that is implementable in polynomial time for polynomial time bounded agents. An approximation to MAXSAT$(\cdot)$ is defined as a social choice rule of the form:

$$c\text{-MAXSAT}(\theta) = \{t : t \text{ satisfies at least } cm(\theta) \text{ clauses}\}$$

where $c \in (0,1)$ and $m(\theta)$ is the maximum number of simultaneously satisfiable clauses in θ. In general, if $F(\cdot)$ is a social choice rule taking the form

$$F(\theta) = \{t \in X : t \in \arg\max_{x \in X} h(x, \theta)\}$$

for some real valued function $h(\cdot)$, a c-approximation for $F(\cdot)$ is a social choice rule $c\text{-}F(\theta)$ such that

$$c\text{-}F(\theta) = \{t \in X : h(t, \theta) \geq c \max_{x \in X} h(x, \theta)\}$$

where c is a constant between 0 and 1.

Definition 16 *A social choice rule $F(\cdot)$ is* **approximately implementable** *if there is some constant $c \in (0, 1)$ such that $c\text{-}F(\cdot)$ is implementable.*

In Sections 4 and 5, we examine whether there exists a constant $c \in (0, 1)$ such that c-MAXSAT$(\cdot)$ is implementable in dominant strategy or Nash equilibrium.

4. DOMINANT STRATEGY IMPLEMENTATION

When dealing with dominant strategy implementation, one must contend with an impossibility theorem known as the Gibbard-Satterthwaite Theorem (Gibbard, 1973; Satterthwaite, 1975) which restricts the set of implementable social choice functions to those that are dictatorial.

Definition 17 *A social choice function $f(\cdot)$ is* **dictatorial** *if there is a single agent i such that, for all type profiles θ, $f(\theta)$ is agent i's most preferred outcome.*

Proposition 4 (The Gibbard-Satterthwaite Theorem) *Let $f(\cdot)$ be any social choice function. Suppose the set of possible outcomes, X, is finite and contains at least three elements and that the range of $f(\cdot)$ is X. Further suppose that the set of possible preference relations over X contains the set of strict preferences over X. Then $f(\cdot)$ is truthfully implementable in dominant strategies if and only if $f(\cdot)$ is dictatorial.*

Since the set of preference relations for MAXSAT does not include the set of strict preference relations, the Gibbard-Satterthwaite Theorem does not apply to multiagent MAXSAT. To see this, observe that if an agent's clause does not include a variable x_i then the agent is indifferent between truth assignments that are identical except in their assignment to x_i. Furthermore, if the agent's clause includes more than one literal then the agent is indifferent between truth assignments that satisfy at least one of the literals. As the following proposition shows, MAXSAT$(\cdot)$ is truthfully implementable in dominant strategies.

Proposition 5 *There is a revelation mechanism with a non-dictatorial outcome function that truthfully implements* MAXSAT(·) *in dominant strategies.*

Proof. See (O'Connell and Stearns, 2000). ∎

We know from Proposition 3 that MAXSAT(·) cannot be implemented in polynomial time so we are interested in determining whether there is some constant c such that c-MAXSAT(·) can be implemented in polynomial time. The question is how to convert one of the many existing approximation algorithms for MAXSAT into a mechanism that truthfully implements c-MAXSAT(·). In the remainder of this section, we show that we can strongly implement $\frac{1}{2}$-MAXSAT(·) in polynomial time using an algorithm we call the complement algorithm.

The following lemma implies that there is a simple 1/2-approximation algorithm for MAXSAT.

Lemma 1 *For a truth assignment t, let $\bar{t}$ denote the truth assignment such that for every variable v, $\bar{t}(v)$ = True if and only if $t(v)$ = False. For all truth assignments t, either t or $\bar{t}$ satisfies at least half the maximum number of satisfiable clauses.*

Proof. Let θ_i be any clause that t does not satisfy. Let l_i be a literal in θ_i. Then $t(l_i)$ = False which implies $\bar{t}(l_i)$ = True. Therefore, $\bar{t}$ satisfies θ_i. ∎

We can use this property to develop a polynomial time mechanism that strongly implements $\frac{1}{2}$-MAXSAT(·). Fix a truth assignment t and define a social choice function $f(\cdot)$ as follows:

$$f(\theta) = \begin{cases} \bar{t} & \text{if } \bar{t} \text{ satisfies more clauses in } \theta \text{ than } t \\ t & \text{otherwise} \end{cases} \quad (1)$$

This social choice function is not dictatorial since for each agent i we can define θ such that $\bar{t}$ does not satisfy θ_i but does satisfy every clause in θ_{-i} while t does not satisfy any clause in θ_{-i}. For example, let $t = x_1\bar{x}_2\bar{x}_3$ and $\theta = (x_1, x_2, x_3)$. We have $f(\theta) = \bar{t}$ which is not agent 1's most preferred outcome. A similar argument applies to the other two agents which implies that $f(\theta)$ is not the same agent's most preferred outcome for every θ. Lemma 1 implies that $f(\theta)$ is a $\frac{1}{2}$-approximation for MAXSAT(·). Therefore, any mechanism that truthfully implements $f(\cdot)$, truthfully implements $\frac{1}{2}$-MAXSAT(·).

Define the following revelation mechanism:

Mechanism 1

1. Given truth assignment t from the definition of $f(\cdot)$ in Equation 1.

2. If t satisfies at least as many of the declared clauses as $\bar{t}$
3. then choose t as the outcome
4. else choose $\bar{t}$.

Theorem 1 *Mechanism 1 above truthfully implements $f(\cdot)$ in dominant strategies in polynomial time for polynomial time bounded agents.*

Proof. Mechanism 1 is polynomial time since, to compute the outcome, it need only compare the number of clauses in θ that are satisfied by two fixed truth assignments. Truth telling is certainly a polynomial time strategy so it suffices to show that truth telling constitutes a dominant strategy equilibrium.

Let t be the truth assignment used in the definition of $f(\cdot)$ in Equation 1. Let θ_i be agent i's true type.

Case 1: t satisfies θ_i and $\bar{t}$ doesn't.

Removing literals from θ_i potentially decreases the number of clauses satisfied by t and does not affect the number of clauses satisfied by $\bar{t}$. Adding literals cannot increase the number of clauses satisfied by t and cannot decrease the number of clauses satisfied by $\bar{t}$. Since lying either leaves the number of clauses satisfied by t the same or decreases it and leaves the number of clauses satisfied by $\bar{t}$ the same or increases it, lying either leaves the outcome unaffected or results in an outcome that is worse for agent i.

Case 2: $\bar{t}$ satisfies θ_i and t doesn't. A symmetric argument applies to this case.

Case 3: Both t and $\bar{t}$ satisfy θ_i. In this case, agent i does not care which of the two truth assignments is chosen so lying cannot be beneficial or harmful.

Since lying is never beneficial, truth telling is a dominant strategy. ▮

Corollary 1 *The social choice rule $\frac{1}{2}$-MAXSAT$(\cdot)$ is strongly implementable in dominant strategies in polynomial time for polynomial time bounded agents.*

Proof. Case 3 is the only case in the proof of Theorem 1 in which lying can affect the outcome and not be harmful to the agent. In (O'Connell and Stearns, 2000), we show that any lie in Case 3 results in an outcome that is in $\frac{1}{2}$-MAXSAT(θ). ▮

5. NASH IMPLEMENTATION

We now consider mechanism design in environments with complete information using Nash equilibrium. There are two properties of social choice rules that are extremely important for Nash implementation – no veto power and monotonicity. No veto power states that one agent cannot override the wishes of all the other agents. Monotonicity says that whenever the type profile changes from θ to θ' and the set of outcomes that an alternative $t \in F(\theta)$ is preferred to remains the same or expands then t must also be in $F(\theta')$. These two properties are formalized in the following definitions.

Definition 18 *A social choice rule $F(\cdot)$ satisfies* **no veto power** *if for any type profile θ such that $I-1$ agents rank an alternative t as their weakly most preferred choice, then $t \in F(\theta)$.*

Let $L_i(t, \theta_i) = \{t' : u_i(t, \theta_i) \geq u_i(t', \theta_i)\}$. $L_i(t, \theta_i)$ is called agent i's *lower contour set* for t.

Definition 19 *A social choice rule $F(\cdot)$ is* **monotonic** *if for all type profiles θ and θ' and all outcomes t such that $t \in F(\theta)$ and $L_i(t, \theta_i) \subseteq L_i(t, \theta_i')$ for all i, we have $t \in F(\theta')$.*

The following result is known as Maskin's Theorem (see (Maskin, 1985)).

Proposition 6 (Maskin's Theorem) *If a social choice rule is fully Nash implementable then it is monotonic. If there are at least three agents then a social choice rule that is monotonic and satisfies no veto power is fully Nash implementable.*

Monotonicity is also required for full Nash implementation in polynomial time for polynomial time bounded agents.

Proposition 7 *If a social choice rule $F(\cdot)$ is fully Nash implementable in polynomial time for polynomial time bounded agents then it is monotonic.*

Proof. See (O'Connell and Stearns, 2000). ∎

Proposition 8 *If there are at least two variables then* MAXSAT$(\cdot)$ *is not monotonic.*

Proof. We provide a proof for the case of two agents and two variables. It can be generalized easily. Let $\theta = (x_1, \bar{x}_1)$. The maximum number of simultaneously satisfiable clauses in θ is 1. Let $t = x_1 x_2$ which is in MAXSAT(θ). Let $\theta' = (x_2, \bar{x}_1)$. Since t satisfies agent 1 when the type profile is either θ or θ', $L_1(t, \theta_1) = L_1(t, \theta_1')$. Since agent 2's type is the same in either case, $L_2(t, \theta_2) = L_2(t, \theta_2')$. However, $t \notin$ MAXSAT(θ') which implies that MAXSAT$(\cdot)$ is not monotonic. ∎

Even if MAXSAT$(\cdot)$ were monotonic, by Proposition 3, it cannot be implemented in polynomial time for polynomial time bounded agents assuming P $\neq$ NP. If the mechanism and the agents are restricted to polynomial time computation, the best one could hope for is to approximately implement MAXSAT($\cdot$). Therefore, we need to find a constant c such that c-MAXSAT($\cdot$) is monotonic even though MAXSAT($\cdot$) is not. We also need to find a polynomial time mechanism that implements c-MAXSAT($\cdot$).

There are several different mechanisms used to prove Maskin's theorem (Maskin, 1985; Repullo, 1987; Saijo, 1988; McKelvey, 1989). For example, in (Saijo, 1988), Saijo defines a mechanism that fully implements a monotonic social choice rule satisfying no veto power as follows. The action sets are defined by $A_i = \Theta_i \times \Theta_{i+1} \times X \times \{1, \ldots, I\}$. Each agent i declares his own type θ_i^i, the type of his neighbor θ_i^{i+1} where we take $I + 1 = 1$, an outcome x_i, and a number k_i between 1 and I. The outcome function $g(\cdot)$ is defined by the following rules:

Mechanism 2 (Saijo's Mechanism)

Let $a = \left[(\theta_i^i, \theta_i^{i+1}, x_i, k_i)\right]_{i=1}^{I}$ *be the action profile.*

Rule I: If $\theta_i^i = \theta_{i-1}^i$ *and* $x_i = x$ *for all* i *and* $x \in F(\theta_1^1, \ldots, \theta_I^I)$ *then* $g(a) = x$.

Rule II: If $\theta_i^i = \theta_{i-1}^i$ *for all* i *except* j *or* $j+1$, *and* $x_i = x$ *for all* i *except* j, *and* $x \in F(\theta_1^1, \ldots, \theta_{j-1}^{j-1}, \theta_{j-1}^j, \theta_{j+1}^{j+1}, \ldots, \theta_I^I)$

$$\text{then } g(a) = \begin{cases} x_j & \text{if } x_j \in L_j(x, \theta_{j-1}^j), \\ x & \text{otherwise} \end{cases}$$

Rule III: If neither Rule I nor Rule II applies then set $g(a) = x_n$ *where* $n = \left(\sum_{i \in I} k_i\right) (\mathrm{mod} I) + 1$.

Notice that Mechanism 2 checks whether a given outcome is socially desirable given the declared type profile. This check, which also appears in the mechanisms used in (Repullo, 1987) and (Maskin, 1985) corresponds to the following decision problem when $F(\cdot)$ is c-MAXSAT($\cdot$).

Definition 20 c**-MAXSAT Membership Problem**
Given a set of Boolean variables $V = \{v_1, \ldots, v_n\}$, *a set of clauses* θ *over* V *and a truth assignment* t, *does* t *satisfy at least* c *times the maximum number of satisfiable clauses in* θ?

Proposition 9 shows that this problem is NP-hard which implies that Saijo's mechanism with $F(\cdot) = c$-MAXSAT($\cdot$) is not a polynomial time mechanism unless P = NP.

Proposition 9 *Let* $c \in (0,1)$ *be such that there is a polynomial time approximation algorithm for c-MAXSAT. The c-MAXSAT Membership problem is NP-hard.*

Proof. See (O'Connell and Stearns, 2000). ∎

Theorem 2 *Let $c \in (0,1)$ be such that there is a polynomial time approximation algorithm for c-MAXSAT. When $F(\cdot) = c$-MAXSAT$(\cdot)$, Saijo's mechanism is not a polynomial time mechanism unless* P = NP.

In Saijo's mechanism, for a strategy profile to be an equilibrium strategy profile, it must always be the case that all but at most one of the agents choose an outcome that is in $F(\theta)$. Theorem 2 implies that if c is such that Saijo's mechanism is polynomial time then the agents cannot find an alternative in c-MAXSAT$(\cdot)$ in polynomial time. Therefore, either the mechanism is not polynomial time or there is no polynomial time equilibrium strategy profile. Whether there is a general mechanism to implement any monotonic social choice rule satisfying no veto power that does not check membership in $F(\cdot)$ is an open question.

As discussed in Section 2, we are willing to settle for strong implementation as opposed to full implementation of c-MAXSAT$(\cdot)$. Thus, it is sufficient for our purposes to find a social choice rule $F(\cdot)$ such that

1. $F(\cdot)$ is monotonic
2. $F(\cdot)$ satisfies no veto power
3. $F(\theta) \subseteq c$-MAXSAT(θ) for all θ
4. checking membership in $F(\cdot)$ is easy.

Such a social choice rule exists for $c = 1/2$.

Theorem 3 *Let $F(\theta)$ be the set of outcomes t such that t satisfies at least $\frac{I}{2}$ clauses in θ. If there are at least three agents, $F(\cdot)$ is fully Nash implementable using Saijo's mechanism.*

Proof. $F(\cdot)$ clearly satisfies no veto power since there are at least three agents. Let θ, θ' and t be such that $t \in F(\theta)$ and $L_i(t, \theta_i) \subseteq L_i(t, \theta'_i)$ for all i. Since every clause has a satisfying truth assignment, if t satisfies θ_i then t must satisfy θ'_i. Therefore, t satisfies at least half of the clauses in θ' so $t \in F(\theta')$. Hence, $F(\cdot)$ is monotonic. Since $F(\cdot)$ is monotonic and satisfies no veto power, Saijo's mechanism fully Nash implements $F(\cdot)$. ∎

Corollary 2 *The social choice rule $\frac{1}{2}$-*MAXSAT$(\cdot)$ *is strongly Nash implementable in polynomial time for polynomial time bounded agents.*

Proof. Let $F(\cdot)$ be the the social choice rule used in Theorem 3. It is easy to check whether a truth assignment t is in $F(\theta)$ so Saijo's mechanism is polynomial time computable. Let θ be the agents' true type profile. For any truth assignment t such that $t \in F(\theta)$ and any $k, 1 \leq k \leq I$, let $s_i(\theta) = (\theta_i, \theta_{i+1}, t, k)$ for all i. The strategy profile $s(\theta) = (s_1(\theta), \ldots, s_I(\theta))$ is a Nash equilibrium.[5] Such a strategy profile is easy for the agents to compute since the agents can use the complement algorithm to find a $t \in F(\theta)$. Therefore, $F(\cdot)$ is strongly Nash implementable in polynomial time for polynomial time bounded agents. Since $F(\theta)$ is a subset of $\frac{1}{2}$-MAXSAT(θ) for all θ, $\frac{1}{2}$-MAXSAT$(\cdot)$ is strongly Nash implementable in polynomial time for polynomial time bounded agents. ∎

6. UPPER BOUNDS ON APPROXIMABILITY

Existing work in mechanism design shows that the set of social choice rules that are implementable in Nash equilibrium is much smaller than the set of social choice rules that are implementable in refinements such as undominated Nash equilibrium[6] (Palfrey and Srivastava, 1991; Jackson et al., 1994) and subgame perfect equilibrium[7] (Moore and Repullo, 1988). In (Palfrey and Srivastava, 1991), it is observed that for a social choice rule to be fully implementable in Nash, undominated Nash, or subgame perfect equilibrium, it must satisfy Property Q defined below. Property Q must also be satisfied for full implementation in dominant strategies. In (O'Connell and Stearns, 2000), we show that Property Q must be satisfied for full implementation in polynomial time for polynomial time bounded agents.

Definition 21 *(Palfrey and Srivastava, 1991) A social choice rule satisfies* **Property Q** *if, whenever θ, θ' and t are such that $t \in F(\theta)$ and $t \notin F(\theta')$, there is an i such that $\theta_i \neq \theta'_i$ and agent i is not completely indifferent under θ'_i.*

Theorem 4 *Let $F(\cdot)$ be a social choice rule such that, for some constant $c \in (0, 1)$, $F(\theta) \subseteq c$-MAXSAT(θ) for all θ. Suppose for some type profile θ there exists a truth assignment $t \in F(\theta)$ that satisfies less than cI clauses. Then $F(\cdot)$ does not satisfy Property Q.*

Proof. Let $F(\cdot)$, θ, c, and t be as defined above. For each i such that t satisfies θ_i, let l_i be any literal in θ_i such that $t(l_i) =$ True. Create θ' from θ by adding $\bar{l}_i$ to θ_i for each clause θ_i that t satisfies. For each i, either $\theta'_i = \theta_i$ or agent i is completely indifferent under type θ'_i. Hence, for $F(\cdot)$ to satisfy Property Q, it must be the case that $t \in F(\theta')$.

Let $m(\theta')$ be the maximum number of clauses that can be satisfiable simultaneously in the type profile θ'. Since $\bar{t}$ must satisfy every clause that t does not satisfy, and $\bar{t}$ must satisfy θ'_i if t satisfies θ_i, $m(\theta') =$

I. Since the clauses in θ that t doesn't satisfy also appear in θ', the number of clauses in θ' that t satisfies is strictly less than $cI = cm(\theta')$. Therefore, $t \notin c$-MAXSAT(θ'). This implies that $t \notin F(\theta')$ since $F(\theta) \subseteq c$-MAXSAT(θ) for all θ. Thus, $F(\cdot)$ does not satisfy Property Q. ∎

Theorem 4 says that the only approximate social choice rules for MAXSAT that can be fully implemented are those that guarantee the number of clauses that the outcomes satisfy is a constant times the total number of clauses rather than a constant times the maximum number of simultaneously satisfiable clauses. The following corollary shows that this is impossible for $c > 1/2$:

Corollary 3 *For any c such that $\lceil cI \rceil > \lceil \frac{I}{2} \rceil$, c-MAXSAT$(\cdot)$ cannot be strongly implemented in dominant strategy, Nash, undominated Nash or subgame perfect equilibrium.*

Proof. Let Γ be a mechanism that strongly implements c-MAXSAT$(\cdot)$ for some constant c such that $\lceil cI \rceil > \lceil \frac{I}{2} \rceil$. Define $E : \Theta_1 \times \cdots \times \Theta_I \to X$ such that $E(\theta)$ is the set of equilibrium outcomes of Γ for a type profile θ. By definition, Γ fully implements $E(\cdot)$. Since Γ strongly implements c-MAXSAT$(\cdot)$, $E(\theta) \subseteq c$-MAXSAT(θ) for all θ.

Let $\theta'_i = x_1$ for $1 \leq i \leq \lfloor \frac{I}{2} \rfloor$. Let $\theta'_i = \bar{x}_1$ for $\lfloor \frac{I}{2} \rfloor + 1 \leq i \leq I$. Since the maximum number of simultaneously satisfiable clauses is $\lceil \frac{I}{2} \rceil$, no truth assignment can satisfy more than $\lceil \frac{I}{2} \rceil$ clauses. Therefore, any $t \in E(\theta')$ satisfies less than cI clauses in θ'. According to Theorem 4, this implies that $E(\cdot)$ does not satisfy Property Q. But then $E(\cdot)$ is not fully implementable which is a contradiction. ∎

7. CONCLUSIONS

We have formalized the computational problem of mechanism design in such a way that classic results from the field can be applied. Using a multiagent version of MAXSAT, we have investigated the difficulties that arise in applying these results to NP-hard optimization problems. We have demonstrated that, despite the impossibility results regarding dominant strategy implementation, it is possible to implement an approximate social choice rule for MAXSAT in dominant strategy equilibrium.

With regard to Nash implementation, we showed that approximation was beneficial in two ways. It enabled strong Nash implementation when the exact social choice rule was not monotonic and it enabled polynomial time implementation when finding a member of the exact rule was NP-hard. In addition, we showed that the standard mechanisms[8] used to prove Maskin's theorem are not polynomial time when the social choice rule is c-MAXSAT$(\cdot)$ for any c such that there is a polynomial time c-approximation algorithm for MAXSAT. This implies that these mechanisms do not fully Nash implement c-MAXSAT$(\cdot)$ for any c in polynomial time for polynomial time bounded agents. We also showed that the best approximate social choice rule for MAXSAT$(\cdot)$ that can be strongly implemented in dominant strategy, Nash, undominated Nash or subgame perfect equilibrium is $\frac{1}{2}$-MAXSAT$(\cdot)$. This is somewhat remarkable given that there are many approximation algorithms for the non-multiagent version of this problem that achieve better lower bounds than 1/2. Several authors (Yannakakis, 1994; Goemans and Williamson, 1995; Asano, 1997) provide algorithms for MAXSAT that achieve lower bounds $\geq \frac{3}{4}$. Our results, therefore, indicate that it can be much more difficult to design good approximation mechanisms than to design good approximation algorithms. It may be the case that these results are peculiar to MAXSAT so future work should consider multiagent versions of other NP-hard problems. However, we have defined multiagent MAXSAT somewhat restrictively with each agent limited to preferences defined by unweighted disjunctions. The fact that such difficulties arise even for this version of the problem suggests that we will be much less successful using approximation to overcome hardness results in settings with self-interested agents than in traditional computational settings.

Acknowledgments

We would like to thank Professor Laurence Kranich of the Department of Economics at the University at Albany for many helpful comments and suggestions.

Notes

1. This example is taken from (Rosenschein and Zlotkin, 1994).

2. We use lower case $f(\cdot)$ when discussing social choice functions and upper case $F(\cdot)$ when discussing social choice rules.

3. See (Johnson, 1974) or (Battiti, 1998).

4. It should also include a description of the mechanism but we will assume that either the description of the mechanism is fixed or that its size is polynomial in the size of the problem's parameters.

5. See (Saijo, 1988) p. 698.

6. An undominated Nash equilibrium is a Nash equilibrium in which no agent plays a weakly dominated strategy.

7. Subgame perfect equilibrium is defined for games that are played in stages. A strategy profile is a subgame perfect equilibrium if it is a Nash equilibrium in every subgame.

8. We are currently evaluating the complexity of the mechanism due to McKelvey (McKelvey, 1989).

References

Asano, T. (1997). Approximation algorithms for MAX SAT: Yannakakis vs. Goemans and Williamson. In *Proceedings of the 3rd Israel Symposium on Theory and Computing Systems*, pages 24–37, Ramat Gan, Israel.

Battiti, R. (1998). Approximation algorithms and heuristics for MAX-SAT. In Zhu, D.-Z. and Pardalos, P., editors, *Handbook of Combinatorial Optimization*, volume 1, pages 77–148. Kluwer Academic Publishers, Norwell, MA.

Gibbard, A. (1973). Manipulation of voting schemes: A general result. *Econometrica*, 41:587–602.

Goemans, M. and Williamson, D. (1995). Improved approximation algorithms for maximum cut and satisfiabilty problems using semidefinite programming. *Journal of the ACM*, 42:1115–1145.

Jackson, M., Palfrey, T., and Srivastava, S. (1994). Undominated Nash implementation in bounded mechanisms. *Games and Economic Behavior*, 6:474–501.

Johnson, D. (1974). Approximation algorithms for combinatorial problems. *Journal of Computer and Systems Sciences*, 9:256–278.

Mas-Colell, A., Whinston, M., and Green, J. (1995). *Microeconomic Theory*. Oxford University Press, New York, NY.

Maskin, E. (1985). The theory of implementation in Nash equilibrium: a survey. In Hurwicz, L., Schmeidler, D., and Sonnenschein, H., editors, *Social goals and social organization: Essays in memory of Elisha Pazner*, pages 173–204. Cambridge University Press, New York, NY.

McKelvey, R. (1989). Game forms for Nash implementation of general social choice correspondences. *Social Choice and Welfare*, 6:139–156.

Moore, J. (1992). Implementation, contracts and renegotiation in environments with complete information. In Laffont, J.-J., editor, *Advances in economic theory: Sixth World Congress*, pages 182–282. Cambridge University Press, Cambridge, UK.

Moore, J. and Repullo, R. (1988). Subgame perfect implementation. *Econometrica*, 56:1191–1220.

Nisan, N. and Ronen, A. (1999). Algorithmic mechanism design. In *Proceedings of the 31st Annual ACM Symposium on Theory of Computing*, pages 129–140, Atlanta, GA.

Nisan, N. and Ronen, A. (2000). Computationally feasible VCG mechanisms. Working Paper, Institute of Computer Science, Hebrew University of Jerusalem.

O'Connell, T. and Stearns, R. (2000). Polynomial time mechanism design. Technical Report SUNYA-CS-TR-00-1, Department of Computer Science, University at Albany, SUNY.

Palfrey, T. and Srivastava, S. (1991). Nash implementation using undominated strategies. *Econometrica*, 59:479–501.

Repullo, R. (1987). A simple proof of Maskin's theorem on Nash implementation. *Social Choice and Welfare*, 4:39–41.

Rosenschein, J. and Zlotkin, G. (1994). *Rules of Encounter: Designing Conventions for Automated Negotiation among Computers.* MIT Press, Cambridge, MA.

Saijo, T. (1988). Strategy space reductions in Maskin's theorem: Sufficient conditions for Nash implementation. *Econometrica*, 56(3):693–700.

Sandholm, T. (1999). Distributed rational decision making. In Weiss, G., editor, *Mutiagent Systems: A Modern Approach to Distributed Artificial Intelligence*, pages 201–258. MIT Press, Cambridge, MA.

Satterthwaite, M. (1975). Strategy-proofness and Arrow's conditions: Existence and correspondence theorems for voting procedures and social welfare functions. *Journal of Economic Theory*, 10:187–217.

Yannakakis, M. (1994). On the approximation of maximum satisfiability. *Journal of Algorithms*, 17:475–502.

Chapter 11

MULTI-AGENT Q-LEARNING AND REGRESSION TREES FOR AUTOMATED PRICING DECISIONS

Manu Sridharan
Dept. of Electrical Engineering and Computer Science
MIT 38-401, Cambridge MA 02139 USA
msridhar@mit.edu

Gerald Tesauro
IBM Thomas J. Watson Research Center
30 Saw Mill River Rd., Hawthorne, NY 10532 USA
tesauro@watson.ibm.com

Abstract We study the use of single-agent and multi-agent Q-learning to learn seller pricing strategies in three different two-seller models of agent economies, using a simple regression tree approximation scheme to represent the Q-functions. Our results are highly encouraging – regression trees match the training times and policy performance of lookup table Q-learning, while offering significant advantages in storage size and amount of training data required, and better expected scaling to large numbers of agents. Clear advantages are seen over neural networks, which yield inferior policies and require much longer training times. Our work is among the first to demonstrate success in combining Q-learning with regression trees. Also, with regression trees, Q-learning appears much more feasible as a practical approach to learning strategies in large multi-agent economies.

Keywords: Agent economies, multiagent learning, regression trees, reinforcement learning

1. INTRODUCTION

Reinforcement learning in multi-agent environments is a challenging forefront of machine learning research that could have immense practical benefit in many real-world problems. Many application domains are envisioned in which teams of software agents cooperate to achieve a global objective. We also foresee signifcant applications for self-interested agents, for example, in electronic marketplaces, where economic software agents can interact with humans and/or other agents to maximize their own individual profits.

This paper studies the use of Reinforcement Learning in multi-agent economies. Specifically, we study the use of Q-learning (Watkins, 1989) to learn pricing strategies in a competitive marketplace. Q-learning is an algorithm for learning to estimate the long-term expected reward for a given state-action pair. It has the nice properties that it does not need a model of the environment, and it can be used for on-line learning. With a lookup table representing the Q-function, the learning procedure is guaranteed to converge to the optimal value function and optimal policy.

The are two main challenges in using Q-learning in multi-agent systems. First, most of the existing convergence proofs only apply to single-agent, stationary Markov Decision Problems. (Important first steps in analyzing the two-agent case have been reported in Littman (1994) and Hu and Wellman (1998).) However, when multiple agents simultaneously adapt, each agent provides an effectively non-stationary environment for the other agents. Hence it is unknown in general whether any global convergence will be obtained in this case, and if so, whether such solutions are optimal. Second, we expect that large multi-agent systems will have states spaces that are too large for lookup tables to be feasible, and hence some sort of function approximation scheme seems necessary to represent the Q-function.

The present work combines single-agent and multi-agent Q-learning with tree-based function approximation in a model two-seller economy. The sellers alternately take turns setting prices, taking into account the other seller's current price. After the price has been set, the consumers choose either seller 1's product or seller 2's product, based on the current price pair. This leads to an instantaneous reward or profit given to the sellers. We assume that both sellers have full knowledge of the expected consumer response for any price pair, and moreover have full knowledge of both reward functions.

Q-learning is one of a variety of ways of endowing agents with "foresight," i.e., an ability to anticipate long-term consequences of actions.

Foresight was found in Tesauro and Kephart (1999, in press) to improve profitability, and to damp out or eliminate the pathological behavior of unending cyclic "price wars," in which long episodes of repeated undercutting by the sellers alternate with large jumps in price. Such price wars were found to be rampant in agent economy models (Kephart, Hanson & Sairamesh, 1998; Sairamesh & Kephart, 1998) when agents use "myopically optimal" or "myoptimal" pricing algorithms that optimize immediate reward, but do not anticipate any longer-term consequences. Q-learning in particular is a principled way to obtain deep lookahead, since the Q-function represents the cumulative discounted reward looking infinitely far ahead in time. In contrast, the prior work of Tesauro and Kephart (in press) was based on shallow finite lookahead.

Q-learning with lookup tables was previously studied in Tesauro and Kephart (1999). A single Q-learner playing against a myoptimal opponent always converged to an optimal policy. In the more interesting case of two agents simultaneously Q-learning against each other, convergence was again obtained in all three models, at least for small-to-moderate values of the discount parameter. In Tesauro (1999), preliminary studies of Q-learning with neural networks found reasonably good policies but excessively long training times. This is a potentially major drawback in an on-line scenario where agents must learn quickly to maintain profitability. The current study of regression trees as an alternate function approximator was motivated by the goal of improving on the training time of neural networks, while generating policies of at least equal profitability.

The remainder of this paper is organized as follows. Section 2 describes the structure and dynamics of the model two-seller economy, and presents three models of seller profit functions[1] that are prone to price wars when agents myopically optimize short-term payoffs. Section 3 describes the implementation of Q-learning in these model economies, and summarizes previous results of lookup table and neural network Q-learning. In Section 4, the algorithms used for constructing regression trees and adapting them to Q-learning in our study are presented. Section 5 presents the results of using Q-learning with regression trees and compares these results to those of section 3. Finally, section 6 summarizes the main conclusions and discusses promising directions and challenges for future work.

2. MODEL AGENT ECONOMIES

Our models make a number of simplifying assumptions relative to the likely complexities of real agent economies. The economy is restricted to two sellers, competing on the basis of price, who offer similar or identical products to a large population of consumer agents. Prices are discretized and lie between a minimum and maximum price; there are typically ~ 100 possible prices. This renders the state space small enough to use lookup tables to represent the agents' pricing policies and expected profits. Time is also discretized; at each time step, the consumers compare the current prices of the sellers, and instantaneously and deterministically choose to buy from at most one seller. Hence at each time step, for each possible pair of prices, the sellers obtain a deterministic profit.

We also assume that the sellers alternately take turns adjusting their prices, rather than simultaneously setting prices. Alternating-turn dynamics has two motivations: (a) It ensures that there will be a deterministic optimal policy (Littman, 1994), and hence normal Q-learning, which yields deterministic policies, can apply. (b) In a realistic many-seller economy, it seems reasonable to assume that the sellers will adjust their prices at different times rather than at the same time (although probably not in a well-defined order).

The three economic models studied here are described in detail elsewhere. In the first model, called the "Price-Quality" model (Sairamesh & Kephart, 1998), the sellers' products have different values of a scalar "quality" parameter, with higher-quality products being perceived as more valuable. At each time step, the consumers buy the lowest-priced product subject to constraints of a maximum allowable price and a minimum allowable quality. The substitutability of seller products enables direct price competition, and the "vertical" differentiation of differing quality values leads to asymmetries in the sellers' profit functions. Such asymmetries can result in unending cyclic price wars when the sellers use myoptimal strategies.

The second model, described in Kephart, Hanson and Sairamesh (1998), is an "Information-Filtering" model in which the two sellers offer news articles in partly overlapping categories. This model contains a "horizontal" differentiation of article categories. To the extent that the categories overlap, there can be direct price competition, and to the extent that they differ, there are asymmetries that again lead to the potential for cyclic price wars.

The third model is the "Shopbot" model, described in Greenwald and Kephart (1999), which models consumers on the Internet who may use a shopbot to compare prices of some or all sellers offering a given product, and select the lowest-priced seller. In this model, the sellers' products are identical, and their profit functions are symmetric. Myoptimal pricing leads the sellers to undercut each other until the minimum price point

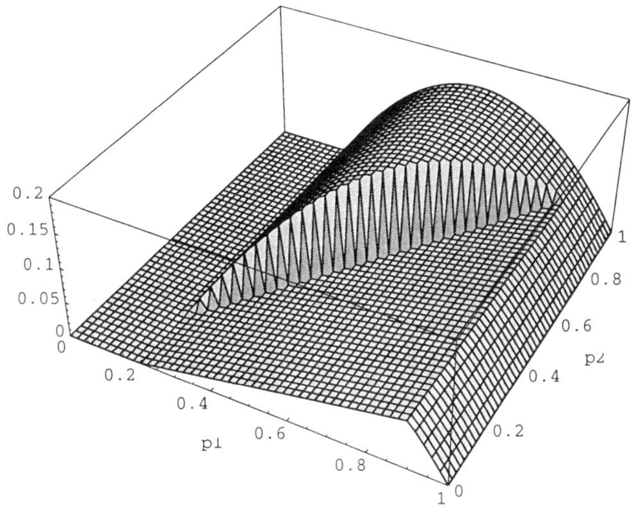

Figure 11.1. Sample profit landscape for seller 1 in Price-Quality model, as a function of seller 1 price p_1 and seller 2 price p_2.

is reached. At that point, a new price war cycle can be launched, due to asymmetric buyer behavior, rather than seller asymmetries. Some buyers choose a random seller rather than bargain-hunt with the shopbot; this makes it preferable to abandon the low-price competition, and instead maximally exploit the random buyers by charging the maximum price.

An example seller profit function, taken from the Price-Quality model, is plotted in Figure 11.1. This shows the instantaneous profit for seller 1, $U_1(p_1, p_2)$. The quality parameters are $q_1 = 1.0$, $q_2 = 0.9$ (i.e. seller 1 is the higher-quality seller). The myoptimal policy for seller 1, $p_1^*(p_2)$, is obtained for each value of p_2 by sweeping across all values of p_1 and choosing the value with the highest profit. For small p_2, the peak profit is obtained at $p_1 = 0.9$. As p_2 increases, the policy eventually jumps to the other peak, which follows the parabolic-shaped ridge along the $p_1 = p_2$ diagonal.

The Information-Filtering and Shopbot models also have similar profit landscapes. In all three models, it is the existence of multiple, disconnected peaks in the landscapes, with varying relative heights depending on the other seller's price, that leads to price wars when the sellers behave myopically.

In these models it is assumed for simplicity that the players can model the consumer behavior perfectly, and they also have perfect knowledge of each other's costs and profit functions. Hence the model is in essence a two-player perfect-information deterministic game, similar to games like chess. The main differences are that the payoffs are not zero-sum, there are no terminating states, and payoffs are given at every time step.

3. SINGLE AND MULTI-AGENT Q-LEARNING

3.1. LEARNING ALGORITHM

Standard Q-learning works as follows: Let $Q(s, a)$ represent the discounted long-term expected reward (with discount parameter γ) to an agent for taking action a in state s. (The value of a reward expected at n time steps in the future is discounted by γ^n.) Assume that $Q(s, a)$ is represented by a lookup table containing a value for every possible state-action pair, and is initialized to arbitrary values. Then to solve for $Q(s, a)$, one infinitely repeats the following two-step loop:

1. Select a particular state-action pair (s, a); observe the immediate reward r and the resulting state s'.

2. Adjust $Q(s, a)$ according to the following equation:

$$\Delta Q(s, a) = \alpha[r + \gamma \max_b Q(s', b) - Q(s, a)] \tag{1}$$

where α is the learning rate parameter. A variety of methods may be used to select state-action pairs in step 1, provided that every state-action pair is visited sufficiently often. When α is decreased over time with an appropriate schedule, the above procedure converges to the correct solution for stationary MDPs.

In these models, the distinction between states and actions is somewhat blurred. It will be assumed that the "state" for each seller is sufficiently described by the other seller's last price, and that the "action" is the current price decision. This should be a sufficient state description, because no other history is needed either for the determination of immediate reward, or for the calculation of the myoptimal price. We have also redefined r and s' for the two-agent case as follows: let s' be the state that is obtained, starting from s, of one action by the Q-learner and a response by the opponent. Likewise, r is defined as the sum of the two rewards obtained after those two actions. These modifications were introduced so that the state s' would have the same player to move as state s. (A possible alternative to this, which has not been investigated, is to include the side-to-move as additional information in the state description.)

3.2. RESULTS OF LOOKUP TABLE LEARNING

Detailed results of lookup table-based Q-learning are presented in Tesauro and Kephart (1999). In brief, single-agent Q-learning in all three models was found to always yield exact convergence to a stationary optimal solution (as expected). The resulting Q-derived policies always outperformed a myopic strategy when tested against myopic opposition, and the expected profit increased montonically with γ. In many cases, Q-learning had the side benefit of also improving the myopic opponent's expected profit. This improvement is due to the Q-learner learning to abandon undercutting behavior more readily as the price decreases. The price-war regime is thus smaller and confined to higher average prices, leading to a closer approximation to collusive behavior, with greater expected profits for both sellers.

For simultaneous Q-learning by both sellers, the procedure utilized was to alternately adjust a random entry in seller 1's Q-function, followed by a random entry in seller 2's Q-function, using the same formalism presented above. As the Q-functions evolved, the policies were correspondingly updated so that they optimized the agents' current Q-function. In modeling the two-step payoff r to a seller in equation 1, the opponent's current policy was used, as implied by its current Q-function.

Simultaneous Q-learning in the Price-Quality model yielded robust convergence to a unique pair of policies, independent of γ, identical to the solution found by shallow lookahead in Tesauro and Kephart (in press). In the Shopbot model, exact convergence of the Q-functions was only found for $\gamma < 0.7$. For $\gamma \geq 0.7$, there was very good approximate convergence, in which the Q-functions converged to stationary solutions to within small random fluctuations. Different solutions were obtained at each value of γ. For small γ, a symmetric solution is generally obtained (in which the shapes of $p_1(p_2)$ and $p_2(p_1)$ are identical), whereas a broken symmetry solution, similar to the Price-Quality solution, is obtained at large γ. There was a range of γ values, between 0.1 and 0.2, where either a symmetric or asymmetric solution could be obtained, depending on initial conditions. The asymmetric solution seems counter-intuitive because one would expect that symmetric profit functions would lead to symmetric policies. Finally, in the Information-Filtering model, simultaneous Q-learning produced exact or good approximate convergence for $0 \leq \gamma \leq 0.5$. For larger γ, no convergence was obtained. The Q-derived policies yielded reduced-amplitude price wars, and montonically increasing profitability for both sellers as a function of γ, up to $\gamma = 0.5$.

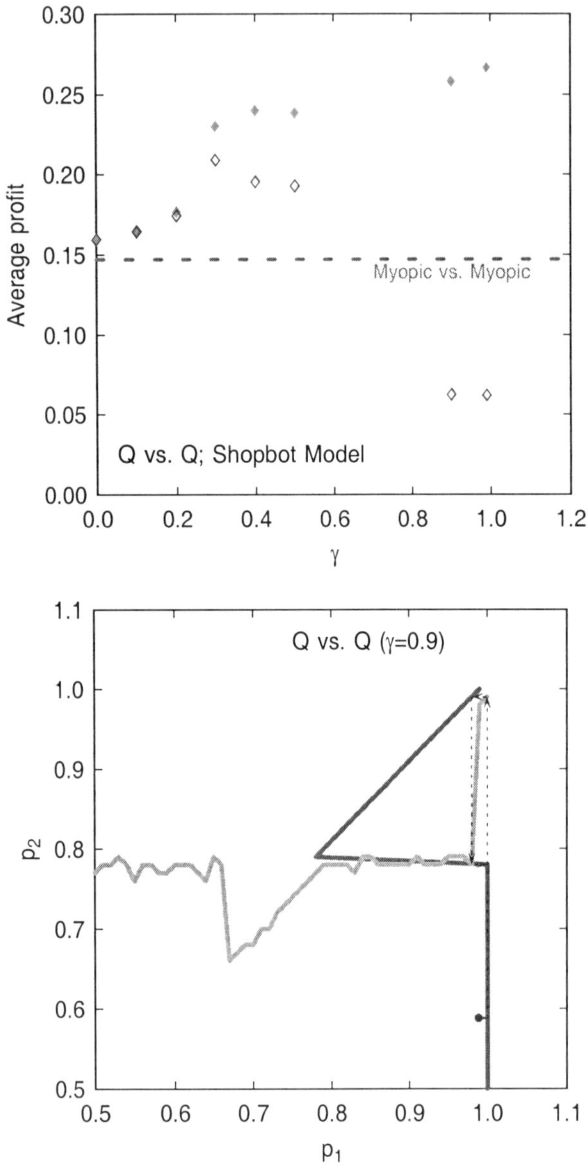

Figure 11.2. Results of simultaneous Q-learning with lookup tables in the Shopbot model. (a) Average utility per time step for seller 1 (solid diamonds) and seller 2 (open diamonds) vs. discount parameter γ. Dashed line indicates baseline myopic vs. myopic expected utility. (b) Cross-plot of Q-derived price curves at $\gamma = 0.9$. Dashed line and arrows indicate a sample price dynamics trajectory.

Figure 11.2 is illustrative of the results of simultaneous Q-learning. The left figure plots the average profit for both sellers in the Shopbot model. The right figure plots Q-derived price curves (at $\gamma = 0.9$) of seller 1 and seller 2 against each other. The system dynamics can be obtained by alternately applying the two pricing policies. This can be done by an iterative graphical construction, in which for any given starting point, one first holds p_2 constant and moves horizontally to the $p_1(p_2)$ curve, and then one holds p_1 constant and moves vertically to the $p_2(p_1)$ curve. For these particular curves, the graphical construction leads to a very short cyclic price war, indicated by the dashed line. The price-war behavior begins at the price pair (1, 1), lasts only a couple of steps, and then drops to $p_2 \sim 0.8$. At this point seller 1 resets its price to $p_1 = 1.0$ and the cycle repeats. The price war amplitude is diminished compared to myopic vs. myopic play, where a long price war would persist all the way to a minimum price point ~ 0.58.

3.3. DIFFICULTIES OF NEURAL NETWORK TRAINING

Some preliminary results of combining Q-learning with neural networks were reported in Tesauro (1999). The neural nets typically appeared to reach peak profitability in a few hundred sweeps through the training cases (corresponding to a few hours of CPU time). The policies were reasonably good at this point, and qualitatively similar to the lookup table policies, but the quality of approximation of the Q-function was poor, as indicated by large Bellman error. With much further training (out to several days of CPU time), the Bellman error improved significantly, but there was no improvement in policy profitability. It is possible that, with enough additional training, further improvements in profitability might be found, but it appeared that the required training times would be prohibitively long.

4. Q-LEARNING WITH REGRESSION TREES

Our regression tree algorithm employs the simplest conceivable heuristics: we use axis-parallel splits, select splits that minimize variance, and approximate the function by constant values in the leaf nodes. A brief description appears below; more details can be found in Breiman et al. (1984).

The trees are constructed in a "batch" mode using a fixed set of training cases. Each training case has d input attribute values, and an associated function value which may be adjusted during training.

Given a training set N, a regression tree is constructed recursively as follows: First, the average a and the variance v of the function values of the cases are calculated. If $|N|$ is less than a given threshold or if v is sufficiently small, the algorithm terminates, returning a leaf node which approximates the function by the constant a. Otherwise, the best axis-parallel split of the cases is found by examining all possible splits on each attribute. The best split is defined as follows: consider a split that divides N into sets N_1 and N_2, with respective variances v_1 and v_2. The split which minimizes the quantity $|N_1|v_1 + |N_2|v_2$ is the best split. A new internal node defined by this split is then created. The training set is separated into two subsets, and two subtrees for the new node are created recursively. Finally, the algorithm returns the node and terminates.

We performed some initial investigations of minimal error-complexity pruning, as described in Breiman et al. (1984). However, it was found that setting a minimum of five cases in each leaf node and not using any pruning gave better results. Minimal error-complexity pruning seems ill-suited to our problem, perhaps because the policies generated by using the tree are more important than the accuracy of the tree's approximation of the function. Other pruning algorithms have not been investigated.

In our problem, the training cases consist of random price pairs (corresponding to uniform random exploration of the state-action space), plus an additional "knowledge-engineered" binary attribute which is set to 1 when the Q-learner's price is less than the opponent's price, and 0 otherwise. This helps the tree represent the critically important "undercutting" discontinuity present in all three models.

The function values for the cases are initialized to the seller's two-step immediate reward for that state and action (as in Section 3). An initial tree is built from these cases. Then, repeated sweeps are performed through the set of training cases, in which the case values Q_i are adjusted according to:

$$\Delta Q_i = \alpha[r + \gamma \max_b Q(s', b) - Q_i] \qquad (2)$$

where the max-Q value for the successor state is found using the current tree. A new tree is built after each sweep through the training set. The algorithm terminates after a fixed number of sweeps. Training runs typically used a fixed learning rate α, which seemed to give good results even though convergence theorems require decreasing α with time.

For multiple Q-learners, tree updates are performed as follows. First, the sweep through the training set of each Q-learner is done, with the most recent policy of the other Q-learners (determined from their most recently constructed tree) used to calculate the immediate reward r. Then, the trees of all the Q-learners are reconstructed. This method promotes consistency of the Q-functions, as all Q-learners have access to equally recent information about other Q-learners.

One slightly different algorithm was also examined, which involved building a tree as above, and then further refining the leaf node values. A random new point in the space is chosen, and the corresponding leaf node is determined. Let a be the current leaf node estimate of the function, k be the number of cases that fell in that leaf node during previous training, and f be the function value of the new point. The leaf node's estimate is then updated according to: $a' = (a*k+f)/(k+1)$. This process continues for a number of new points, after which a new tree is built from a newly constructed training set. This algorithm worked well in the case of a single Q-learner, but for multiple Q-learners was not as good at producing stable policies as the batch updating algorithm.

5. RESULTS

In the case of single-agent Q-learning, the batch training algorithm converged rapidly, within a few dozen iterations, taking $\sim$ 1 minute or less of CPU time on a fast RS/6000 workstation. This represents a huge improvement over neural network training, which ranged from a few hours to several days of CPU time. Furthermore, in all three economic models, regression tree learning was consistently able to match the exact optimal policies that were obtained from lookup table Q-learning. (Again, this is superior to neural nets, which usually were not able to find the optimal policies.)

These optimal policies were found with training set sizes that were relatively small fractions of the total state space size. Typically we find that data sets $\sim$ 20-25% of the state space size in the Price-Quality and Information-Filtering models, and $\sim$ 40-50% in the Shopbot model, are sufficient to generate exact or near-optimal policies. As shown in Figure 11.3, the algorithm's behavior as a function of training set size is quite good. The policy profitability quickly asymptotes at the level of the lookup table policy, and for smaller training sets, there is graceful degradation: in most cases a near-optimal policy is found, however, there is an increasing chance that a poor solution will be obtained as the training set size is reduced.

We can also see in Figure 11.3 that the trees grow to reasonable sizes (much smaller than the corresponding lookup tables) as a function of training set size. For example, in the Shopbot model, trees with ap-

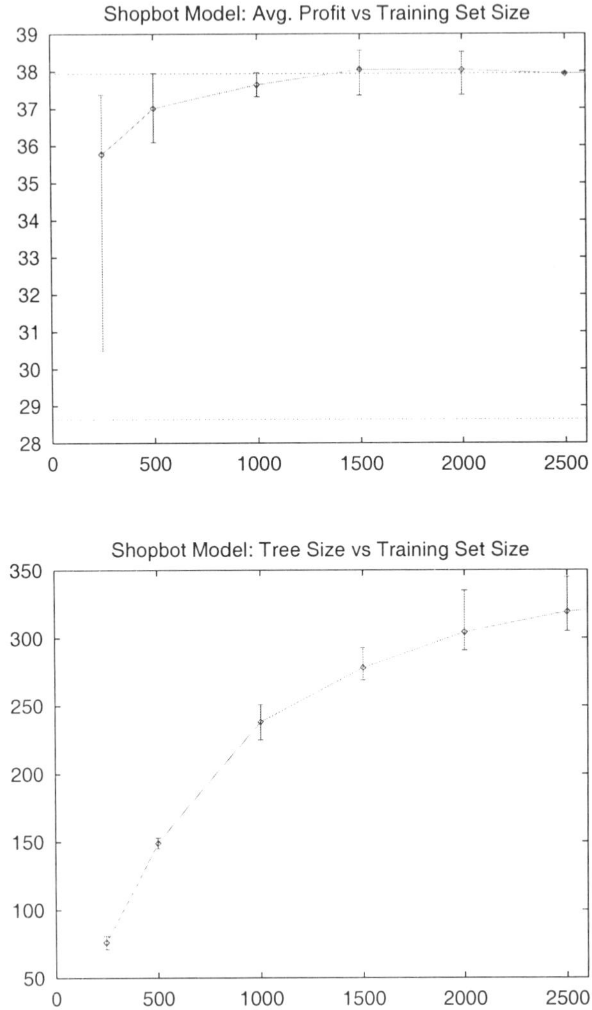

Figure 11.3. Results of single-agent Q-learning with regression trees in the Shopbot model, $\gamma = 0.5$. (a) Average Q-learner profit per time step as function of training set size. Error bars represent min and max profit over 10 runs with different randomly-generated training sets. The lower dashed line represents average profit in myopic vs. myopic play. The upper dashed line indicates average profit of exact optimal policy obtained by lookup table Q-learning. (b) Average tree size as a function of training set size.

proximately 300 nodes (or 150 leaf nodes) were able to generate optimal policies. Trees with approximately 600 nodes generated optimal policies in the Price-Quality model, with the higher quality seller as the Q-learner.

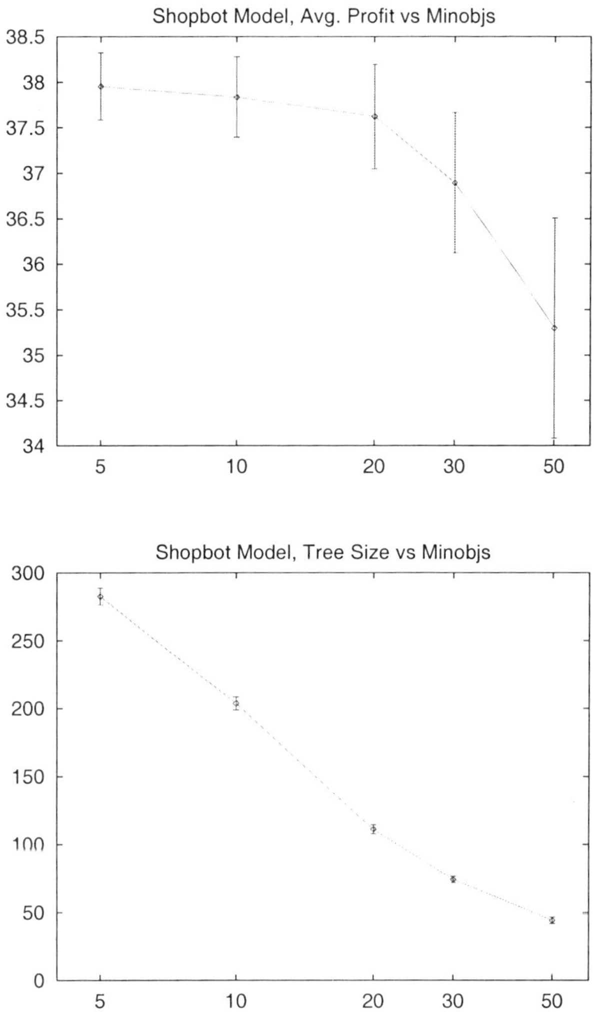

Figure 11.4. Results of varying Minobjs, the minimum no. of cases per leaf node. Single-agent Q-learning with regression trees in the Shopbot model, $\gamma = 0.5$, 1500 training cases. (a) Average Q-learner profit per time step as function of Minobjs. Error bars represent min and max profit over 10 runs with different randomly-generated training sets. (b) Average tree size as a function of Minobjs.

Most of our results were obtained with the minimum number of training cases per leaf node ("Minobjs") set to five. This lenient criterion resulted in a large number of leaf nodes, with relatively good constant-value approximation within each leaf node. We have also investigated more stringent stopping criteria by increasing Minobjs. This results in

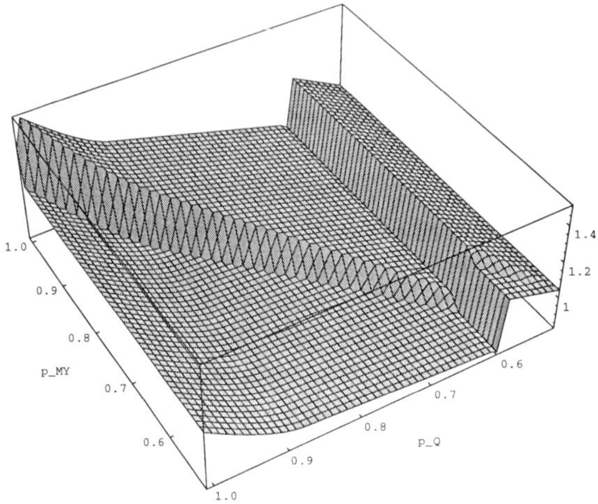

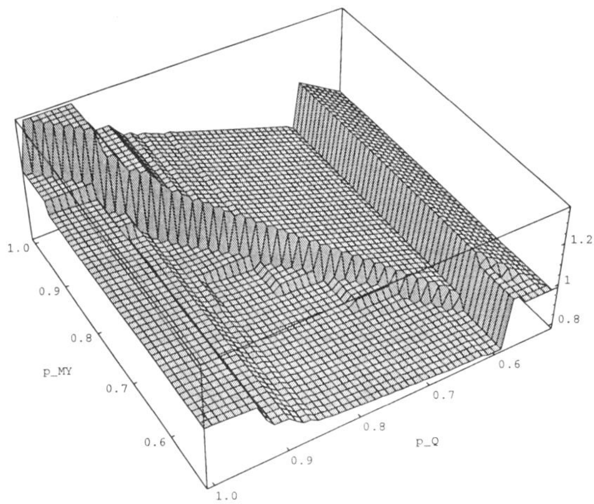

Figure 11.5. Plot of the Q-function for single-agent Q-learning vs. a myopic opponent in the Shopbot model at $\gamma = 0.7$. p_Q is the Q-learner's price and p_MY is the myopic player's price. (a) Exact optimal Q-function obtained by lookup table Q-learning. (b) Regression tree approximation to the Q-function. The tree was trained on 1500 cases and contained at least ten cases per leaf node.

smaller trees, with potentially cruder function approximation within the leaf nodes. Generally we find robust behavior with increasing Minobjs, as illustrated in Figure 11.4. Increasing Minobjs from 5 to 20 was found to reduce the tree size by nearly a factor of 3, with only a slight decrease in policy profitability.

An illustration of the quality of regression tree function approximation compared to an exact lookup table solution is shown in Figure 11.5. The tree correctly fits the diagonal discontinuity along the undercutting line $p_Q < p_{MY}$ (aided greatly by the "knowledge-engineered" input attribute), and unlike neural networks, it also correctly fits the second discontinuity in the small p_Q regime. This is important in obtaining the optimal policy. Neural nets fit the flat and smoothly curving portions of the landscape better, but this is irrelevant to finding the optimal policy.

The results of multi-agent Q-learning using regression trees were also promising. In the Price-Quality model, results were identical to those generated with lookup tables: robust convergence was found to a self-consistent pair of policies independent of γ. In the Information-Filtering model, convergence was obtained for γ up to 0.5, with cumulative profits for both sellers increasing with higher γ, exactly as in the lookup table case. Results for the Shopbot model were mixed. For γ up to 0.2, convergence to symmetric policies was obtained. Using the leaf-node updating algorithm described briefly in Section 4, the same asymmetric solutions were obtained for $\gamma > 0.2$. Using the batch training algorithm with large γ, the policies of the sellers remain symmetric for some time, after which they suddenly become asymmetric and no convergence is obtained. Although these results are not completely understood, they seem to correspond well with the lookup table results, where it was also seen that the symmetric solution for multi-agent Q-learning was highly unstable.

Training times in the multi-agent case increased, but convergence was still usually obtained in under 5 minutes. Partly this was due to having to use a smaller value of $\alpha \sim 0.1$, whereas in the single agent case, it was often possible to go as high as $\alpha \sim 1$ and still obtain convergence. Tree sizes increased slightly in multi-agent Q-learning tests, which is reasonable since the profit landscapes are more complicated. In the Shopbot model, trees that generated optimal policies ranged from 350-400 nodes, and trees with approximately 700 nodes generated optimal policies for the higher-quality seller in the Price-Quality model.

6. CONCLUSIONS

We have studied a simple combination of tree-based function approximation with single-agent and multi-agent Q-learning, and have found highly encouraging results in three different models of agent economies. Each model is of interest because it captures a different mechanism believed likely to be important in future agent-based e-commerce (asymmetric buyer information, vertical differentiation, and horizontal differentiation of seller products), and each has a different mechanism for generating price wars. Empirically, we also find different multi-agent learning phenomena in each model. While we are far from having modeled all important aspects of real economies, we are encouraged by our early results. The success of our approach in each model suggests that it is not due to specific details of any single model, and hence may be more generally applicable in agent economies.

The main contributions of this work are twofold. First, an important open question in Reinforcement Learning research is how to combine RL methods such as Q-learning with nonlinear function approximation. Most of the empirical research on this topic has involved neural networks. Our work, along with Wang and Dietterich (1999), is among the first to demonstrate success in combining Q-learning with regression trees. Our results show clear advantages over the neural net approach, and thus provide impetus for further studies of tree-based methods.

The second contribution is that with regression trees, Q-learning appears much more feasible as a practical approach to learning strategies in multi-agent economies. With lookup tables, Q-learning generally performed well in terms of training time and policy profitability, both in the single-agent and multi-agent cases, but is expected to scale poorly with the number of agents. Regression trees are expected to offer much better scaling. We find that they train quickly, achieve profits equal or very close to the lookup table policies, and offer significant advantages in terms of tree size and amount of training data required.

A potential limitation of the tree-based approach may eventually occur in very high dimensional spaces (i.e. hundreds of inputs). Trees do not generalize as aggressively as neural nets, and thus tend to have more curse-of-dimensionality problems. Neural nets might turn out to be preferable in such problems (provided that a solution to the slow training time could be found).

One clearly important direction of future research is to examine how well regression trees perform in economic models with more agents and more realistic details. (Preliminary results indicate that in a three-seller case, regression trees still seem to perform well with a manageable increase in the size of the trees.) It also seems likely that more sophisticated methods for generating splits and leaf-node function approximation, such as oblique splits and linear function approximation, could be of clear benefit. It will also be important to resolve how to adapt our batch training methodology to online learning, which is inherently incremental.

Finally, we have observed empirical behavior of multi-agent Q-learning, such as approximate but not exact convergence, that has no analog in ordinary Q-learning, where the Bellman equation is known to be a contraction mapping, leading to a unique global attractor with zero residual error. The empirical results here, together with those of Sandholm and Crites (1995) for Prisoner's Dilemma, suggest that more interesting theoretical results may be available in the multi-agent case. Depending on the payoff functions and on γ, there may be a single global attractor, or multiple basins of attraction, or no attractor dynamics at all. Also, the attractors may be small finite regions or limit cycles rather than points, with non-zero asymptotic residual error.

Notes

1. Detailed descriptions and prior studies of these economic models are available on the Web at: www.research.ibm.com/infoecon/researchpapers.html.

References

Breiman, L., et al. (1984). *Classification and Regression Trees.* Monterey CA: Wadsworth.

Greenwald, A., & Kephart, J. O. (1999). Shopbots and pricebots. *Proceedings of the Sixteenth International Joint Conference on Artificial Intelligence* (pp. 506-511). San Francisco: Morgan Kaufmann.

Hu, J., & Wellman, M. P. (1998). Multiagent reinforcement learning: theoretical framework and an algorithm. *Proceedings of the Fifteenth International Conference on Machine Learning* (pp. 242–250). San Francisco: Morgan Kaufmann.

Kephart, J. O., Hanson, J. E., & Sairamesh, J. (1998). Price-war dynamics in a free-market economy of software agents. *Proceedings of ALIFE-VI.* Cambridge MA: MIT Press.

Littman, M. L. (1994). Markov games as a framework for multi-agent reinforcement learning. *Proceedings of the Eleventh International Conference on Machine Learning* (pp. 157–163). San Francisco: Morgan Kaufmann.

Sairamesh, J., & Kephart, J. O. (1998). Dynamics of price and quality differentiation in information and computational markets. *Proceedings of the First International Conference on Information and Computation Economics* (pp. 28–36). New York: ACM Press.

Sandholm, T. W., & Crites, R. H. (1995). On multiagent Q-Learning in a semi-competitive domain. *Proceedings of the IJCAI-95 Workshop on Adaptation and Learning in Multiagent Systems* (pp. 71–77).

Tesauro, G. J., & Kephart, J. O. (in press). Foresight-based pricing algorithms in agent economies. *Decision Support Sciences.*

Tesauro, G., & Kephart, J. O. (1999). Pricing in agent economies using multi-agent Q-learning. *Proceedings of the Workshop on Decision Theoretic and Game Theoretic Agents*, London, England, 5-6 July 1999.

Tesauro, G. (1999). Pricing in agent economies using neural networks and multi-agent Q-learning. *Proceedings of the IJCAI-99 Workshop on Learning About, From and With Other Agents*, Stockholm, Sweden.

Wang, X., & Dietterich, T. G. (1999). Efficient value function approximation using regression trees. *Proceedings of the IJCAI-99 Workshop on Statistical Machine Learning for Large-Scale Optimization*, Stockholm, Sweden.

Watkins, C. J. C. H. (1989). Learning from delayed rewards. Doctoral dissertation, Cambridge University, Cambridge UK.

Chapter 12

SATISFICING EQUILIBRIA

A Non-Classical Theory of Games and Decisions

Wynn C. Stirling
Electrical and Computer Engineering Department
Brigham Young University, Provo, Utah, 84602, USA
wynn@ee.byu.edu

Michael A. Goodrich
Computer Science Department
Brigham Young University, Provo, Utah, 84602, USA
mike@cs.byu.edu

Dennis J. Packard
Philosophy Department
Brigham Young University, Provo, Utah, 84602, USA
dennis_packard@byu.edu

Abstract Satisficing, or being "good enough," is the fundamental obligation of rational decision makers. An operational definition of what it means to be good enough is a prerequisite for making rational choices. In a departure from the traditional notion of satisficing as a species of bounded rationality, satisficing is here redefined in terms of a notion of intrinsic rationality. Epistemic utility theory serves as the philosophical foundation of a new praxeological decision-making paradigm of satisficing equilibria that is applicable to both single- and multiple-agent scenarios. All interagent relationships are modeled by an interdependence function that explicitly accommodates both self and group interest, from which multilateral and unilateral selectability and rejectability mass functions can be derived and compared via the praxeic likelihood ratio test.

Keywords: Decision theory, game theory, satisficing, equilibria, epistemology, praxeology

1. INTRODUCTION

When Simon introduced the notion of "bounded rationality," he appropriated the term *satisficing* to describe a process of constructing expectations, or aspiration levels, of how good a solution might reasonably be achieved, and halting search when these expectations are met (Simon, 1955; Simon, 1956; Simon, 1990; Simon, 1996). Subsequently, other notions of bounded rationality were generated by relaxing one or more assumptions of standard optimization approaches. Procedures such as augmenting the utility function with computational costs are species of constrained optimization (see, for example, (Sandholm and Lesser, 1997; Zilberstein, 1998; Kaufman, 1990)), and yield optimal solutions according to modified criteria. Regardless of the details of how a boundedly rational decision is obtained, however, the ultimate reason for adopting it is that it represents a compromise between performance and cost that is, by definition if for no other reason, "good enough."

The notion of being good enough is an underlying issue in all decision problems. While it is tacitly understood that a truly optimal solution to any well-framed problem is certainly good enough, the issue becomes problematic when other-than-optimal procedures are implemented. A characteristic of boundedly rational approaches *à la* Simon is that the aspiration levels are arbitrary; if set too low, performance may be sacrificed needlessly, and if set too high, the solution set may be empty. Although aspiration levels at least superficially establish minimum requirements, the approach provides no guidance as to how these levels are to be specified, and relies instead upon experience-derived expectations. But it is difficult to establish a good and practically attainable aspiration level without first exploring the limits of what is possible, that is, without first identifying optimal solutions—the very procedure satisficing is designed to circumvent. For simple, low-dimensional problems, specifying the aspiration levels may be noncontroversial. But, particularly with multi-agent systems, interdependence between decision makers becomes complex, and aspiration levels become conditional (what is satisfactory for me may depend upon what is satisfactory for you). The current state of affairs regarding aspiration levels does not appear to address completely the problem of specifying minimum requirements in multi-agent contexts. Perhaps what is needed is a notion of minimum requirements that depends on intrinsic properties of the decision problem rather than upon aspiration levels.

A characteristic of boundedly rational approaches *à la* constrained optimization is that they employ criteria (such as computational costs) that are independent of actual performance. It is possible, even for single-agent decision problems, that slightly modifying one or more assumptions of a bounded optimality-based solution methodology can lead to solutions that would not otherwise have been good enough. This issue becomes even more delicate for multi-agent systems, since the interconnections between agents can be extremely sensitive, and slightly modifying an assumption can have a substantial effect on the outcome.

In this paper we propose a new definition for satisficing, or being good enough, that applies to both single- and multi-agent situations. As we define it, satisficing does not rely on notions of bounded rationality either *à la* aspiration levels or *à la* constrained optimization. Instead, our definition hinges on a new notion of rationality that is constructive, in that it leads to a practical framework for decision making. In Section 2 we motivate and formalize this definition, and establish a new concept for equilibrium. In Section 3 we develop the mathematical structure of our decision framework, in Section 4 we extend it to the multi-agent case, and in Section 5 we apply our theory to the Prisoner's Dilemma game. Finally, in Section 6 we summarize our results, discuss some practical applications, suggest some research directions, and finish with reflections on the relationship between our engineering need to act and our scientific need to know.

2. A COMPARATIVE PARADIGM

If the notion of "good enough" is to be made precise in a way analogous to, but distinct from, the way "best" is made precise, we must rely on a notion of rationality that does not depend upon optimization. At the same time, we must not retreat into *ad hocism,* that is, decision making based on notions of desirability or convenience, but without a facility for self-criticism. Heuristics may foster good, even very good, outcomes, but we cannot know if they are good enough. In this section, we first illustrate by way of example two extreme views of rational decision making, then we introduce an alternative concept of rationality and establish the notion of satisficing equilibria.

2.1. AN EXAMPLE

Strict optimality and pure heuristics represent two extreme views of decision making. We illustrate this situation by means of an example. One of the most widely studied of all games is the Prisoner's Dilemma. This game involves two players, X_1 and X_2, who have been charged with a serious crime, arrested, and incarcerated in a way that precludes any communication between them. The prosecution has evidence sufficient only to convict them of a lesser crime with a short jail sentence. To

Table 12.1 . Payoff array in ordinal form for traditional Prisoner's Dilemma game.

	Player X_2	
Player X_1	Confess (C)	Defect (D)
Confess (C)	(3,3)	(1,4)
Defect (D)	(4,1)	(2,2)

Key: 4 = best; 3 = next best; 2 = next worst; 1 = worst

get at least one conviction on the more serious crime, the prosecution entices each prisoner to give evidence against the other, that is, to defect (D). Otherwise, each prisoner may attempt to cooperate (C) with the other by not supplying evidence. Defection yields dropped charges if the other cooperates; cooperation yields the maximum sentence if the other defects. If both cooperate, both receive short sentences; if both defect, both receive intermediate-length sentences.

The standard approach to this problem is obtained via von Neumann-Morgenstern (VNM) game theory (von Neumann and Morgenstern, 1944), and assumes that each player is determined to choose a strategy that is best for itself. VNM game theory requires the formation of a payoff array, which is a table of juxtaposed individual expected utility functions that provide a total ordering of the strategies available to each player. This array is illustrated in ordinal form in Table 12.1.

The most well known notion of what constitutes a viable solution to games such as this is the concept of equilibria. An *equilibrium* for a single-stage game such as Prisoner's Dilemma is a vector of options (one element for each player), or joint option, such that each player's individual option is acceptable to it according to some criterion. Three of the most widely used equilibrium concepts are dominance, Nash equilibria, and Pareto equilibria (see, for example, (Rasmusen, 1989) for an expanded discussion of game-theoretic equilibria). A joint option is a *dominant equilibrium* if each individual option is best for the corresponding player, no matter what options the other players choose. Dominance is a very strong condition, and dominant equilibria do not always exist. When they do not, an alternative is to seek a Nash equilibrium. A joint option is a *Nash equilibrium* if, were any single agent to change its decision, it would reduce its level of satisfaction. The concept of Nash equilibria is consistent with an attitude of exclusive self-interest. However, if the players are disposed to cooperate, they may seek a Pareto equilibrium. A joint option is a *Pareto equilibrium* if no single agent, by changing its decision, can increase its level of satisfaction without lowering the satisfaction level of at least one other agent. An exclusively self-interested player, however, would have no incentive to choose a Pareto equilibrium if it were not also a Nash equilibrium.

By inspection, playing (D, D) is a dominant equilibrium. Furthermore, (D, D) is also the unique Nash equilibrium. Unfortunately, both of these equilibria are inferior to playing the Pareto equilibrium, (C, C), as they result in next-worst, rather than next-best, play. Hence, the dilemma.

One of the main characteristics of the VNM approach is that it abstracts the game from its context, or story-line—all information relevant to devising a strategy is captured by the expected utility functions. Under this view, the players are assumed to be absolutely certain that self-interest is the only issue. The VNM approach does not countenance mixed motives. Accounting for any dispositions for coordinated behavior would change the payoff array and, hence, the game.[1] There is no room for equivocation. This is a powerful, but necessary, assumption under the VNM approach to this game.

This game has also been analyzed in the context of repeated play. Of course, the standard VNM approach must return the Nash solution, no matter how many times the game is played, but other approaches have revealed quite different behavior. A very interesting approach was introduced by Rapoport, who proposed a simple *tit-for-tat* rule of repeated play: start by cooperating, thereafter play what the other player chose in the previous round. This purely heuristic rule won the Axelrod Tournament (Axelrod, 1984). One of its main characteristics is that it does not abstract the game from the story—the actions are taken completely within the game context.

The VNM and Rapoport approaches to the Prisoner's Dilemma represent two extremes. With the VNM approach, all relevant knowledge is assumed to be compactly and precisely encoded into mathematical expressions, self interest is the only issue, and expected utility maximization is the operative solution concept. With the Rapoport approach, the players' dispositions enter into the decision, the solution concept is an *ad hoc* rule to be followed, and there are no attempts to rank-order options or maximize performance.[2]

2.2. INTRINSIC RATIONALITY

The need for an alternative to the extremes of strict optimality on the one hand and pure heuristics on the other was recognized by Kreps, who observed, in the context of general two-player games, that

> ...the real accomplishment will come in finding an interesting middle ground between hyperrational behaviour and too much dependence on *ad hoc* notions of similarity and strategic expectations. When and if such a middle ground is found, then we may have useful theories for dealing with situations in which the rules are somewhat ambiguous. (Kreps, 1990, p. 184).

Our objective is to explore such middle ground. In doing so, we find it instructive to appeal to the three degrees of comparison in grammar. The highest, or superlative, degree, is founded on the notion of being "best," which requires the total-ordering of preferences for the consequences associated with the options. The evaluation of an option under the superlative paradigm requires comparing preferences for its consequences with the preferences for the consequences of all other possible options. Such evaluations are extrinsic, since they depend upon qualities that are external to the option under evaluation. In the literature, decision makers who operate under the superlative paradigm are said to be *substantively rational* (Simon, 1986). To reflect the relativity of this rationality notion, we will refer to substantive rationality also as *extrinsic rationality.*

The lowest, or positive, degree of comparison is founded on the notion of being "good," which requires no explicit preference orderings or comparisons. Decision makers who operate under the positive paradigm are often characterized as being *procedurally rational* (Simon, 1986). Procedures are developed by experts and derive their validity from that authority. Substantive rationality tells us where to go, but not how to get there; procedural rationality tells us how to get there, but not where to go. Substantive rationality is viewed in terms of the outcomes it produces; procedural rationality is viewed in terms of the methods it employs.

There is a logical gap between the *superlative* paradigm and the *positive* paradigm. This gap is filled by the *comparative* paradigm of being "better." Literature involving the superlative paradigm, particularly that of statistical decision theory, game theory, optimal control theory, and operations research, is overwhelmingly vast, reflecting many decades of serious research and development. The positive paradigm, manifest in the form of heuristics, rule-based decision systems, and various other *ad hoc* techniques, has also been well-represented in the computer science, social science, and engineering literatures. A formally stated comparative paradigm, however, has not yet been well represented in the literature as a basis for a viable decision-making concept for general application.

As a first step toward the development of a comparative paradigm, we must formalize a notion of rationality that does not depend upon rank-ordering, yet is amenable to self policing. Rank-ordering is an extrinsic exercise involving inter-option comparisons; that is, comparing a given attribute of one option to the same attribute of another option or to a fixed standard. This is not, however, the only way to make comparisons. We may also consider making intra-option, or intrinsic, comparisons; that is, comparisons between different attributes of a given option.

The approach we advocate is very simple and well-precedes any formal theories regarding its use: a common way people evaluate personal and business propositions is to compare potential gains with potential losses. Forming dichotomies provides a convenient way of accounting for the desirable and undesirable properties associated with options under consideration and conforms with common praxeological[3] behavior. We will say that a decision is *intrinsically rational* if the gains achieved by making it equal or exceed the loss incurred, provided the gains and losses can be expressed in commensurable units. This is perhaps the most primitive form of self policing—it is local, rather than global, and appeals only to intrinsic properties of the option.

As a formalized means of decision making, the concept appears in at least two very different contexts: economics and epistemology—the former intensely practical and concrete, the latter intensely theoretical and abstract. Economists implemented cost-benefit analysis in the 1930's. The usual procedure is to express all costs and benefits in monetary units, and to sanction a proposition if "the benefits to whomsoever they accrue are in excess of the estimated costs." (Pearce, 1983). Cost-benefit analysis is a useful way to reduce a complex problem to a simpler, more manageable one. One of its chief virtues is its fundamental simplicity.

Epistemic logic, or the classification of propositions on the basis of knowledge and belief regarding their content, has also made use of a comparative paradigm. Building upon the American epistemological tradition of pragmatism as developed by Peirce, James, and Dewey, Levi (Levi, 1980) has developed a distinctive school of thought regarding the evolution of knowledge corpora. Unlike the conventional epistemological doctrine of expanding one's knowledge corpus by adding information that has been justified as unequivocally true, Levi proposes the more modest goal of avoiding error. This theory has been detailed elsewhere (see (Stirling and Morrell, 1991; Stirling and Goodrich, 1999b; Goodrich et al., 1998)), and we provide here only a brief summary. The gist of it is that, given the task of determining which, if any, of a set of propositions should be retained in an agent's knowledge corpus, the agent should evaluate each proposition on the basis of two distinct criteria—first, the credal, or subjective, probability of it being true, and second, the informational value[4] of rejecting it, that is, the degree to which discarding the proposition results in a corpus with the kind of information that is demanded by the question. Thus, for a proposition to be admissible, it must be both believable and informative—all implausible or uninformative options should be rejected. Levi proposes a very simple test, based on the mathematical structure of probability theory, to perform such evaluations. He constructs an *expected epistemic utility function,* defined as the difference between credal probability and a certain constant (called the index of boldness) times another probability function, termed the informational-value-of-rejection probability. The set of options that maximizes this difference is the admissible set—all others are rejected. A key feature of this procedure is that the admissible set is generally not a singleton.

2.3. SATISFICING EQUILIBRIA

Following the tradition in economics and epistemology, we define a *satisficing option as any option that is intrinsically rational.* We retain the "satisficing" terminology because we are driven by exactly the same issue that motivated Simon's original usage of the term—to find options that are good enough by directly comparing attributes of options. We differ only in the standards used for comparison. Whereas Simon's approach is extrinsic and compares attributes to fixed aspiration levels, our approach is intrinsic and compares the positive attributes to the negative attributes of each option.

Under the superlative paradigm, once the utilities are in place, all actual decision making ceases—expected utility maximization is simply a matter of searching. Under the comparative paradigm, the utilities are used to provide rankings of attributes for each option. Thus, instead of making one global decision with respect to the entire collection of options, the comparative paradigm requires a separate local decision to be made for each option. Each option is thus subjected to individual scrutiny, and a decision is made either to retain it as an admissible choice or to reject it.

The comparative paradigm enables the formation of a precise definition of satisficing equilibria without the need to invoke any notions of optimality. We will say that an option is in a state of *satisficing equilibrium* if

S-1 The benefits derived from adopting it at least compensate for the costs incurred.

S-2 No other option provides more benefit without also costing more, or costs less without also providing less benefit.

Condition S-1 provides a weak notion of adequacy, and is a valid praxeological concept. Condition S-2 applies the domination principle to the cost-benefit framework to eliminate options that needlessly either sacrifice performance or incur expense.

If a sequential-search decision problem is resource-constrained (for example, with computational or time limitations), the decision maker may not be able to afford to perform an exhaustive search. In such cases, it is not necessary to evaluate every option. In the spirit of Simon's aspirations concept, sequential search can cease once the first intrinsically rational option is found. In general, however, the set of satisficing equilibrium options will not be a singleton, and further refinement will be required before action can be taken. Each element of this set enjoys the property that it results in a tradeoff that favors benefits over costs—there are no bad choices. The final choice depends upon the disposition of the decision maker. If the budget is tight, then the results may be commensurately low, but if cost is no object, then performance need not be sacrificed.[5] Whereas the essence of the superlative paradigm is "Nothing but the very best will do," and the essence of the positive paradigm is "It has always worked before," the essence of the comparative paradigm is "You get what you pay for." In this sense, satisficing equilibria provide an operational definition of being "good enough."

By weakening the principles of rationality we hope better to accommodate both individual and group preferences. Our philosophy is consistent with Levi's dictum that

> ... principles of coherent or consistent choice, belief, desire, etc. will have to be weak enough to accommodate a wide spectrum of potential changes in point of view. We may not be able to avoid some fixed principles, but they should be as weak as we can make them while still accommodating the demand for a systematic account. (Levi, 1997, p. 24).

In particular, our goal is to develop a satisficing notion that is equally applicable to both group and individual points of view.

We do not view intrinsic rationality, and satisficing equilibrium in particular, as a replacement for substantive rationality, or for conventional game theory in general. We do, however, invite consideration of intrinsic rationality for situations where complexity and uncertainty make it difficult or inappropriate to apply classical superlative approaches. In the final analysis, all decision methods are subjective; they are simply tools to be used with judgment and skill.

3. PRAXEIC UTILITY

Our notion of satisficing requires us to obtain dichotomous mathematical descriptions of cost and benefit, and then to use the resulting cost-benefit pair to identify members of the satisficing equilibrium set. Our approach is to adapt the mathematics of epistemic utility theory, originally developed for epistemological decision making (committing to beliefs), to the praxeological domain (taking action). Our interest in this theory is motivated by its striking compatibility with the comparative paradigm, since it permits costs and benefits to be characterized via probability measures that can be evaluated by the expected epistemic utility test discussed in (Levi, 1980).

Epistemic utility theory offers a number of significant features. First, it addresses an important issue of insuring comparable units by ascribing a unit of mass to both belief and informational value; second, it permits the evaluation of non-singleton sets of options; and third, it permits the extension of such evaluations to the multivariate case (that is, situations involving more than one decision maker or multiple attribute decision problems for a single agent).

Epistemic utility theory captures the essence of Popper's imperative: "Yet we must also stress that *truth is not the only aim of science.* We want more than truth: what we look for is *interesting truth.*" (Popper, 1963, p. 229). Although truth and informational value are natural semantic notions for cognitive decision making, they are not always natural for practical decision making. To apply the ideas of epistemic utility theory to practical decision making we must formulate praxeological analogs to the epistemological notions of truth and informational value.

A natural analog for *truth* is *success,* in the sense of achieving the fundamental goal of taking action. To formulate an analog for informational value, we observe that, just as the management of a finite amount of relevant information is important when inquiring after truth in the epistemological context, taking effective action requires the management of a finite amount of resource, such as conserving wealth, materials, energy, safety, or other assets, in the praxeological context. Thus, an apt praxeological analog to *the informational value of rejection* is *the conservational value of rejection.* We thereby rephrase Popper's injunction to become: *we want more than success—what we look for is efficient success.* (Note that this is a softer goal than the superlative imperative of minimizing cost or maximizing payoff, which is tantamount to finding the one and only truly successful option.) Thus, we change the context of the decision problem from one of acquiring information while avoiding error to one of conserving resources while avoiding failure. With the context shift from the epistemological issue of belief to the praxeological issue of action, we refer to the resulting utility function as *praxeic utility,* rather than epistemic utility.

Our development mathematically parallels Levi's original development of epistemic utility theory. We will refer to the degree of resource consumption as *rejectability,* and require rejectability to be expressed in terms of a function that conforms with the axioms of probability. We use this new terminology to emphasize the semantic distinction of using the mathematics of probability in a non-conventional way. Thus, for a finite action space, U, rejectability is expressed in terms of a mass function, p_R: $U \to [0,1]$ such that $p_R(u) \geq 0$ for all $u \in U$ and $\sum_{u \in U} p_R(u) = 1$. Inefficient options (those with high resource consumption) should be highly rejectable; that is, if considerations of success are ignored, one should be prone to reject options that result in large costs, high energy consumption, exposure to hazard, etc. Normalizing p_R to be a mass function, which we will term the *rejectability mass function,* insures that the agent will have a unit of resource consumption to apportion among the elements of U. We may view p_R as the inutility of consuming resources. If $u \in U$ is rejected, then the agent conserves $1 - p_R(u)$ worth of its unit of resources which is therefore available to be applied to other options. We will usually assume that $p_R(u) > 0$ for all $u \in U$ (that is, there are no completely cost- or risk-free options).

Failure is avoided if potentially successful options are not rejected, but efficiency, as well as success, must also be considered. Our approach to evaluating candidate sets of options for retention is to define the utility of not rejecting them in the interest of both success and resource conservation, and to retain the set that maximizes this utility. Suppose that implementing $u \in U$ would lead to success and let $A \subset U$ be a set we are considering for retention. The utility of not rejecting A in the interest of avoiding failure is the *indicator function,*

$$I_A(u) = \begin{cases} 1 & \text{if } u \in A \\ 0 & \text{otherwise} \end{cases} . \tag{1}$$

This utility function is very even-handed; it is concerned only with the value of not rejecting A when u is successful—it does not count one option to be more desirable than another. Matters of desirability are deferred to the conservation-determining function, p_R, and not with accounting for the seriousness of one failure relative to another. We define the *praxeic utility* of not rejecting A when u is successful as the convex combination of the utility of avoiding failure and the utility of conserving resources:

$$\phi(A, u) = \alpha I_A(u) + (1 - \alpha)\left(1 - \sum_{v \in A} p_R(v)\right), \tag{2}$$

where $\alpha \in [0,1]$ is chosen to reflect the agent's personal weighting of these two desiderata—setting $\alpha = \frac{1}{2}$ associates equal concern for avoiding failure and conserving resources.

Clearly, praxeic utility is maximized when $A = \{u\}$ (the singleton set). Unfortunately, it generally is not known precisely which u will lead to success (or that only one u will do so), so we cannot simply reject $U \setminus \{u\}$, the complement of $\{u\}$. We may, however, possess information regarding the degree of success support possessed by each u. Let $p_S \colon U \to [0,1]$ be a mass function that evaluates each option with respect to the degree to which it accomplishes the objective of the decision problem, independently of how much resource is consumed by implementing it. We will refer to the degree of success support as *selectability,* and we will term p_S the *selectability mass function.*

In the tradition of VNM theory, we may then calculate expected praxeic utility for any set $A \subset U$ by weighting the utility by the degree of success support associated with each u and summing over all $u \in U$. The *expected praxeic utility* is then

$$\begin{aligned}\bar{\phi}(A) &= \sum_{u\in U}\left[\alpha I_A(u) + (1-\alpha)\left(1 - \sum_{v\in A} p_R(v)\right)\right] p_S(u) \\ &= \alpha \sum_{u\in U} I_A(u) p_S(u) - (1-\alpha)\sum_{v\in A} p_R(v) \sum_{u\in U} p_S(u) \\ &\quad + (1-\alpha)\sum_{u\in U} p_S(u) \\ &= \alpha \sum_{v\in A} p_S(v) - (1-\alpha)\sum_{v\in A} p_R(v) + (1-\alpha), \qquad (3)\end{aligned}$$

Since $\sum_{u\in U} p_S(u) = 1$. Because VNM utility functions are invariant with respect to positive affine transformations, dividing by α and ignoring the constant term yields a more convenient but equivalent form for expected praxeic utility:

$$\bar{\varphi}(A) = \sum_{u\in A} \left[p_S(u) - b p_R(u)\right], \qquad (4)$$

where $b = \frac{1-\alpha}{\alpha}$.

We may obtain the largest set of options for which the selectability is greater than or equal to b times the rejectability by choosing the set that maximizes[6] expected praxeic utility, resulting in the *satisficing set*

$$\Sigma_b = \arg\max_{A\subset U} \bar{\varphi}(A) = \{u \in U \colon p_S(u) \geq b p_R(u)\} = \left\{u \in U \colon \frac{p_S(u)}{p_R(u)} \geq b\right\}. \qquad (5)$$

Σ_b is the set of all options for which the benefits outweigh the costs, as scaled by b. The parameter, b, is the *boldness* of the decision problem and parameterizes the degree to which the agent is willing to risk rejecting successful options in the interest of conserving resources. Setting $b > 1$ attributes more weight to resource conservation than to success. Setting $b \leq 1$ ensures that $\Sigma_b \neq \emptyset$, since otherwise $p_S(u) < bp_R(u)$ for all $u \in U$, which would imply $1 = \sum_{u \in U} p_S(u) < b \sum_{u \in U} p_R(u) = b$, a contradiction. We will refer to (5) as the Praxeic Likelihood Ratio Test (PLRT).

The set Σ_b contains all options such that the benefits of adoption outweigh the cost, and therefore is consistent with condition S-1, but it may include options that needlessly sacrifice benefit or incur cost. We may eliminate such options as follows. For every $u \in U$ let

$$\begin{aligned} B_S(u) &= \{v \in U\colon p_R(v) < p_R(u) \text{ and } p_S(v) \geq p_S(u)\} \\ B_R(u) &= \{v \in U\colon p_R(v) \leq p_R(u) \text{ and } p_S(v) > p_S(u)\}, \end{aligned} \tag{6}$$

and define the set of options that are *strictly better* than u:

$$B(u) = B_S(u) \cup B_R(u); \tag{7}$$

that is, $B(u)$ consists of all possible options that are either less rejectable and not less selectable than u, or are not more rejectable and more selectable than u. If $B(u) = \emptyset$, then no options can be preferred to u on the basis of both selectability and rejectability. The set of equilibrium options is consistent with condition S-2, and is defined as

$$\mathcal{E} = \{u \in U\colon B(u) = \emptyset\}. \tag{8}$$

The set of *satisficing equilibrium* options is the intersection of the satisficing and equilibrium sets:

$$\mathfrak{S}_b = \mathcal{E} \cap \Sigma_b. \tag{9}$$

4. MULTI-AGENT SYSTEMS

4.1. CONVENTIONAL INTERAGENT MODELING

Perhaps the most intensely studied multi-agent decision formalism is game theory. VNM Game theory is built on one basic principle: individual self-interest. Under this view, when a player is faced with uncertainty, the only justifiable course of action is to choose an action that maximizes expected utility as constrained by the expected actions of other players. This is called the *principle of individual rationality.*

For two-person constant-sum games, individual self-interest is the only possible non-vacuous principle—what one player wins, the other loses. Game theory insists, however, that this same principle applies to the general case. Thus, even in situations where there is the opportunity for group, as well as individual interest, only individually rational actions are viable. If a joint solution (that is, for the group) is not individually rational for some agent, that exclusively self-interested agent would not be a party to such a joint action.

Consider the VNM version of the Prisoner's Dilemma game. The only information X_1 can use regarding X_2 is X_2's payoff values. Even if X_1 had some knowledge of X_2's preferences, there would be no way, under the VNM regime, to factor that information into the payoff array, since X_1's payoff values are determined only by the actions that it and X_2 jointly make. Perhaps the actual "dilemma" of this game is as much an artifact of the structure of the utility functions used to model the story-line as it is an intrinsic property of the story-line itself.

This limitation of game theory has been observed by Luce and Raiffa:

> ...general game theory seems to be in part a sociological theory which does not include any sociological assumptions ... it may be too much to ask that any sociology be derived from the single assumption of individual rationality. (Luce and Raiffa, 1957, p. 196).

With conventional game theory, attitudes such as cooperation and conflict are not built into the expected utility functions. Expected utilities are functions only of the actions, or strategies, of the player, and not of their preferences. It is not until the expected utilities of all players are juxtaposed that the "game" aspect of the decision problem emerges.

Social choice theory is another multi-agent formalism that is widely studied. Like game theory, social choice theory has built largely on the foundation of individual rationality. For example, under Harsanyi's formulation, a social-welfare function is a positive linear combination of individual utilities. Each individual utility in this combination is a mapping of group options to individual utility, then a weighted combination of these individual utilities defines the utility to society. Each player then proceeds under the substantively rational paradigm by maximizing its expected social-welfare function subject to any constraints that are relevant (Harsanyi, 1977; Hogg and Jennings, 1999).

This social-welfare function modifies the decision maker's stance from consideration of purely selfish preferences to consideration of what are termed *moral* (or social) preferences and gives weight to the legitimate interests of each participant. However, the sequence of mappings from group options to individual utilities and then from individual utilities to a social utility provides a very constrained linkage between one decision-maker's preferences (for itself or for the group) and another decision-maker's utilities, and may not deal adequately with the rich diversity of interconnections that can exist in multi-agent societies.

4.2. SATISFICING INTERAGENT MODELING

We suggest that interagent modeling should provide for explicit linkage between the preferences of decision makers and their evaluations of their options. These preference linkages, however, may be quite different from the type of preference orderings that are the substance of conventional utility theory. An act by any single member X_i of a multi-agent system, $\{X_1, \ldots, X_N\}$, has possible ramifications throughout the entire community. Some of the agents may be benefited, some may be damaged, and some may be indifferent. Furthermore, although the single agent may perform the act for its own benefit, or for the benefit of other agents, or for the benefit of the entire system, the act is usually not implemented free of cost. Resources are expended, or risk is taken, or some other penalty or unpleasant consequence is incurred, perhaps by the single agent itself, perhaps by other agents, and perhaps by the entire community. Although these undesirable consequences may be defined independently from the benefits, the measures associated with benefits and costs cannot be specified independently of each other, due to the possibility of interaction. A critical aspect of modeling the behavior of such a collection, therefore, is the means of representing the interdependence of both positive and negative consequences of all possible joint options that could be undertaken.

Definition 22 *A* mixture[7] *is any subset of agents considered in terms of their interaction with each other, exclusively of possible interactions with other agents not in the subset.*

A selectability mixture, *denoted $\mathcal{S} = S_{i_1} \ldots S_{i_k}$, is a mixture consisting of agents $X_{i_1}, \ldots X_{i_k}$ being considered from the point of view of success. The* joint selectability mixture *is the selectability mixture consisting of all agents in the system, denoted $\mathbf{S} = S_1 \ldots S_N$. A* myopic *selectability mixture is a mixture of the form $\mathcal{S} = S_i$; that is, when an agent views success as though it were functioning with complete disregard for all other agents.*

A rejectability mixture, *denoted $\mathcal{R} = R_{j_1} \ldots R_{j_\ell}$, is a mixture consisting of agents $X_{j_1}, \ldots X_{j_\ell}$ being considered from the point of view of resource consumption. The* joint rejectability mixture *is the rejectability mixture consisting of all agents in the system, denoted $\mathbf{R} = R_1 \ldots R_N$. A* myopic *rejectability mixture is a mixture of the form $\mathcal{R} = R_i$.*

An intermixture *is the concatenation of a selectability mixture and a rejectability mixture, and is denoted* $\mathcal{SR} = S_{i_1} \ldots S_{i_k} R_{j_1} \ldots R_{j_\ell}$. *The* joint intermixture *is the concatenation of the joint selectability and joint rejectability mixtures, and is denoted* $\mathbf{SR} = S_1 \ldots S_N R_1 \ldots R_N$. *A* myopic *intermixture is a mixture of the form* $\mathcal{SR} = S_i R_i$.

Definition 23 *Given arbitrary action spaces* U_i, $i = 1, \ldots, N$, *the* product action space, *denoted* $\mathbf{U} = U_1 \times \cdots \times U_N$ *is the set of all* N*-tuples* $\mathbf{u} = (u_1, \ldots, u_N)$ *where* $u_i \in U_i$. *The* selectability action space *associated with a selectability mixture* $\mathcal{S} = S_{i_1} \ldots S_{i_k}$ *is the product space* $\mathbf{U}_\mathcal{S} = U_{i_1} \times \cdots \times U_{i_k}$. *The* rejectability action space *associated with a rejectability mixture* $\mathcal{R} = R_{j_1} \ldots R_{j_\ell}$ *is the product space* $\mathbf{U}_\mathcal{R} = U_{j_1} \times \cdots \times U_{j_\ell}$. *The* interaction space *associated with an intermixture* $\mathcal{SR} = S_{i_1} \cdots S_{i_k} R_{j_1} \cdots R_{j_\ell}$ *is the product space* $\mathbf{U}_{\mathcal{SR}} = U_{i_1} \times \cdots \times U_{i_k} \times U_{j_1} \times \cdots \times U_{j_\ell}$. *The* joint interaction space *is* $\mathbf{U}_{\mathbf{SR}} = \mathbf{U} \times \mathbf{U}$.

Definition 24 *A* selectability mass function *(smf) for the mixture* $\mathcal{S} = \{S_{i_1} \ldots S_{i_k}\}$ *is a mass function denoted* $p_\mathcal{S} = p_{S_{i_1} \ldots S_{i_k}} \colon \mathbf{U}_\mathcal{S} \to [0, 1]$. *The* joint smf *is an smf for* $\mathbf{S}$, *denoted* $p_\mathbf{S}$.

A rejectability mass function *(rmf) for the mixture* $\mathcal{R} = \{R_{j_1} \ldots, R_{j_\ell}\}$ *is a mass function denoted* $p_\mathcal{R} = p_{R_{j_1} \ldots R_{j_\ell}} \colon \mathbf{U}_\mathcal{R} \to [0, 1]$. *The* joint rmf *is a rmf for* $\mathbf{R}$, *denoted* $p_\mathbf{R}$.

An interdependence mass function *(imf) for the intermixture* $\mathcal{SR} = \{S_{i_1} \ldots S_{i_k} R_{j_1} \ldots R_{j_\ell}\}$ *is a mass function denoted* $p_{\mathcal{SR}} = p_{S_{i_1} \ldots S_{i_k} R_{j_1} \ldots R_{j_\ell}} \colon \mathbf{U}_\mathcal{S} \times \mathbf{U}_\mathcal{R} \to [0, 1]$. *The* joint imf *is an imf for* $\mathbf{SR}$, *denoted* $p_{\mathbf{SR}}$.

Let $\mathbf{v} \in \mathbf{U}_\mathcal{S}$ and $\mathbf{w} \in \mathbf{U}_\mathcal{R}$ be two option vectors. Then $p_{\mathcal{SR}}(\mathbf{v}, \mathbf{w})$ is a representation of the success support associated with $\mathbf{v}$ and the resource consumption associated with $\mathbf{w}$ when the two option vectors are viewed simultaneously. In other words, $p_{\mathcal{SR}}(\mathbf{v}, \mathbf{w})$ is the mass associated with adopting $\mathbf{v}$ in the interest of success and rejecting $\mathbf{w}$ in the interest of conserving resources.

The joint imf provides a complete description of the individual and interagent relationships in terms of their positive and negative consequences, but specifying the imf can be complex, especially for high-dimensional systems. Fortunately, the structure of the imf makes it possible to generate global behavior (the joint imf) from interdependence mass functions for intermixtures. To develop this theory we require some additional definitions.

Definition 25 *Given an intermixture* $\mathcal{SR} = S_{i_1} \ldots S_{i_k} R_{j_1} \ldots R_{j_\ell}$, *a* subintermixture *of* $\mathcal{SR}$ *is an intermixture formed by concatenating subsets of* $\mathcal{S}$ *and* $\mathcal{R}$: $\mathcal{S}_1\mathcal{R}_1 = S_{i_{p_1}} \ldots S_{i_{p_q}} R_{j_{r_1}} \ldots R_{j_{r_s}}$, *where* $\{i_{p_1}, \ldots, i_{p_q}\} \subset$

$\{i_1, \ldots, i_k\}$ and $\{j_{r_1}, \ldots, j_{r_s}\} \subset \{j_1, \ldots, j_\ell\}$. We shall use the notation $\mathcal{S}_1\mathcal{R}_1 \subset \mathcal{SR}$ to indicate that $\mathcal{S}_1\mathcal{R}_1$ is a subintermixture of $\mathcal{SR}$.

The $\mathcal{SR}$-complementary subintermixture *associated with a subintermixture $\mathcal{S}_1\mathcal{R}_1$ of an intermixture $\mathcal{SR}$, denoted $\mathcal{SR}\backslash\mathcal{S}_1\mathcal{R}_1$, is an intermixture created by concatenating the selectability and rejectability mixtures formed by the relative compliments of $\mathcal{S}_1$ and $\mathcal{R}_1$. Clearly, $\mathcal{SR}\backslash\mathcal{S}_1\mathcal{R}_1 \subset \mathcal{SR}$. We will say that $\mathcal{SR}$ is the union of $\mathcal{SR}\backslash\mathcal{S}_1\mathcal{R}_1$ and $\mathcal{S}_1\mathcal{R}_1$, denoted $\mathcal{SR} = \mathcal{SR}\backslash\mathcal{S}_1\mathcal{R}_1 \cup \mathcal{S}_1\mathcal{R}_1$.*

Definition 26 *Let $\mathcal{SR}$ be an intermixture with subintermixture $\mathcal{S}_1\mathcal{R}_1$. A* conditional interdependence mass function, *denoted $p_{\mathcal{SR}\backslash\mathcal{S}_1\mathcal{R}_1|\mathcal{S}_1\mathcal{R}_1}$, is a mapping of $(\mathbf{U}_{\mathcal{SR}\backslash\mathcal{S}_1\mathcal{R}_1} \times \mathbf{U}_{\mathcal{S}_1\mathcal{R}_1})$ into $[0,1]$ such that, for every $\mathbf{v} \in \mathbf{U}_{\mathcal{S}_1\mathcal{R}_1}$, $p_{\mathcal{SR}\backslash\mathcal{S}_1\mathcal{R}_1|\mathcal{S}_1\mathcal{R}_1}(\cdot|\mathbf{v})$ is a mass function on $\mathbf{U}_{\mathcal{SR}\backslash\mathcal{S}_1\mathcal{R}_1}$.*

We require all conditional interdependence mass functions to be consistent with interdependence mass functions, that is, for $\mathcal{SR}$ an arbitrary intermixture with subintermixture $\mathcal{S}_1\mathcal{R}_1$ with $\mathbf{w} \in \mathcal{SR}\backslash\mathcal{S}_1\mathcal{R}_1$ and $\mathbf{v} \in \mathcal{S}_1\mathcal{R}_1$, we have

$$p_{\mathcal{SR}}(\mathbf{v}, \mathbf{w}) = p_{\mathcal{SR}\backslash\mathcal{S}_1\mathcal{R}_1|\mathcal{S}_1\mathcal{R}_1}(\mathbf{w}|\mathbf{v}) \cdot p_{\mathcal{S}_1\mathcal{R}_1}(\mathbf{v}). \tag{10}$$

To illustrate conditioning, let $\{X_1, X_2, X_3\}$ be a multi-agent system and let $\mathcal{S} = S_1S_2$ and $\mathcal{R} = R_3$. Then $\mathcal{SR} = S_1S_2R_3$ and $\mathbf{SR}\backslash\mathcal{SR} = S_3R_1R_2$. The imf is

$$\begin{aligned} p_{S_1S_2S_3R_1R_2R_3}(v_1, v_2, v_3, w_1, w_2, w_3) = \\ p_{S_3R_1R_2|S_1S_2R_3}(v_3, w_1, w_2|v_1, v_2, w_3) \\ \cdot p_{S_1|S_2R_3}(v_1|v_2, w_3) \cdot p_{S_2R_3}(v_2, w_3). \end{aligned} \tag{11}$$

$p_{S_3R_1R_2|S_1S_2R_3}(v_3, w_1, w_2|v_1, v_2, w_3)$ is the conditional selectability and rejectability associated with X_3 selecting action v_3, X_1 rejecting action w_1, and X_2 rejecting action w_2, conditioned on X_1 ascribing its entire unit of selectability mass to v_1, X_2 ascribing its entire unit of selectability mass to v_2, and X_3 ascribing its entire unit of rejectability mass to w_3. The other two conditional interdependence mass functions are interpreted similarly.

An advantage of factoring the imf into conditional components is that preferences conditioned on various situations make it possible to characterize global preferences in terms of conditional local preferences. As noted by Pearl (Pearl, 1988), it is often easier to specify conditional local characteristics rather than unconditional joint global characteristics. Conditional probabilities permit local, or specific responses to be characterized; they possess modularity features similar to production rules.

Factorizations make it possible to simplify the structure of the interdependence function when agents or agent groups are indifferent to each other, in which case the interdependence function will factor into products of marginal interdependence functions. When agents are conditionally independent of each other, various Markovian structures may be present, which also results in a significant simplification of the interdependence function. The interdependence function thus provides a parsimonious way of accounting for interagent relationships. This is important, because as the dimensionality of the agent system becomes large, the number of possible inter-agent relationships grows combinatorially. Thus, it is necessary to exploit any simplifying structure that may exist, while at the same time not ignoring dependencies that are critical. Multivariate probability theory is ideally suited to this task. While it can be complex, it is not more complex than it needs to be. It fulfills the dictum offered by Palmer:

> ... complexity is no argument against a theoretical approach if the complexity arises not out of the theory itself but out of the material which any theory ought to handle. (Palmer, 1971, p. 176).

5. SATISFICING GAMES

5.1. MULTILATERAL AND UNILATERAL DECISIONS

The role of the joint imf is to characterize all of the logically possible interagent preference relationships that exist between the members of the multi-agent system. Once this function is defined, dichotomies may be extracted from it by computing appropriate marginals. We examine two perspectives. The first is appropriate for *multilateral,* or group decision making, and the second is appropriate for *unilateral,* or individual decision making. We then expose some relationships between group decision making and individual decision making.

To form a multilateral decision, we compute the joint selectability and rejectability marginals as

$$p_{\mathbf{S}}(\mathbf{u}) = \sum_{\mathbf{v} \in \mathbf{U}} p_{\mathbf{SR}}(\mathbf{u}, \mathbf{v}) \tag{12}$$

$$p_{\mathbf{R}}(\mathbf{v}) = \sum_{\mathbf{u} \in \mathbf{U}} p_{\mathbf{SR}}(\mathbf{u}, \mathbf{v}) \tag{13}$$

for all $(\mathbf{u}, \mathbf{v}) \in \mathbf{U} \times \mathbf{U}$. These marginals may then be used to generate the *multilateral satisficing set* according to a straightforward extension of the PLRT:

$$\Sigma_b = \{\mathbf{u} \in \mathbf{U}\colon p_{\mathbf{S}}(\mathbf{u}) \geq b p_{\mathbf{R}}(\mathbf{u})\}. \tag{14}$$

The *multilateral satisficing equilibrium set* is

$$\mathfrak{S}_b = \boldsymbol{\mathcal{E}} \cap \mathbf{\Sigma}_b, \tag{15}$$

where $\boldsymbol{\mathcal{E}}$ is the set of multilateral satisficing equilibrium options defined in accordance with (8).

The multilateral satisficing set given by (14) provides the set of jointly satisficing options as viewed from the group perspective. To obtain the individual perspective, we must compute the univariate satisficing sets for each agent. We first compute the univariate selectability and rejectability marginals, given, for agent X_i, by

$$p_{S_i}(u_i) = \sum_{j \neq i} p_{S_1 \cdots S_N}(u_1, \ldots, u_{i-1}, u_i, u_{i+1} \ldots, u_N) \tag{16}$$

$$p_{R_i}(u_i) = \sum_{j \neq i} p_{R_1 \cdots R_N}(u_1, \ldots, u_{i-1}, u_i, u_{i+1} \ldots, u_N). \tag{17}$$

The *univariate satisficing set* for X_i is

$$\Sigma_b^i = \{u \in U_i \colon p_{S_i}(u) \geq b p_{R_i}(u)\}, \tag{18}$$

and the *unilateral satisficing equilibrium set* is

$$\mathfrak{S}_b^i = \mathcal{E}_i \cap \Sigma_b^i. \tag{19}$$

We may examine the relationships between group-level satisficing and individual-level satisficing, that is, between $\mathbf{\Sigma}_b$ and the family of individually satisficing sets $\{\Sigma_b^i, i = 1, \ldots, N\}$. Under the special circumstance of all agents acting completely independently of each other, the joint distributions will factor into the product of the marginals, yielding

$$p_{\mathbf{S}}(\mathbf{u}) = p_{S_1}(u_1) \cdots p_{S_N}(u_N) \tag{20}$$

$$p_{\mathbf{R}}(\mathbf{u}) = p_{R_1}(u_1) \cdots p_{R_N}(u_N), \tag{21}$$

in which case

$$\mathbf{\Sigma}_b = \Sigma_b^1 \times \cdots \times \Sigma_b^N. \tag{22}$$

In general, however, the interdependencies among agents will render (22) invalid, and jointly satisficing actions will not necessarily be individually satisficing for all agents. Jointly satisficing and individually satisficing solutions must, however, obey a weak consistency relationship in that, if u_i is individually satisficing for X_i, then it must be an element of some jointly satisficing vector. To establish this claim, we prove the contrapositive: if u_i is not the ith element of any $\mathbf{u} \in \mathbf{\Sigma}_b$, then $u_i \notin \Sigma_b^i$. Without loss of generality, let $i = 1$. By hypothesis, $p_{\mathbf{S}}(u_1, \mathbf{v}) <$

Table 12.2 . Selectability and rejectability functions.

$p_{S_1S_2}$	v	v'
u	0.28	0.40
u'	0.30	0.02

$p_{R_1R_2}$	v	v'
u	0.30	0.30
u'	0.20	0.20

$bp_{\mathbf{R}}(u_1, \mathbf{v})$ for all $\mathbf{v} \in U_2 \times \cdots \times U_N$, so $p_{S_1}(u_1) = \sum_{\mathbf{v}} p_{\mathbf{S}}(u_1, \mathbf{v}) < b \sum_{\mathbf{v}} p_{\mathbf{R}}(u_1, \mathbf{v}) = bp_{R_1}(u_1)$, hence $u_1 \notin \Sigma_b^1$. This result says that no one is ever completely frozen out of a deal—every individual has a seat at the negotiating table. This is perhaps the weakest condition under which meaningful negotiations are possible. It is not true, however, that there always exists a jointly satisficing option vector such that each component is simultaneously individually satisficing—it is possible that $\mathbf{\Sigma}_b \cap (\Sigma_b^1 \times \cdots \times \Sigma_b^N) = \emptyset$, as evidenced by the following simple example. Let $U_1 = (u, u')$, $U_2 = (v, v')$ and let $p_{S_1S_2}$ and $p_{R_1R_2}$ be given in Table 12.2.
From this we easily obtain, for $b = 1$, that $\mathbf{\Sigma}_b = \{(u, v'), (u', v)\}$ and $\Sigma_b^1 \times \Sigma_b^2 = \{(u, v)\}$. As b is decreased, however, these decision sets change. At the value $b = 0.933$, we have $\mathbf{\Sigma}_b = \{(u, v), (u, v'), (u', v)\}$ and $\Sigma_b^1 \times \Sigma_b^2 = \{(u, v)\}$, so $\mathbf{\Sigma}_b \cap (\Sigma_b^1 \times \Sigma_b^2) = \{(u, v)\}$ and there does indeed exist, at this level of boldness, a jointly satisficing option vector whose components are both individually satisficing. Reducing the boldness, b, is a controlled way to relax the standards of intrinsic rationality, which may be necessary in difficult situations if a compromise is to be reached. The amount b must be reduced below unity is a measure of the amount of compromising needed to reach a mutually acceptable solution.

5.2. THE PRISONER'S DILEMMA

Let us now apply satisficing game theory to the Prisoner's Dilemma game. Our task is to define the imf, from which the smf and rmf can be obtained and compared for each joint option. To generate the imf, we must first define operational notions for success and resource consumption.

The fundamental objective of the players, as expressed by their utility functions, is to get out of jail quickly—self-interest is the primary consideration. Laboratory experiments with randomly selected humans, however, result in the cooperative solution being chosen relatively frequently with single play and no communication between the participants (Wolf and Shubik, 1974). This evidence suggests that for the game to be a model of human behavior, there may exist motives in addition to self interest. An obvious possibility is group interest, which apparently emerges in repeated play as a result of learned cooperation.[8] We view the players of this game as individuals who are concerned primarily with their own task, but at the same time have a degree of consideration for other players' difficulties and consider it a cost to them if an action they take makes it difficult for others. Such players are *enlightened liberals,* who are intent upon pursuing their own self interest but give some deference to the interests of the group in general. Accordingly, we proceed by associating success with reducing individual jail time and associating resource consumption with increasing group jail time. Thus, short individual sentences will have high selectability and long group sentences will have high rejectability.

The notion of group interest can have significance only if the players each acknowledge some form of dependence on the other, however weak it may be. To the extent that these dependencies reinforce each other, the players implicitly forge a joint opinion regarding the relative merits of cooperation and conflict. The success of Rapoport's approach suggests that there may be (at least) two attitudes in the minds of the players that may affect their decisions: (a) a propensity for *dissociation,* that is, for the agents to go their separate ways without regard for coordination, and (b) a propensity for *vulnerability,* that is, for the agents to expose themselves to individual risk in the hope of improving the joint outcome.

Let $\alpha \in [0, 1]$ be a measure of the joint value the players place on rejecting the joint option (C, C). We may identify α as a *dissociation index:* if $\alpha \approx 1$, the agents are completely dissociated and coordination is unlikely. Also, let $\beta \in [0, 1]$ be a measure of the joint value placed on rejecting the joint option (D, D). β may be viewed as a *vulnerability index:* $\beta \approx 1$ means the agents are each willing to risk a long jail sentence in the hope of both obtaining a shorter one. A condition of high dissociation and high vulnerability would indicate a contradictory unconcern for possible cooperation while implying a hope for cooperation. We may prohibit this situation by imposing the constraint that $\alpha + \beta \leq 1$. If, for example, $\alpha = 1$ and $\beta = 0$, then self-interest is the only consideration. If, however, $\alpha = 0$ and $\beta = 1$, then the players are willing to assume high risk to achieve cooperation.

If the sentence lengths are independent of the identity of the agents, then the joint rejectability of (D, C) should be equal to the joint rejectability of (C, D). With this assumption and the constraints on α

and β, we define the joint rmf:

$$p_{R_1R_2}(C,C)=\alpha, \qquad p_{R_1R_2}(C,D)=\frac{1-\alpha-\beta}{2}$$
$$p_{R_1R_2}(D,C)=\frac{1-\alpha-\beta}{2} \qquad p_{R_1R_2}(D,D)=\beta. \tag{23}$$

Although selectability deals with individual objectives, it is a joint consideration, since the consequences of the agents' decisions are not independent, and therefore preferences cannot be independent. A convenient way to express this dependency is to exploit the probabilistic structure of the imf and to compute joint selectability conditioned on joint rejectability, $p_{S_1S_2|R_1R_2}(v_1,v_2|w_1,w_2)$ for all (v_1,v_2) and (w_1,w_2) in $U\times U=\mathbf{U}=\{(C,C),(C,D),(D,C),(D,D)\}$, from which the imf may be obtained by the product rule:

$$p_{S_1S_2R_1R_2}(v_1,v_2,w_1,w_2)=p_{S_1S_2|R_1R_2}(v_1,v_2|w_1,w_2)\cdot p_{R_1R_2}(w_1,w_2). \tag{24}$$

The conditional selectability mass function, $p_{S_1S_2|R_1R_2}(v_1,v_2|w_1,w_2)$, characterizes the selectability of the joint option (v_1,v_2) given that the agents jointly place all of their rejectability mass on (w_1,w_2). We may compute the conditional selectability by invoking straightforward and intuitive rules of the form: "If X_1 and X_2 jointly reject (w_1,w_2), then they should jointly select (v_1,v_2)." Let, say, $(w_1,w_2)=(C,C)$, that is, the agents jointly reject cooperating. Given this situation, it is trivially obvious that the preferred joint option is for both to defect.[9] We may encode this rule into the conditional selectability mass function by placing all of the mass of on the joint option (D,D), that is, $p_{S_1S_2|R_1R_2}(D,D|C,C)=1$. If exactly one agent rejects cooperation, then it is obvious that, in this case as well, the preferred joint option is for both to defect, consequently, $p_{S_1S_2|R_1R_2}(D,D|C,D)=1$ and $p_{S_1S_2|R_1R_2}(D,D|D,C)=1$. Finally, if both agents reject defecting, then it is clear that $p_{S_1S_2|R_1R_2}(C,C|D,D)=1$. Table 12.3 summarizes the structure of this conditional selectability function.

Substituting the conditional selectability interdependence function given by Table 12.3 and the joint rejectability given by (23) into (24) and applying (12) yields the joint selectability function:

$$p_{S_1S_2}(C,C)=\beta \qquad p_{S_1S_2}(C,D)=0$$
$$p_{S_1S_2}(D,C)=0 \qquad p_{S_1S_2}(D,D)=1-\beta.$$

Comparing the bilateral selectability, given by (25), with the bilateral rejectability, given by (23), we obtain the bilateral satisficing set, which consists of all decision pairs, $(w_1,w_2)\in\mathbf{U}$, such that $p_{S_1S_2}(w_1,w_2)\geq$

Table 12.3 . Conditional selectability for the satisficing Prisoner's Dilemma.

	$p_{C_1C_2\|R_1R_2}(v_1, v_2\|w_1, w_2)$			
		(w_1, w_2)		
(v_1, v_2)	(C, C)	(C, D)	(D, C)	(D, D)
(C, C)	0	0	0	1
(C, D)	0	0	0	0
(D, C)	0	0	0	0
(D, D)	1	1	1	0

$bp_{R_1R_2}(w_1, w_2)$. Thus, parameterized by α and β, this set is, for the special case $b = 1$,

$$\mathfrak{S}_b = \begin{cases} \{(C, C)\} & \text{for} \quad \beta \geq \frac{1}{2} \\ \{(D, D)\} & \text{for} \quad \beta \leq \alpha \\ \{(C, C), (D, D)\} & \text{for} \quad \alpha < \beta < \frac{1}{2} \end{cases} . \tag{26}$$

These regions are depicted in Figure 12.1(a). The bilateral satisficing set coincides with the Pareto equilibrium solution, (C, C), when the vulnerability index is at least as large as $\frac{1}{2}$. It coincides with the Nash solution when the vulnerability index is less than the dissociation index. If the vulnerability index is greater than dissociation index but less than $\frac{1}{2}$, then the bilateral satisficing set contains both (C, C) and (D, D). To take action in this situation requires the invocation of a tie-breaker. For example, (D, D) is the satisficing option placing higher emphasis on individual interest (higher selectability) and (C, C) is the satisficing option placing higher emphasis on group interest (lower rejectability).

We may compute the unilateral satisficing equilibrium set by computing the univariate selectability and rejectability marginals in accordance with (16) and (17), from which the univariate satisficing set for either agent is

$$\mathfrak{S}_b = \begin{cases} \{C\} & \text{for} \quad \beta > \frac{1+\alpha}{3} \\ \{D\} & \text{for} \quad \beta < \frac{1+\alpha}{3} \\ \{C, D\} & \text{for} \quad \beta = \frac{1+\alpha}{3} \end{cases} . \tag{27}$$

The unilateral decision regions are illustrated in Figure 12.1(b). Note that the set is a singleton except in the special situation of $\beta = \frac{1+\alpha}{3}$.

In contrast to the VNM solution to this game, the satisficing solution accounts for the inclinations of the players. The VNM solution emerges as a special case (e.g., $\alpha = 1$ and $\beta = 0$), but the satisficing solution

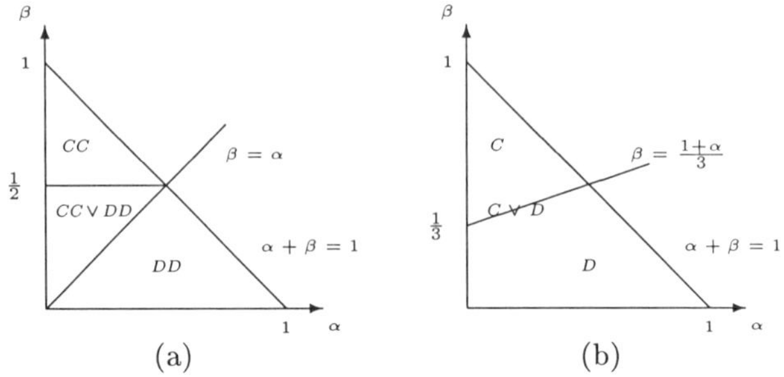

Figure 12.1. Decision regions for (a) bilateral decisions and (b) unilateral decisions.

gives the solution for all admissible (α, β) pairs. Uncertainty regarding these parameters may be handled in several ways. An agent is free to regard them as (a) random variables with known or interval-valued distributions and compute expectations, (b) deterministic interval-valued parameters and perform worst-case analyses, or (c) unknown parameters for which only an ordinal relationship is assumed. Thus, under fairly general circumstances, a decision can be rendered even though there may be considerable uncertainty regarding the values of the dispositional indices. In repeated-play scenarios, the players may refine their estimates of these parameters by invoking various learning mechanisms.

6. SUMMARY AND DISCUSSION

6.1. SUMMARY

Current decision-making methodologies have two extreme categories: (a) superlative approaches, such as VNM game theory, that rely on expected utility maximization; and (b) positive, or heuristic approaches, such as searching, planning, and scheduling procedures, that rely on the knowledge and experience of recognized experts. Much current work is devoted to finding hybrid approaches, such as bounded rationality, which blend optimality and heuristics by modifying the performance or stopping criteria. In this paper we do not rely upon optimality, or heuristics, or even a hybrid. Instead, we invoke a comparative paradigm and establish a concept for satisficing based on the logical middle ground between the superlative and positive paradigms. Using this concept, we define satisficing equilibria and establish it as a basis for making decisions that can be systematically justified as "good enough." Table 12.4 provides a summary of the two extreme paradigms and ours, and compares them in terms of the solution concept and knowledge requirements.

Substantive rationality—seeking the best and only the best—is perhaps the strongest possible notion of rationality, but it is more important

Table 12.4 . Frameworks for decision making.

RATIONALITY CONCEPT	DECISION PARADIGM	SOLUTION CONCEPT	KNOWLEDGE REQUIREMENTS
Substantive (extrinsic) rationality	Superlative degree (optimal)	Maximal expectations	Global mathematical models
Intrinsic rationality	Comparative degree (dichotomous)	Acceptable tradeoffs	Local mathematical models
Procedural rationality	Positive degree (heuristic)	Authoritative procedures	Local behavioral rules

to be good enough than to be best. Intrinsic rationality—getting what you pay for—is a weaker notion, but it is also more fundamental. Stated another way, evaluating dichotomies is a more elementary activity than searching for maxima. It is local, rather than global; it is an internal, rather than an external, evaluation.

A useful mechanism for identifying intrinsically rational options is epistemic utility theory. This theory is a cognitive decision-making procedure designed to acquire information and avoid error. Using the same mathematical structure, we provide new semantics for the design of a practical decision-making procedure designed to conserve resources and avoid failure. To distinguish this practical context from the cognitive context, we term this approach praxeic utility theory. This approach applies to both univariate (single-agent) and multivariate (multi-agent) decision making. It provides a compatible interface with the notion of satisficing we use, since it is designed to accommodate dichotomies. Furthermore, the mathematical structure of the interdependence function (as a mass function) permits the characterization of global interdependence in terms of conditional local interdependence and provides a viable means of constructing the interdependence function to account for whatever dependencies exist among the agents.

In complex and uncertain environments, it is essential that decision-making procedures be parsimonious and robust. The middle ground offered by the comparative paradigm accommodates both of these desiderata in a natural way. By softening the demands of strict optimality, the door is opened for a flexible and economical way of characterizing multi-agent behavior while not abandoning demands for acceptable performance. The comparative paradigm uses whatever modeling information and performance principles are available to formulate the interdependence function and apply the praxeic likelihood ratio test. The resulting satisficing solutions are justified as being good enough in the sense that they result in tradeoffs that favor benefits over costs. This line of research is still in its early stages of development, but substantive results are beginning to emerge. In the single-agent context, (Goodrich et al., 1999; Goodrich et al., 1998; Stirling et al., 1996a; Stirling, 1993) demonstrate its applicability to the control of nonlinear dynamic systems that have proven to be difficult to solve using conventional means. In (Stirling and Goodrich, 1999a; Stirling et al., 1996b; Stirling and Goodrich, 1999b) the approach is applied to the multi-agent case, resulting in the formalization of the notion of satisficing games.

6.2. SOME QUESTIONS AND REFLECTIONS

The approach developed in this paper provides a contrast to conventional game theory, which tells us about outcomes we can expect from substantively rational agents. Game theory has been used extensively as a means of modeling human behavior, but there is considerable evidence that people often do not behave in ways consistent with substantive rationality; that is, they are not optimizers, or even constrained optimizers (Bazerman, 1983; Bazerman and Neale, 1992; Rapoport and Orwant, 1962; Slote, 1989; Gigerenzer and Todd, 1999) An important open question is whether or not, and under what conditions, our notion of satisficing based on intrinsic rationality provides a valid model for human behavior; that is, are people satisficers as we have defined the term? An answer to this question may be provided by appropriately designed psychological testing and evaluation.

Another characteristic of conventional game theory is that it employs rationality postulates that are imposed at the individual level, rather than on the group. Game theory does not easily accommodate group interests, since the preferences of each agent are expressed as functions only of the choices of other agents and are not conditioned on their preferences. With conventional game theory, attitudes such as cooperation and conflict are not built into the utility functions, but become evident only when the utilities are juxtaposed—the linkages are external. A characteristic of our approach, however, is that preference relationships between agents can be expressed via the interdependence function—the linkages are internal. This feature invites further investigation into its significance. Does the explicit linkage of interagent preferences provide a basis upon which to construct an artificial society that captures important aspects of human behavior?

Satisficing decision theory may provide a convenient framework for negotiation. One of the problems with VNM game theory as a framework for negotiation is that it is not constructive—it may identify a best compromise, but does not provide a procedure for reaching it. The main trouble is dealing with the dynamic nature of coalition formation. Consequently, the strategic form is used extensively, and much of classical game-theoretic attention has been focused on situations where the process of negotiation can be presumed irrelevant to actual play. In other words, all of the deals, promises, threats, etc., are presumed to take place before the first move is actually played (Shubik, 1982). Heuristic approaches to negotiation are naturally amenable to the development of processes, but they lack the capacity for self-policing and quality cannot be assured. The principal run-time "decision-making" activity under substantive rationality is searching for an option that possesses the externally defined attribute of "optimality," and under procedural rationality it is rule-following. But under intrinsic rationality, the principal run-time activity is evaluating dichotomies and actually making choices dynamically and interactively according to internal assessments of both group and individual preferences. This feature is potentially a great advantage when designing negotiatory processes. Is seeking a good enough compromise when negotiating a more robust and flexible posture than directly seeking a best compromise?

No realistic decision problem can account for all logically possible options. All decision problems are framed against a background of knowledge and assumptions that result in a subset of options that are deemed relevant by the decision maker. This set may or may not be adequate for the task at hand, and one of the most difficult of all decision-theoretic issues is to decide whether or not this set of options should be enlarged, and if so, how to go about expanding it. Rank-ordering-based techniques, by their very nature, provide only relative information and cannot be used to address this concern. Dichotomy-based techniques, such as praxeic utility theory, may stand a better chance of addressing this issue, since they are grounded in the fundamental properties of the options and permit self policing.

For example, being unable to find a good enough option in a situation may lead an agent to reconsider what it is willing or unwilling to do. If, for another example, there are no options for which the benefit-to-cost ratio provides a clear choice, this is evidence that the decision problem is a tense one for the agent. This is not to say that the agent cannot make a good decision. Rather, it is merely evidence that it may not be well-suited, at present, for the task. This realization may trigger the expansion of the set of options. In practical situations this may require activating additional sensors, applying more computational power, interrogating information sources, exploring unfamiliar states of the world, or other means of acquiring additional information, perhaps at significant cost, in an attempt better to equip the agent to deal with its environment. In this way, the practical effects of actions may influence the design of the set of options that are available. Once again the praxeic has influenced the epistemic. Our need to act has led our need to know.

Notes

1. It is interesting to note that the story-line that gives the Prisoner's Dilemma its name was invented post factum to conform, for pedagogical illustration, to the payoff array with the given structure, rather than the other way around (Straffin, 1980). This illustrates the mind-set of many game theorists: the actual "game" is the payoff array, not the story-line.

2. Traditional game theorists are quick to point out that Rapoport's solution is not optimal. In fact, however, Axelrod proves (Axelrod, 1984) that, for repeated-play games where future payoffs are important, there does not exist an optimal strategy that is independent of the strategies used by other players.

3. Praxeology: the branch of knowledge that deals with practical activity and human conduct; the science of efficient action (Kotarbiński, 1965).

4. Informational value, as used in this context, is distinct from the notion of "value of information" of conventional decision theory (Raiffa, 1968), which deals with the change in expected utility that obtains if uncertainty is reduced or eliminated from a decision problem.

5. Such tradeoffs are also made under the superlative paradigm, but in a different way. As every control engineer well knows, specifying the performance criterion and solving for the optimum is only the first step in a subjective exercise of controller design. Consider, for example, the elementary control problem of designing an optimal linear quadratic regulator, where the goal is to maintain the state of a linear dynamical system within acceptable deviations from a desired set point while keeping the cost of control at an acceptable level. The tradeoff between performance and cost is accomplished by adjusting the relative weights of these two desiderata. Each weighting ratio leads to a different optimal control policy, and the task for the designer is to tune the weights iteratively to achieve an acceptable balance. In reality, there is no universally optimal solution—the "optimal" solution technique is nothing more than a convenient and systematic design procedure to achieve a solution that can be defended, at the end of the day, as being "good enough."

6. It is in this sense only that the superlative paradigm is involved in our usage of satisficing; the members of this set do not necessarily possess any superlative attributes.

7. Not to be confused with a mixture of distributions, which is a convex combination of probability distributions.

8. Although other psychological factors, such as expected behavior, may contribute to the results of repeated play games, for simplicity we restrict attention to the notion of group interest.

9. This apparent triviality is a consequence of each agent having only two options, since rejecting one implies selecting the other, but the situation is not so trivial when there are more than two options.

References

Axelrod, R. (1984). *The Evolution of Cooperation.* Basic Books, New York.

Bazerman, M. (1983). A critical look at the rationality of negotiator judgement. *Behavorial Science*, 27:211–228.

Bazerman, M. H. and Neale, M. A. (1992). Negotiator rationality and negotiator cognition: The interactive roles of prescriptive and descriptive research. In Young, P. H., editor, *Negotation Analysis*, pages 109–129. Univ. of Michigan Press, Ann Arbor, MI.

Gigerenzer, G. and Todd, P. M. (1999). *Simple Heuristics That Make Us Smart.* Oxford Univ Press, New York.

Goodrich, M. A., Stirling, W. C., and Frost, R. L. (1998). A theory of satisficing decision and control. *IEEE Trans. Systems, Man, Cybernet.*, 28(6):763–779.

Goodrich, M. A., Stirling, W. C., and Frost, R. L. (1999). Model predictive satisficing fuzzy logic control. *IEEE Trans. on Fuzzy Systems*, 7(3):319–332.

Harsanyi, J. (1977). *Rational Behavior and Bargaining Equilibrium in Games and Social Situations.* Cambridge Univ. Press, Cambridge.

Hogg, L. and Jennings, N. R. (1999). Variable sociability in agent-based decision making. In Parsons, S. and Wooldridge, M. J., editors, *Workshop on Decision Theoretic and Game Theoretic Agents*, pages 29–42, University College, London, United Kingdom, 5 July.

Kaufman, B. E. (1990). A new theory of satisficing. *The Journal of Behavorial Economics*, 19(1):35–51.

Kotarbiński, T. (1965). *Praxiology: An Introduction to the Sciences of Efficient Action.* Pergamon Press, Oxford. Translated by Olgierd Wojtasiewicz.

Kreps, D. M. (1990). *Game Theory and Economic Modelling.* Clarendon Press, Oxford.

Levi, I. (1980). *The Enterprise of Knowledge.* MIT Press, Cambridge, MA.

Levi, I. (1997). *The Covenant of Reason.* Cambridge Univ. Press, Cambridge.

Luce, R. D. and Raiffa, H. (1957). *Games and Decisions.* John Wiley, New York.

Palmer, F. R. (1971). *Grammar.* Harmondsworth, Penguin, Harmondsworth, Middlesex.

Pearce, D. W. (1983). *Cost-Benefit Analysis.* Macmillan, London, second edition.

Pearl, J. (1988). *Probabilistic Reasoning in Intelligent Systems.* Morgan Kaufmann, San Mateo, CA.

Popper, K. R. (1963). *Conjectures and Refutations: The Growth of Scientific Knowledge.* Harper & Row, New York.

Raiffa, H. (1968). *Decision Analysis.* Addison-Wesley, Reading, MA.

Rapoport, A. and Orwant, C. (1962). Experimental games: a review. *Behavorial Science*, 7:1–36.

Rasmusen, E. (1989). *Games and Information.* Basil Blackwell, Oxford.

Sandholm, T. and Lesser, V. (1997). Coalitions among computationally bounded agents. *Artificial Intelligence*, 94(1):99–137.

Shubik, M. (1982). *Game Theory in the Social Sciences.* MIT Press, Cambridge, MA.

Simon, H. A. (1955). A behavioral model of rational choice. *Quart. J. Econ.*, 59:99–118.

Simon, H. A. (1956). Rational choice and the structure of the environment. *Psychological Review*, 63(2):129–138.

Simon, H. A. (1986). Rationality in psychology and economics. In Hogarth, R. M. and Reder, M. W., editors, *Rational Choice.* Univ. Chicago Press, Chicago.

Simon, H. A. (1990). Invariants of human behavior. *Annu. Rev. Psychol.*, 41:1–19.

Simon, H. A. (1996). *The Sciences of the Artificial.* MIT Press, Cambridge, MA, third edition.

Slote, M. (1989). *Beyond Optimizing.* Harvard Univ. Press, Cambridge, MA.

Stirling, W. C. (1993). Coordinated intelligent control via epistemic utility theory. *IEEE Control Systems Magazine*, 13(5):21–29.

Stirling, W. C. and Goodrich, M. A. (1999a). Satisficing equilibria: A non-classical approach to games and decisions. In Parsons, S. and Wooldridge, M. J., editors, *Workshop on Decision Theoretic and Game Theoretic Agents*, pages 56–70, University College, London, United Kingdom, 5 July.

Stirling, W. C. and Goodrich, M. A. (1999b). Satisficing games. *Information Sciences*, 114:255–280.

Stirling, W. C., Goodrich, M. A., and Frost, R. L. (1996a). Procedurally rational decision-making and control. *IEEE Control Systems Magazine*, 16(5):66–75.

Stirling, W. C., Goodrich, M. A., and Frost, R. L. (1996b). Toward a theoretical foundation for multi-agent coordinated decisons. In *Proceedings of The Second International Conference on Multi-Agent Systems*, pages 345–352, Kyoto, Japan.

Stirling, W. C. and Morrell, D. R. (1991). Convex bayes decision theory. *IEEE Trans. Systems, Man, Cybernet.*, 21(1):173–183.

Straffin, P. (1980). The prisoner's dilemma. *UMAP Journal*, 1:101–103.

von Neumann, J. and Morgenstern, O. (1944). *The Theory of Games and Economic Behavior.* Princeton Univ. Press, Princeton, NJ. (2nd ed., 1947).

Wolf, G. and Shubik, M. (1974). Concepts, theories and techniques: Solution concepts and psychological motivations in prisoner's dilemma games. *Decision Sciences*, 5:153–163.

Zilberstein, S. (1998). Satisficing and bounded optimality. In *Proceedings of the 1998 AAAI Symposium*, pages 91–94. March 23–25, Stanford California. Technical Report SS-98-05.

Chapter 13

INVESTIGATING COMMITMENT FLEXIBILITY IN MULTI-AGENT CONTRACTS

Vanessa Teague
Department of Computer Science, Stanford University
vteague@cs.stanford.edu

Liz Sonenberg
Department of Information Systems, The University of Melbourne
l.sonenberg@dis.unimelb.edu.au

Abstract Reputation and commitment are important issues for automated contracting. Leveled commitment contracts, i.e. contracts where each party can decommit by paying a predetermined penalty, were introduced to allow self interested agents to accommodate events that unfolded since the contract was entered into. Various approaches to modelling reputation have been explored, allowing an agent to make decisions after considering not only current opportunities, but also possible longer term (social) implications of maintaining a reputation for reliability in contract fulfilment. An agent may consider both its own reputation for honesty and the past behaviour of others. We explore a combination of these issues in the context of a simple contracting scenario. Analytical and simulation results are presented.

1. INTRODUCTION

The advent of widespread Electronic Commerce promises to have enormous benefits for consumers. Due to the proliferation of on-line auction houses such as eBay and on-line retailers such as Amazon.com, we now have a vast array of options for selling or purchasing almost anything. The resulting environment is well suited to intelligent autonomous agents, since some reasonably intelligent behaviour is required for bidding or bargaining but there are often too many possibilities for a human to investigate manually.

Doing business on the internet also presents significant new problems. Possibly the most important of these is the problem of deciding when another entity on the web is trustworthy. Trustworthiness is a particularly important issue for organisers of on-line markets and auction houses, since the credibility of their market is undermined if participants frequently succeed in cheating others. The main question is whether market organisers should try to intervene and guarantee the trustworthiness of participants (as stock markets do, for example, by enforcing laws such as regular auditing of the traded companies), or adopt a *laissez faire* approach and assume that dishonourable parties will eventually acquire a bad reputation, as do eBay and MIT Media Lab's Kasbah system (Maes et al., 1999).

The Internet is an environment that involves a large degree of uncertainty and changes quickly over time. In such environments, allowing agents to renegue on contracts sometimes can greatly improve performance, since agents can agree to contracts that seem beneficial and then decommit if circumstances change adversely.

Trust of other agents is difficult to establish when some decommitment from contracts is permitted. If an agent enters into a contract, it needs to be confident that it will not be damaged by the dishonesty of other contract participants. In real-world (human) contracts, compliance can be enforced by an authoritative body such as a law court, but in artificial agent scenarios it is difficult or impossible to trace the behavior of an artificial agent to its human "owner". Agents need some way of knowing how to evaluate the "trustworthiness" of others.

The model of trading we adopt is a two-round process. The first round involves only partial commitment; in the second round agents have an opportunity to leave before they commit fully. This has some intuitive appeal for an agent-mediated electronic commerce setting, since a normal pattern of behaviour could be to send an agent out on to the web to make tentative arrangements, but only spend large amounts of money with explicit confirmation from the human who sent it (Nwana et al., 1998). It also provides a precise way to measure "trustworthiness", since for any agent it is possible to count the number (or percentage) of times it honours its initial commitment.

Sandholm and colleagues introduce "Leveled Commitment Contracts" which allow agents to choose their level of commitment at the beginning (Sandholm and Lesser, 1996; Sandholm et al., 1999; Sandholm and Zhou, 1999). Agents are allowed to decommit, but only if they pay the other participant(s) a predetermined penalty. The payment is assumed to be enforced by the market.

An example of a *laissez faire* market is is the work of Sen *et al.* (Sen, 1996; Sen and Biswas, 1998). They assume no market intervention, but that agents choose which others to work with according to a system of reciprocity. They remember which agents have reciprocated the favours they have done for them, gradually learning which others can be trusted. A stable community of cooperating (though self-interested) agents emerges.

The "brownie point" model of Grosz, Kraus and colleagues (Glass and Grosz, 1999; Sullivan et al., 1999) represents a moderate level of intervention, where there are both market-imposed penalties for decommitting and agent-imposed reluctance to do so.

1.1. OUR PROBLEM

Our main interest was investigating different approaches to market intervention, the primary aims being to explore ones that elicit cooperative behaviour from the agents and to consider identifying characteristics of the environment and agent interactions that may make one approach more suitable than another.

We have devised a scenario general enough to incorporate all three of the decision factors described (decommitment penalties, brownie points and reliability) and flexible enough to adopt various combinations of their features. See Section 2.

Some parameterisations that involved only decommitment penalties could be analysed mathematically. In most cases, the analysis showed that the agents behaved optimally, in the sense that they always ended up maximising the total value of the targets they caught. In one case, agents were shown to behave "fairly", meaning that the payoff of the second-lowest-paid agent was maximised. Section 3 describes the analytical results.

Since it was not possible to analyse the "brownie point" and reciprocation based markets, we constructed a simulation in Java and performed experiments on the efficiency of the agents under different market conditions. The first experiments were mainly checks that the simulation reproduced the results contained in (Andersson and Sandholm, 1998; Andersson, 1998) and (Glass and Grosz, 1999). These were generally successful, finding that a moderate emphasis on "brownie points" and small to medium decommitment penalties were optimal. Section 4 contains empirical investigations. The most interesting result was the formation of stable teams using the reliability-based approach (section 4.2). Other results confirmed the intuition that "brownie points" and decommitment penalties perform similar roles — section 4.4 shows that combining them does not increase performance compared to "brownie points" alone, and section 4.3 shows that "brownie points" perform similarly to decommitment penalties but slightly better. However, the reliability strategy was significantly different.

2. THE SCENARIO

The theme is that agents band together to capture "targets". This corresponds to operating successfully together as a team. Agents are self-interested, though they may cooperate when they believe it is in their long-term interests.

There are a number of rounds, each of which has four stages:

1. Each agent in turn chooses an initial ("first-stage") target to try to capture. It moves to that target, and may make a commitment to stay there until the end of the run.

2. New targets appear.

3. Each agent in turn has one opportunity to leave its current target, pay any decommitment penalty required, and move to another.

4. (Clearing) The number of agents at each target is counted. If a target has "enough" agents it is considered captured and its value is counted towards the final total.

At the end of each round all targets disappear, all agents are released from all contracts and removed from the board, then the first-stage targets reappear.

Each target has three important properties: the number of agents required to capture it, the amount of "money" it will pay each of them, and the decommitment penalties it will demand if they renegue. Suppose target k pays the first agent to arrive x_1^k, the second x_2^k and so on (of course it only pays if it is captured). We will assume that $x_1^k \geq x_2^k \geq \cdots$.

The agents play in turns, in the same order in each round. Agent α_n is the n-th to play. We chose a consistent (rather than random) playing order so that all the uncertainty in the game was controlled

at one point, the second-stage targets. This is somewhat less realistic, but it made it much easier to design and evaluate agent strategies. In some cases, we have commented on how the game would have differed if playing order had been random.

The following are common knowledge:

- The payoffs offered, number of agents required and decommitment penalties demanded by each of the first-stage targets.
- The number of second-stage targets that will appear and the number of agents required to capture each.
- The number of agents.
- That agents are perfect reasoners, each one aiming to maximise its own payoff.
- The probability distributions of the payoffs offered by each second-stage target (common knowledge at the start of stage one).
- The payoffs offered by each second-stage target (common knowledge at the end of stage two).

Each simulation consists of a number of rounds, each of which has the same first-stage targets and the same agents. A new set of second-stage targets is generated in each round, according to a fixed probability distribution. We recorded the average value of the targets captured per round.

Agents use two different brownie-point-like numbers in their calculations. The first one (which we will refer to as "brownie points") is almost exactly like that in (Glass and Grosz, 1999) — each agent keeps a record of its own behaviour and is uninclined to decommit if its "brownie points" will suffer much. The second *(reliability)* involves keeping a record of other agents' behaviour, and preferring to work with agents who have been reliable in the past. Each agent α updates its record of "brownie points" and reliability at the end of each round. Denoting α's "brownie points" by BP and writing $\text{Rel}(\alpha_i)$ for α's evaluation of agent α_i's reliability (for all other agents α_i), the update formulas are:

$$\text{BP} = \begin{cases} \text{BP} + 1 & \text{if } \alpha \text{ stayed committed} \\ \text{BP} - o_o/o_n & \text{if } \alpha \text{ decommitted from a target offering } o_o \text{ to} \\ & \text{one offering } o_n \end{cases}$$

$$\text{Rel}(\alpha_i) = \begin{cases} \text{Rel}(\alpha_i) + 1 & \text{if agent } \alpha_i \text{ was at the same target as } \alpha \\ & \text{and stayed committed} \\ \text{Rel}(\alpha_i) - o_o/o_n & \text{if } \alpha_i \text{ decommitted from a target where} \\ & \alpha \text{ was offered } o_o \text{ and moved to a target} \\ & \text{where it was offered } o_n. \end{cases}$$

The justification for deducting the ratio of offers from reliability when an agent decommits is that the agent who has been decommitted against is more likely to be forgiving if it didn't lose very much due to the decommitment or if the decommitting agent was making a large gain by leaving. Similarly, in the case of BP, agents are more likely to forgive themselves if they decommitted from a low-paying target to a high-paying one.

When deciding how to play, agents use a weighted sum of the three main considerations: the expected payoff, the effect on their own "brownie points" and the past behaviour of others (reliability). The decommitment penalties are automatically included in the expected payoff.

Estimating the probability that a target will be captured requires some approximations, since agents can't predict exactly what others will do if the others are considering "brownie points" or reliability as well as their expected payoff (see the discussion in section 4).

3. SOME ANALYSIS OF SPECIAL CASES

The most interesting results in this section are the ones that show that the self-interested agents succeed in achieving the optimal payoff for the group (sections 3.2 and 3.3). Section 3.3 shows that this happens even for quite general assumptions about the decommitment penalties. The first result, in the case where agents have to work in pairs but do not have any decommitment penalties, demonstrates that the agents' behaviour maximises the payoff of the second-least paid agent. This is surprising for a group of agents that are self-interested and do not even consider the payoffs of others.

Note about phrasing: When we write "plays in the first (second) place," we mean "becomes the first (second) agent at a target."

3.1. CASE 1: AGENTS WORKING IN PAIRS BEHAVE FAIRLY

Simplifying assumptions:

- No targets appear or disappear after the first stage. This means that agents will not change targets in the second stage because no new information has become available to them. Hence this analysis is about how they should choose their first target correctly.

- There are no decommitment penalties.

- Each target requires exactly two agents to capture it.

- If a target is captured, it pays strictly positive amounts to each agent at it.

- There are less than twice as many agents as there are targets. If there were more than that, all the agents would know that all the targets would be captured, so their strategy would simply be to take the largest available spot.
- There are only finitely many targets.
- Being the first agent to arrive at any target is always better than being the second to arrive at any target, as long as both targets are captured. That is, $\min_k x_1^k > \max_k x_2^k$.

It is not the case that the first few agents can simply choose to be the first agents at lucrative targets, because they need to worry that those targets may not be successfully captured.

Let there be m agents. Since they are all perfect reasoners, every agent that has an opportunity to guarantee a positive payoff will achieve a positive payoff.

Claim 1 *Every agent, except the last one if m is odd, can guarantee a positive payoff.*

Proof If, at agent α's turn, there is a target k with one agent already at it, then α can choose to be the second agent there and guarantee itself a payoff $x_2^k > 0$.

If all the targets are either free of agents or occupied by two, and if there is at least one more agent left to play after α, then α can choose to be the first agent at the target k' with k' chosen so that $x_2^{k'}$ is maximised. If it does this, another agent will surely play at the second place on the same target because at least one of the agents yet to play must play in second place (the last agent always does if it can), and an agent playing in second place can maximise its payoff by playing at target k'.

If m is even, one of the two cases above always holds. If m is odd, one of the cases holds for every agent except the last one. Therefore α can always achieve a positive payoff unless it is the last of an odd number of agents. $\diamond$

Corollary At the end of the round, every agent (except α_m when m is odd) will successfully capture a target, though not necessarily the one described by the above algorithm.

A simple optimal strategy Since each agent has only finitely many possible moves, they could each perform full lookahead to determine which targets would be captured, then choose the most lucrative. Therefore, each of the first $\lfloor \frac{m}{2} \rfloor$ agents to play can work out which targets will eventually be captured, depending on what move it makes. They will always play in the first place at some target, since $\min_k x_1^k > \max_k x_2^k$.

Consider agent $\alpha_{\lfloor \frac{m}{2} \rfloor + 1}$. Playing in the first place at any target must result in a zero payoff, because if a second agent cooperated with $\alpha_{\lfloor \frac{m}{2} \rfloor + 1}$ then one of the preceding agents would end up with a zero payoff, violating the corollary. Even if $\alpha_{\lfloor \frac{m}{2} \rfloor + 1}$ plays as the first agent at a vacant target k such that x_2^k is maximised, no other agent will cooperate — all of the $m - \lfloor \frac{m}{2} \rfloor - 1$ agents that play afterwards will prefer to play in second place at one of the first $\lfloor \frac{m}{2} \rfloor$ established targets. Therefore, the first $\lfloor \frac{m}{2} \rfloor$ agents have chosen (in first place) all the targets offering the best $m - \lfloor \frac{m}{2} \rfloor - 1$ second places.

If m is odd, $m - \lfloor \frac{m}{2} \rfloor - 1 = \lfloor \frac{m}{2} \rfloor$, so the first $\lfloor \frac{m}{2} \rfloor$ agents occupied exactly those targets that had the $\lfloor \frac{m}{2} \rfloor$ highest values of x_2^k.

Therefore the best algorithm for odd m (supposing this is the jth agent to play) is:

If ($j \geq \lfloor \frac{m}{2} \rfloor + 1$) play in second place at an already-occupied target k such that x_2^k is the highest available.

else sort the targets by their values of x_2^k. Choose a target that has one of the highest $\lfloor \frac{m}{2} \rfloor$ values of x_2^k, such that the corresponding value of x_1^k is the highest available.

Theorem 1 *The result always maximises the payoff of the second-last (and hence second-lowest-paid) agent.*

The $m - 1$-th agent occupies the $\lfloor \frac{m}{2} \rfloor$-th best x_2^k position. $\diamond$

The case when there are an even number of targets is very similar, except that the last agent always has only one possible move, so one first-playing agent can get away with choosing a target that has a small reward for the second agent.

If second-round playing order is random, new information (the playing order) is revealed. If m is odd and the agents have executed the algorithm above, no agent will benefit from moving in round two because the lucrative places will already be taken. If there is an even number of agents, this is not the case — an agent which played in second place can benefit by moving to the first place at another target and relying on the last agent having only one possible move.

3.2. CASE 2: UNCOMMITTED AGENTS WORKING ALONE BEHAVE OPTIMALLY

Assumptions:

- There are no decommitment penalties.

- Each target requires exactly one agent to capture it.
- If a target is captured, it pays a strictly positive amount to capturing agent.
- There are at least as many first-round targets as there are agents.

Let the values of the first-stage targets be $x_1^1, \cdots, x_1^{n'}$ where $n' \geq n$. Without loss of generality, $x_1^1 \geq x_1^2 \geq \cdots \geq x_1^{n'}$. Let the values of the second-stage targets (known probabilistically) be represented by the random variables $X_{(1)} \cdots X_{(n'')}$, where $X_{(1)} \geq X_{(2)} \geq \cdots \geq X_{(n'')}$. The random variable $X_{(i)}$ is the i-th order statistic of the second-stage targets' values. Suppose it is given by p.d.f. p_i.

Agents have nothing to lose by committing in the first round, since there are no decommitment penalties. If an agent (apart from the last one) does not commit it may reduce its payoff because another agent playing later may "take its place". Only if agent α_k is certain that at least one second-stage target will pay more than x_1^k does it have nothing to lose by not committing. We will assume such agents commit. All the others will commit because they have nothing to lose by doing so, but the potential to lose if they fail to.

Each agent will chose the most lucrative available target in the first round. When second-stage targets appear, the agents will shift into any more lucrative positions in the order they first played. A first-stage target that is vacated may be re-occupied by a later agent. Therefore the agents always capture the n most lucrative targets, with their payoffs decreasing in the order of their playing turns.

Agent α_i captures first-stage target $i-k$ iff there are exactly k second-stage targets more lucrative than it. That is, if $X_{(k)} > x_1^{i-k} > X_{(k+1)}$. The agent captures second-stage target $i-l$ iff there are exactly l first-stage targets greater than it. That is, if $x_1^l \geq X_{(i-l)} \geq x_1^{l+1}$.

Therefore, at the start of round one, agent α_i's expected payoff is

$$\sum_{k=0}^{i-1} x_1^{i-k} \Pr\left(X_{(k)} > x_1^{i-k} > X_{(k+1)}\right) + \sum_{l=0}^{i-1} \int_{x_1^{l+1}}^{x_1^l} x p_{i-l}(x)\, dx$$

where $X_{(0)} = x_1^0 = \infty$ and p_{i-l} is the p.d.f. of the $(i-l)$-th best second-stage target.

If second-round playing order is random, the analysis is similar except that each agent has to consider its expected payoff in each of the $n!$ equiprobable permutations of players. This makes computation much more intensive and also considerably increases the uncertainty of the agent's payoff. The same observation is true for the following case.

3.3. CASE 3: MODERATELY-COMMITTED AGENTS WORKING ALONE STILL BEHAVE OPTIMALLY

Assumptions: The same as case 2, except that now there are decommitment penalties. The assumptions about them are:

- Decommitment penalties are always non-negative.
- The decommitment penalty for a target is a function ϕ of the target price, with the property that $id + \phi$ is monotonic increasing.
- If a target is decommitted from, it offers a higher payoff in the second round. Target τ_i offers $x_1^i + \phi(x_1^i)$.

The second condition asserts that the ordering of the targets is the same after adding on the decommitment penalties. This will be true whenever the offers are far apart and the decommitment penalties are relatively small.

If all the agents certainly committed in the first round, the second-round analysis would be the same as case 2 with $x_1^j + \phi(x_1^j)$ substituted for x_1^j for all j. However, since the agents now have something to lose by decommitting, they might not always commit in the first round.

It is possible to arrange a decommitment scheme so that agents are too reluctant to decommit, even when decommiting would be extremely beneficial.

This would defeat the purpose of leveled commitment contracts. In this case we can show that such a problem does not occur.

Theorem 2 *In this case (where there are decommitment penalties) it is still true that the n agents occupy the n most lucrative targets at the end of the second round (where the value of first-stage target τ_i is taken to be $x_1^i + \phi(x_1^i)$).*

Proof (By contradiction). Suppose there is an agent α_k which, at the end of round 2, occupies a target τ_i that is not among the n most lucrative ones. Then there is at least one unoccupied target in the n most lucrative ones. Therefore when α_k took its turn in the second round, there was at least one unoccupied target τ_j in the n most lucrative ones — the number of unoccupied targets in that set can't have increased since no agents would have left those targets to move to lower-paying ones. If α_k chose τ_i rather than τ_j in round 2, it must have been because

$$O_2(\tau_i) > O_2(\tau_j) \quad \text{if } \alpha_k \text{ was not already committed to } \tau_i$$

or

$$x_1^i > O_2(\tau_j) - \phi(x_1^i) \quad \text{if } \alpha_k \text{ was already committed to } \tau_i$$

where $O_2(\tau_k)$ is target k's second-round offer.

These (equivalent) conditions both violate the assumption that τ_j is one of the n most lucrative while τ_i is not (because of the assumption that $id + \phi$ is monotonic increasing. ◇

In the last two cases, the targets collected when there are decommitment penalties are exactly the targets collected when there are no penalties. Hence decommitment penalties neither improve nor detract from the group's payoff in this case. It is also interesting to ask whether adding decommitment penalties alters which agents capture which targets. In the one- or two-agent cases, there is no change.

Theorem 3 *If there are two or fewer agents, the final state in case 3 (where there are non-zero decommitment penalties) is the same as that in case 2 (where there are none).*

Proof In the one-agent case, the best strategy is to wait until the second round and then pick the best target. This produces the same final state as case 2.

If there are exactly two agents, suppose there are two first-stage targets τ_1 and τ_2 and two second-stage targets σ_1 and σ_2. Assuming two second-stage targets does not result in a loss of generality because if there were more targets, all but the first two would be ignored by both agents and if there were fewer targets, this could be modelled by setting the probability distributions of the two so that one or both of them were guaranteed to be so low that they would not be considered. Assuming two first-stage targets is also not a loss of generality, because we can assign one or both of the targets a value near zero and no decommitment penalty.

Since the final state can only be different from case 2 when in the first round α_2 commits to target 1 while α_1 either doesn't commit or commits to target 2, it suffices to show that it is never individually rational for the agents to behave in this way.

The agents' expected payoffs for all possible stage one actions are given in table 13.1. In the table, "NC" indicates no commitment, "C1" indicates committing to target τ_1 and "C2" to target τ_2 in the first stage.

$X_{(1)}$ and $X_{(2)}$ (where $X_{(1)} > X_{(2)}$) are the payoffs of the second-stage targets, given by p.d.fs p_1 and p_2 respectively.

It can be seen from Table 1 that α_2's expected payoff depends only on its action, not on that of α_1, although of course α_1's commitments restrict the possible places that α_2 can choose to commit. However, α_1's payoff is dependent on the behaviour of α_2, since it matters whether α_2 commits to target τ_1 possibly forcing α_1 to accept a lower payoff if the second-stage targets turn out to be of lower value than expected.

To prove the theorem, we will show that cases 2 and 7 never occur, because α_1 can always predict α_2's first-stage behaviour, and if it knows that α_2 would choose to commit to target τ_1 then α_1 will commit to that target instead.

Suppose that α_1 doesn't commit to target τ_1. Agent α_2 only needs to compare committing to τ_1 against not committing, since committing to τ_2 (cases 3 and 5) is always worse than not committing (cases 1, 4 and 6). It will commit to τ_1 iff

$$\begin{aligned}
&\int_{x_1^1+\phi(x_1^1)}^{\infty} p_2(x)[x-\phi(x_1^1)]\,dx \\
&\qquad + \Pr\left(X_{(2)} \le x_1^1+\phi(x_1^1)\right)x_1^1 \\
&> \int_{x_1^1+\phi(x_1^1)}^{\infty} x p_2(x)\,dx \\
&\qquad + \Pr\left(X_{(1)} > x_1^1+\phi(x_1^1) \ge X_{(2)}\right)(x_1^1+\phi(x_1^1)) \\
&\qquad + \int_{x_1^2+\phi(x_1^2)}^{x_1^1+\phi(x_1^1)} x p_1(x)\,dx \\
&\qquad + \Pr\left(X_{(1)} \le x_1^2+\phi(x_1^2)\right)(x_1^2+\phi(x_1^2))
\end{aligned}$$

This implies

$$\begin{aligned}
&\int_{x_1^1+\phi(x_1^1)}^{\infty} [x-\phi(x_1^1)]p_1(x)\,dx \\
&\qquad + \Pr\left(X_{(1)} \le x_1^1+\phi(x_1^1)\right)x_1^1 \\
&> \int_{x_1^2+\phi(x_1^2)}^{\infty} x p_1(x)\,dx \\
&\qquad + \Pr\left(X_{(1)} \le x_1^2+\phi(x_1^2)\right)(x_1^2+\phi(x_1^2)),
\end{aligned}$$

which is exactly the requirement for α_1 to choose case 4 or 5 rather than case 2. That is, it will always commit to target τ_1 when it thinks

Case		*Payoffs*
1	α_1(NC):	$\int_{x_1^1+\phi(x_1^1)}^{\infty} xp_1(x)\,dx$
		$+\Pr(X_{(1)} \leq x_1^1 + \phi(x_1^1))(x_1^1 + \phi(x_1^1))$
	α_2(NC):	$\int_{x_1^1+\phi(x_1^1)}^{\infty} xp_2(x)\,dx$
		$+\Pr(X_{(1)} > x_1^1 + \phi(x_1^1) \geq X_{(2)})(x_1^1 + \phi(x_1^1))$
		$+\int_{x_1^2+\phi(x_1^2)}^{x_1^1+\phi(x_1^1)} xp_1(x)\,dx$
		$+\Pr(X_{(1)} \leq x_1^2 + \phi(x_1^2))(x_1^2 + \phi(x_1^2))$
2	α_1(NC):	$\int_{x_1^2+\phi(x_1^2)}^{\infty} xp_1(x)\,dx$
		$+\Pr(X_{(1)} \leq x_1^2 + \phi(x_1^2))(x_1^2 + \phi(x_1^2))$
	α_2(C1):	$\int_{x_1^1+\phi(x_1^1)}^{\infty} p_2(x)[x - \phi(x_1^1)]\,dx$
		$+\Pr(X_{(2)} \leq x_1^1 + \phi(x_1^1))(x_1^1)$
3	α_1(NC):	Same as state 1
	α_2(C2):	$\int_{x_1^1+\phi(x_1^1)}^{\infty} p_2(x)[x - \phi(x_1^2)]\,dx$
		$+P \times (x_1^1 + \phi(x_1^1) - \phi(x_1^2))$
		$+\int_{x_1^2+\phi(x_1^2)}^{x_1^1+\phi(x_1^1)} p_1(x)[x - \phi(x_1^2)]\,dx$
		$+\Pr(X_{(1)} \leq x_1^2 + \phi(x_1^2))(x_1^2)$
	where	$P = \Pr(X_{(1)} > x_1^1 + \phi(x_1^1) \geq X_{(2)})$
4	α_1(C1):	$\int_{x_1^1+\phi(x_1^1)}^{\infty} p_1(x)[x - \phi(x_1^1)]\,dx$
		$+\Pr(X_{(1)} \leq x_1^1 + \phi(x_1^1))(x_1^1)$
	α_2(NC):	Same as states 1 and 6
5	α_1(C1):	Same as state 4
	α_2(C2):	Same as state 3
6	α_1(C2):	$\int_{x_1^1+\phi(x_1^1)}^{\infty} p_1(x)[x - \phi(x_1^2)]\,dx$
		$+\Pr(X_{(1)} \leq x_1^1 + \phi(x_1^1))(x_1^1 - \phi(x_1^2))$
	α_2(NC):	Same as states 4 and 1
7	α_1(C2):	$\int_{x_1^2+\phi(x_1^2)}^{\infty} p_1(x)[x - \phi(x_1^2)]\,dx$
		$+\Pr(X_{(1)} \leq x_1^2 + \phi(x_1^2))(x_1^2)$
		(which is always worse than state 2).
	α_2(C1):	Same as state 2

Table 13.1 . Agents' payoffs for the two-player game.

that α_2 will do so otherwise. Therefore α_1 always captures the most lucrative target and α_2 the second most lucrative. ◇

4. EMPIRICAL WORK

When agents are considering "brownie points" or reliability, their behaviour becomes difficult to predict, both for us and for the other agents in the simulation. Since it was no longer feasible to do analysis, we investigated it using empirical methods.

Because the agents cannot predict exactly the behaviour of others, they need to make some estimates of probabilities such as the probability that a particular target will be captured or that a better offer will appear. These estimates are based partly on discrete order statistics (Kendall and Stuart, 1977) and partly on less rigorous methods[1].

Our first experiments were designed to check that our system behaved reasonably when presented with situations similar to those in (Andersson and Sandholm, 1998; Andersson, 1998) and (Glass and Grosz, 1999). These experiments generally replicated the results reported in those papers, showing that a moderate level of decommitment inhibiting (that is a moderate "brownie point" weight in the former, and small-to-moderate decommitment penalties in the latter) were optimal. In both cases, they were better than allowing agents to decommit freely or enforcing full commitment.

The first experiment reported in detail here demonstrates an interesting property of the reliability considerations. It shows that this method can be used to form stable teams that can be more effective overall than ad hoc alliances. In the next experiment, "brownie points" and decommitment penalties are compared on the same scenarios. "Brownie points" are shown to be more effective in both cases, mainly because high decommitment penalties tend to discourage agents from comitting initially. The third experiment shows that having both decommitment penalties and "brownie points" is not better than having only one. The final experiment shows that, although reliability can often be very effective, decommitment penalties are sometimes better.

4.1. BACKGROUND EXPERIMENTS: CHECKING THAT "BROWNIE POINTS" AND DECOMMITMENT PENALTIES WORK AS EXPECTED

Figure 13.1 shows the agents' average effectiveness in a particular scenario as a function of the weight they attach to "brownie points". There were no decommitment penalites in this experiment. As expected,

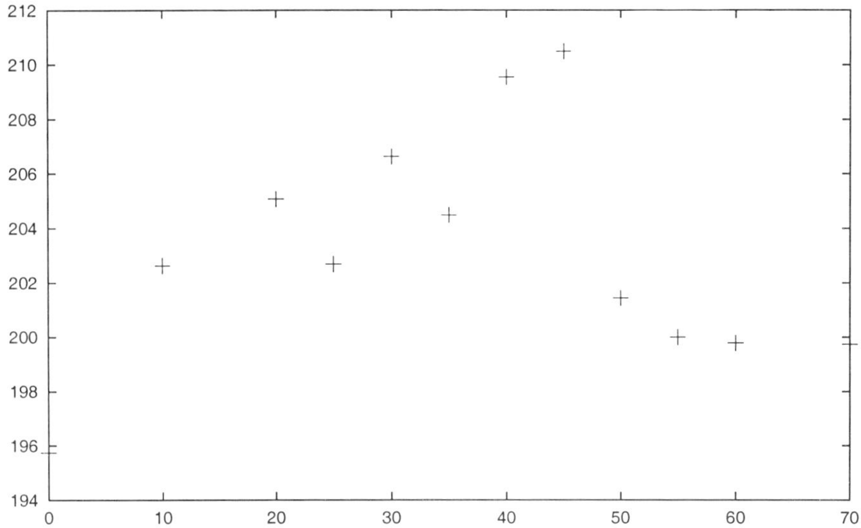

Figure 13.1. Average group income vs. "brownie point" weight.

a medium value (about 40) is the most effective, since then agents tend to decommit exactly when it is particularly beneficial to do so.

Figure 13.2 shows the results for agents paying no attention to "bro-nie points", but having to pay decommitment penalties. All decommitment penalties were equal (since all first-round offers were approximately equal). As expected, a small but non-zero decommitment penalty (about 30) was optimal. This corresponds to 15% of the target's offered value, which is slightly higher than the percentage noted elsewhere (Andersson and Sandholm, 1998; Andersson, 1998).

4.2. EXPERIMENT 1: USING RELIABILITY TO FORM STABLE TEAMS

In this case there were no targets in the first round (so an approach based on decommitment penalties or brownie points would have had no effect), then a large number of equally sized targets in the second round. The number of agents was far too small to capture all the targets, so cooperation was extremely important. Each agent kept a record of every other agent's past decommitting behaviour, summarised by a reliability

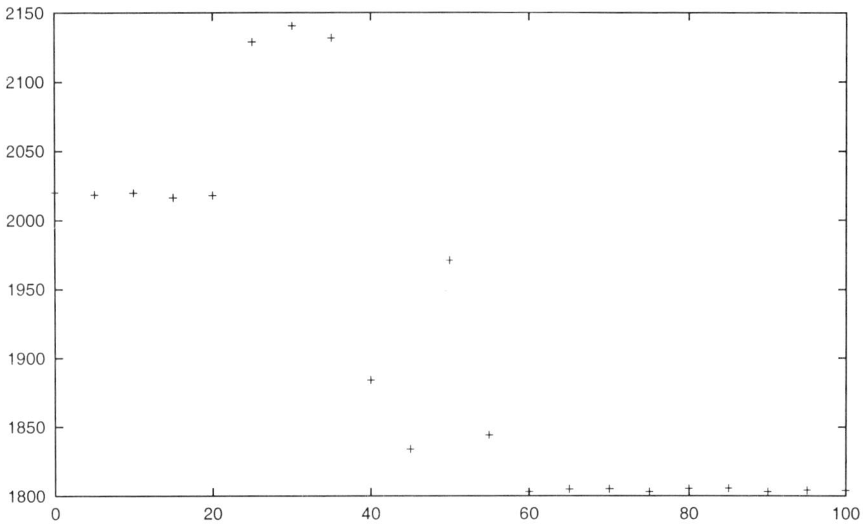

Figure 13.2. Average group income vs. decommitment penalty.

value that was increased when the other agent successfully captured a target with the first, and decreased when it decommitted against it.

A graph of the agents' average effectiveness over time is shown in Figure 13.3, for a fixed reliability weighting. Agents were generally ineffective in the first few rounds because later agents ignored low-paying places on targets that were already partially occupied, preferring instead to take higher-paying places on targets that did not yet have other agents. The result was that at the end of a round, there were many targets with not quite enough agents to capture them. As the simulation progressed, those agents who had cooperated successfully in the past (i.e. caught a target together) gradually coalesced into stable teams that always succeed in capturing a target. The teams were much more effective than the disorganised agents, though their behaviour was still below optimal because they often did not capture the best target, just the one that offered the first agent the highest payoff. For reference, the expected optimal payoff (if the agents had some central planner to choose their targets for them) would be approximately 34, about double what the agents achieved.

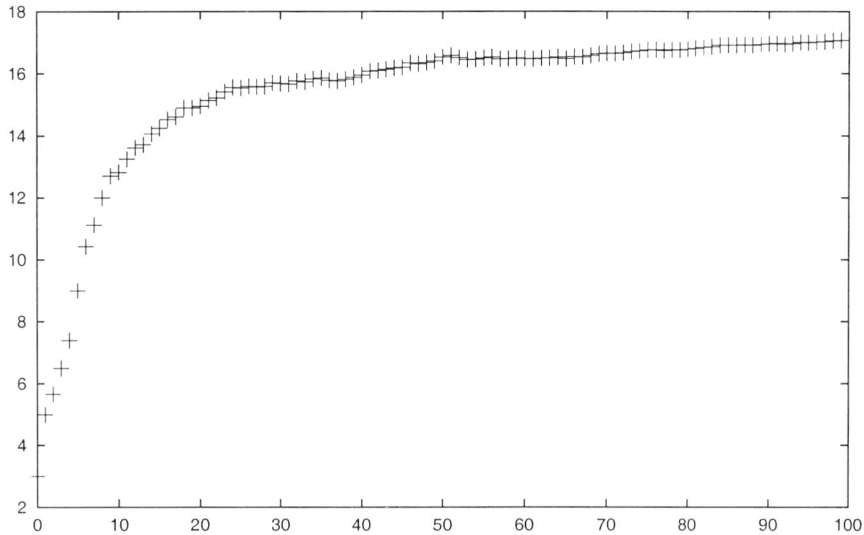

Figure 13.3. Running average group income over time for fixed reliability weight.

The exact value of the weight attached to reliability seems unimportant, as long as it is non-zero. See Figure 13.4 for a graph of effectiveness as a function of this weight. The intuition is that after a short while an agent's commitment to its small team becomes an overriding concern, so its exact weight doesn't matter greatly. This is actually a bad thing — if the agents were a little less committed to their team they would be better able to take advantage of the best targets, though they would be less certain of successfully capturing them.

A random playing order would probably disrupt this team formation, because it relies upon agents just happening to appear in the same teams repeatedly, early on in the simulation. It would be interesting to investigate whether teams still formed under random playing order — they would certainly take much longer to do so.

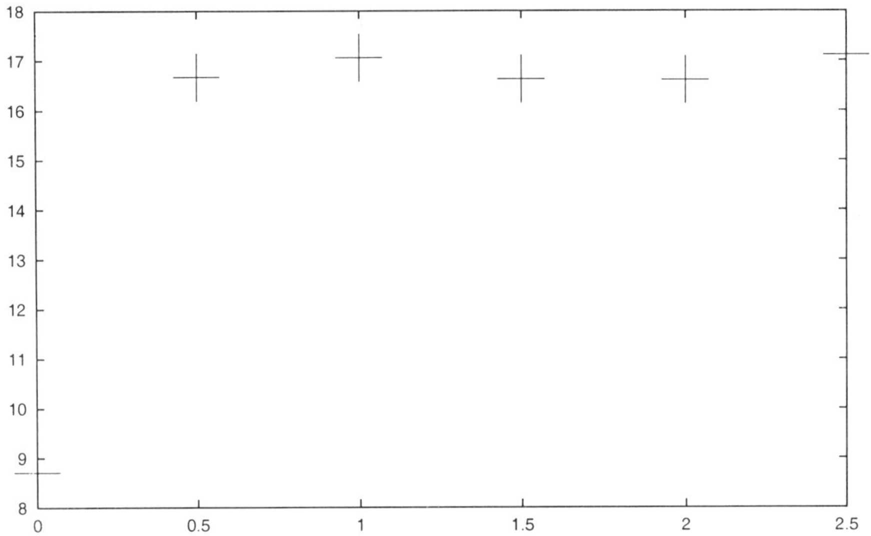

Figure 13.4. Average group income vs. reliability weight.

4.3. EXPERIMENTS 2: COMPARISONS OF THE DIFFERENT METHODS ON THE SAME SCENARIOS

Figure 13.5 shows the effect of using decommitment penalties rather than "brownie points" on the same scenario as Figure 13.1. The main difference is that the greatest value attained by decommitment penalties is lower than that achieved using brownie points (about 205, compared to 210 in Figure 13.1). Figure 13.6 shows the effectiveness of "brownie points" for the scenario in Figure 13.2. Again the "brownie points" experiment has a higher maximum group payoff.

In the former scenario, where there was one large first-stage group target and many smaller second-stage ones, decommitment penalties were unsuccessful because they tended to deter some agents from decommitting in the first place. (This was clear from watching the simulation). Agents considering committing in the first stage factored the expected loss due to decommitment penalty into their calculations. If the decommitment penalty was high and the agent was likely to get a better offer from the second-stage targets, it did not commit. The result was that

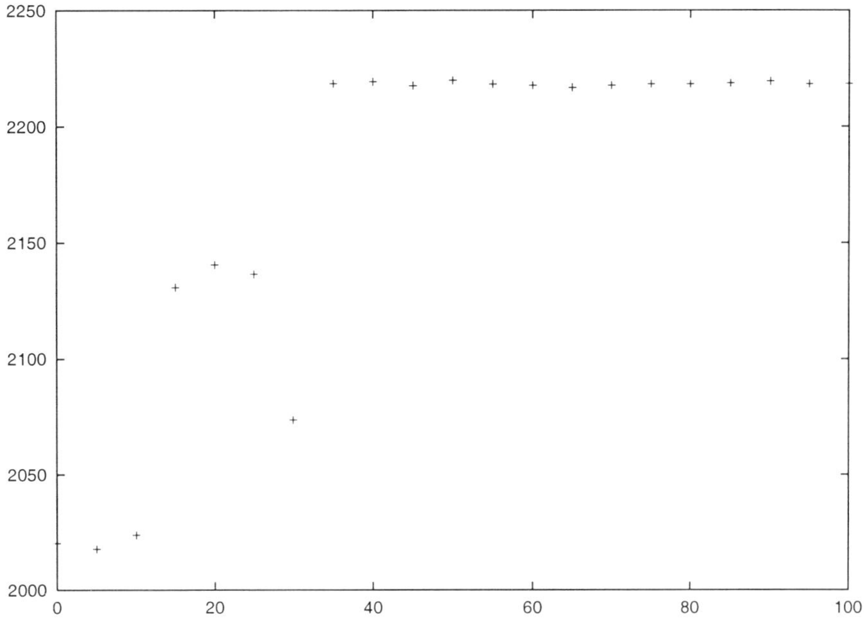

Figure 13.5. Average group income vs. decommitment penalty, for Figure 13.1 scenario.

the group target was rarely captured, so the average group payoff was not much higher than the payoff of an always-decommitting group. Losing "brownie points" however, was not a disincentive to commit in the first stage, just a disincentive to decommit in the second. Hence agents tended to commit to the group target in the first stage and then only sometimes decommit to chase the smaller targets. Their group payoff was consequently higher than an always-decommitting group's.

A similar effect occurred in the latter scenario, though the results were more pronounced because of the scenario itself. Again high decommitment penalties deterred earlier agents from committing in the first stage, so the final result was less lucrative than it was for the "brownie-point" oriented agents. In this case there was a large discrepancy because the usual final arrangement of the agents was different for the different cases — with decommitment penalties, agents tended to capture one first-stage and four second-stage targets (because of their reluctance to commit to first-stage targets initially), but using "brownie points" agents tended to capture three first-stage and one second-stage target, which was generally more lucrative.

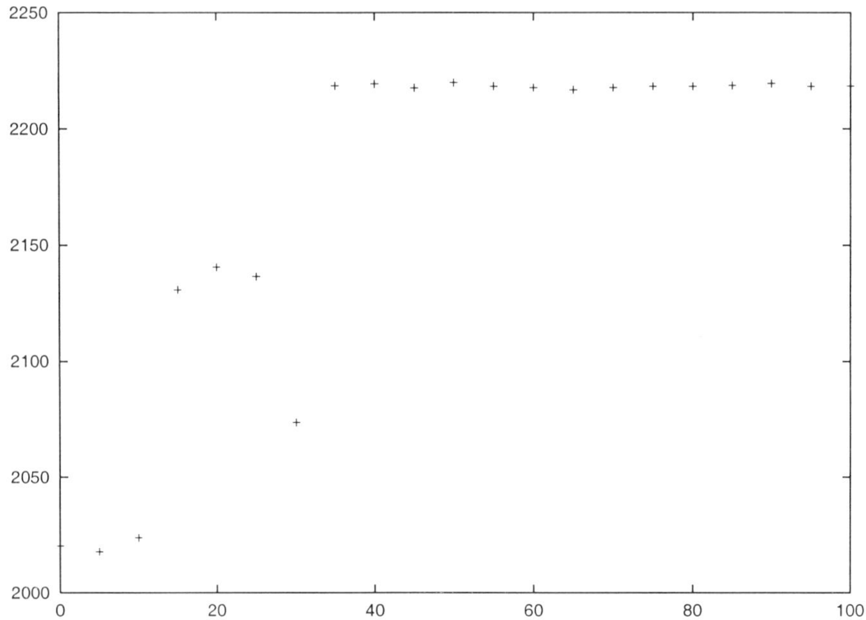

Figure 13.6. Average group income vs. "brownie point" weight, for Figure 13.2 scenario.

"Brownie points" have the benefits of decommitment penalties, since they gently discourage agents from decommitting, but they don't have the problem of acting as a deterrent to initial commitment.

4.4. EXPERIMENT 3: COMBINING "BROWNIE POINTS" AND DECOMMITMENT PENALTIES

So far in the experiments described, either there were no decommitment penalties or the weights attached to "brownie points" were always zero. "Brownie points" and decommitment penalties work in very similar ways, by discouraging decommitment just enough that agents decommit only when it is very lucrative. Therefore, intuitively we would expect that combining decommitment penalties and "brownie points" would result in little or no overall improvement, relative to using "brownie points" alone.

This is exactly the result shown in Figure 13.7, where group effectiveness is plotted against decommitment penalty, for several different

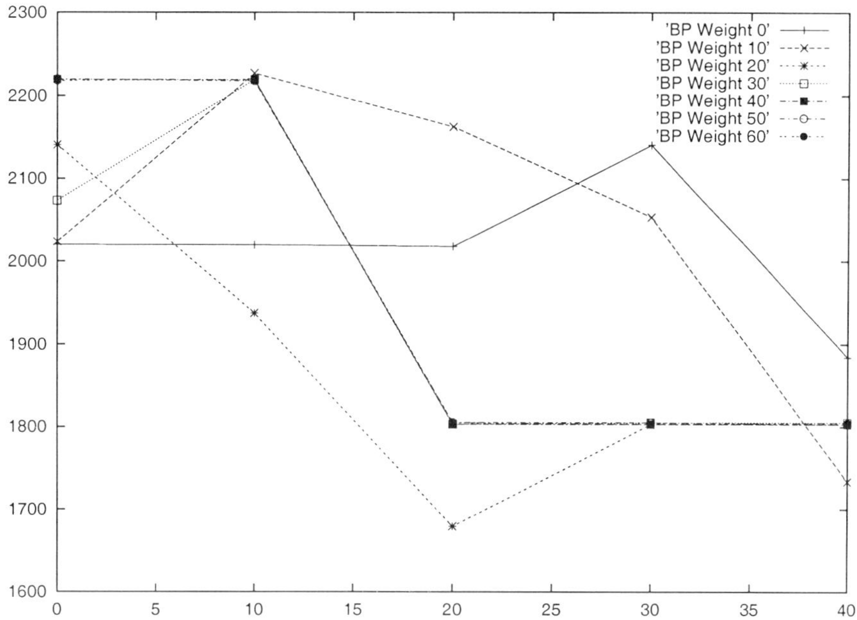

Figure 13.7. Average group income vs. decommitment penalty, for different "brownie point" weights.

"brownie-point" weights. The overall maximum (about 2227 for decommitment penalty 10 and "brownie-point" weight 10) is not significantly higher than the maximum for "brownie points" only (about 2220 for "brownie point" weight 50). As the agents' "brownie point" weight increases, the optimum decommitment penalty becomes smaller. This reflects the fact that decommitment penalties and "brownie points" combine to limit decommitting.

4.5. EXPERIMENT 4: WHEN DECOMMITMENT PENALTIES WORK BETTER THAN RELIABILITY

One problem with reliability is that it may take agents some time to form into teams. By contrast, decommitment penalties work immediately. Another issue is that agents using only reliability can sometimes form teams that are too strong, from which they will never decommit no matter how good the outside opportunities. Both of these effects could be seen from the agents' behaviour in this experiment — Figure 13.8

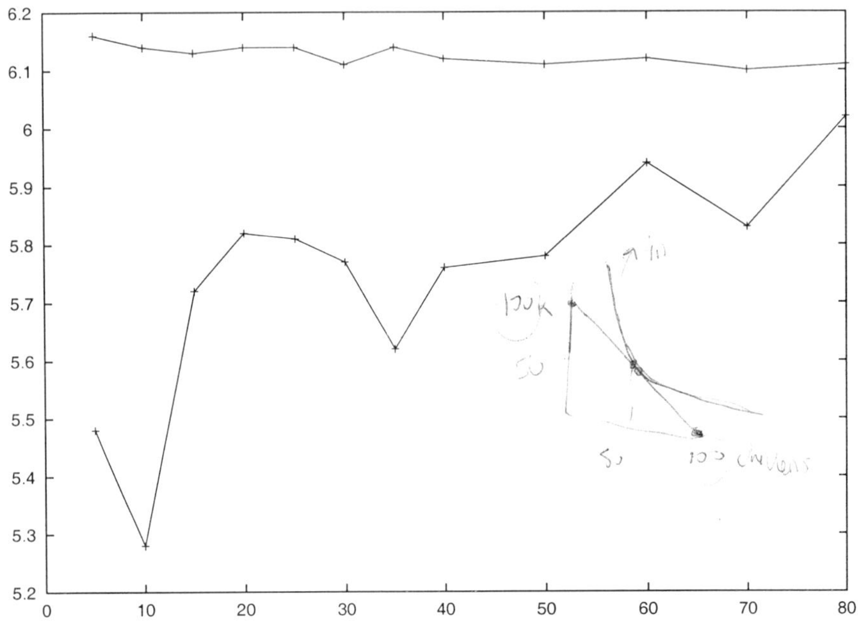

Figure 13.8. Average group income vs. number of runs, for agents using only decommitment penalties and agents using only reliability. Those using reliability improve more slowly and reach a lower maximum than those using decommitment penalties.

shows that agents using reliability were initially much less effective than those using decommitment penalties, and even when they formed into teams were on average slighly less effective.

The scenario consisted of one two-agent target in the first round, with each place offering 3 units, and one one-agent second-round target, offering between 0 and 7 units. Hence the optimum behaviour has expected group payoff $6\frac{1}{8}$, which is about what the decommitment-based group achieved.

5. CONCLUSIONS AND FURTHER WORK

The agents in this scenario displayed some quite complicated and interesting behaviours, including forming stable teams (section 4.2), obtaining optimal group payoff (sections 3.2 and 3.3) and behaving fairly to the less fortunate (section 3.1). The results of the experiments are consistent with other research when the scenarios coincided (Andersson and Sandholm, 1998; Andersson, 1998; Glass and Grosz, 1999), showing that a moderate "brownie-point" weight or small-to-moderate decommitment penalties produced optimal behaviour.

The most interesting aspect of this work is in further investigations of the team-forming that occurs when reliability is considered. The aims would be to discover a general description of when this happens, and investigate whether it is possible to make the resulting teams more efficient overall.

More generally it would be interesting to try to evaluate more systematically the performance of different methods under many different circumstances, and try to characterise methods that seem to suit particular types of scenario.

The interaction between methods is another direction for research. Intuitively, we would have expected the result in section 4.4 that combining decommitment penalties with consideration of an agent's "brownie points" results in little or no improvement, since the purpose of either method is simply to restrict agent decommitting to those occasions when it is very beneficial. However, adding reliability weights to another method could be very interesting, since reliability has a significantly different effect.

Another interesting question is whether the reliability count is unnecessarily complicated. An alternative approach would be that agents simply remember which agents had decommitted against them and then avoid them completely. This would make the computations significantly simpler, since it would effectively involve a boolean value for reliability. It also coincides more closely with human attitudes to defaulters, at least in some cultures and contexts (Kumar et al., 1998). It would be interesting to see whether this method was as effective as the more complicated version in causing teams to form.

This scenario could also be considered from the perspective of agent design. So far we have considered only homogeneous agents, but this is not very realistic. One interesting question is what kind of agent is best suited to interact with a given set of (possibly diverse) other agents. Suppose that the designer knows the distribution and types of other agents that will occur. What kind of agent (i.e. what weights for each criterion) should he or she choose? It would be possible to evaluate different types of agents in different contexts using the simulation.

There is still a lot of work to be done before these results can realistically be applied to building intelligent agents for e-commerce. However, the scenario does provide a flexible framework for investigating trust and reliability. Its results so far are consistent with intuition about how a simplified market should behave.

Acknowledgments

This work was completed while both authors were in the Department of Computer Science and Software Engineering at The University of Melbourne. A preliminary version of the work was reported at the Workshop on Artificial Intelligence and Electronic Commerce held in conjunction with the Twelfth Australian Joint Conference on Artificial Intelligence held in Sydney, December 1999.

Notes

1. The main idea is to assume that all the other agents who are not committed will choose targets at random, then to count the various ways in which this can be done. The assumption of random choice is the non-rigorous part, but it seems a reasonable model for behaviour that the agent can't predict

References

Andersson, M. (1998). Performance of leveled commitment protocols for automated negotiation: an empirical study. Master's thesis, Royal Institute of Technology, Stockholm.

Andersson, M. and Sandholm, T. (1998). Leveled commitment contracting among myopic individually rational agents. In *Proceedings of the Third International Conference on Multiagent Systems (ICMAS-98)*, pages 26–33, Paris, France.

Glass, A. and Grosz, B. (1999). Socially conscious decision-making. In *proceedings of the Bar-Ilan Symposium on Foundations of Artificial Intelligence (BISFAI-99)*.

Kendall, M. and Stuart, A. (1977). *The advanced theory of statistics, Volume 1: distribution theory*. Charles Griffin & Co. Ltd., London, 4 edition.

Kumar, K., van Dissel, H., and Bielli, P. (1998). The merchant of prato–revisited: Toward a third rationality of information systems. *MIS Quarterly*, pages 199–226.

Maes, P., Guttman, R., and Moukas, A. (1999). Agents that buy and sell: Transforming commerce as we know it. *Communications of the ACM*.

Nwana, H. et al. (1998). Agent-mediated electronic commerce: Issues, challenges and some viewpoints. In *Second International Conference on Autonomous Agents (AGENTS-98)*, pages 189–196. ACM press.

Sandholm, T. and Lesser, V. (1996). Advantages of a leveled commitment contracting protocol. In *Thirteenth National Conference on Artificial Intelligence (AAAI-96)*, pages 126–133, Portland, Oregon.

Sandholm, T., Sikka, S., and Norden, S. (1999). Algorithms for optimizing leveled commitment contracts. In *International Joint Conference on Artificial Intelligence (IJCAI)*, pages 535–540, Stockholm, Sweden.

Sandholm, T. and Zhou, Y. (1999). Revenue equivalence of leveled commitment contracts. In *AAAI Workshop on Negotiation: Settling Conflicts and Identifying Opportunities*, pages 38–43, Orlando, Florida. AAAI Workshop Technical Report WS-99-12.

Sen, S. (1996). Reciprocity: a foundational principle for promoting cooperative behavior among self-interested agents. In *Proceedings of the First International Conference on Multiagent Systems (ICMAS-96)*, pages 315–321. AAAI Press.

Sen, S. and Biswas, A. (1998). Effects of misconception on reciprocative agents. In *Proceedings of the Second International Conference on Autonomous Agents (AGENTS-98)*, pages 430–435. ACM Press.

Sullivan, D. et al. (1999). Intention reconciliation in the context of teamwork: an initial empirical investigation. In *Proceedings of the 10th International Conference and Workshop on Database and Expert Systems Applications (DEXA-99)*.

Chapter 14

PRICING IN AGENT ECONOMIES USING MULTI-AGENT Q-LEARNING

Gerald Tesauro
IBM Thomas J. Watson Research Center,
30 Saw Mill River Rd., Hawthorne, NY 10532 USA
tesauro@watson.ibm.com

Jeffrey O. Kephart
IBM Thomas J. Watson Research Center,
30 Saw Mill River Rd., Hawthorne, NY 10532 USA
kephart@watson.ibm.com

Abstract This paper investigates how adaptive software agents may utilize reinforcement learning algorithms such as Q-learning to make economic decisions such as setting prices in a competitive marketplace. For a single adaptive agent facing fixed-strategy opponents, ordinary Q-learning is guaranteed to find the optimal policy. However, for a population of agents each trying to adapt in the presence of other adaptive agents, the problem becomes non-stationary and history dependent, and it is not known whether any global convergence will be obtained, and if so, whether such solutions will be optimal. In this paper, we study simultaneous Q-learning by two competing seller agents in three moderately realistic economic models. This is the simplest case in which interesting multi-agent phenomena can occur, and the state space is small enough so that lookup tables can be used to represent the Q-functions. We find that, despite the lack of theoretical guarantees, simultaneous convergence to self-consistent optimal solutions is obtained in each model, at least for small values of the discount parameter. In some cases, exact or approximate convergence is also found even at large discount parameters. We show how the Q-derived policies increase profitability and damp out or eliminate cyclic price "wars" compared to simpler policies based on zero lookahead or short-term lookahead. In one of the models (the "Shopbot" model) where the sellers' profit functions are symmetric, we find that Q-learning can produce either symmetric or broken-symmetry policies, depending on the discount parameter and on initial conditions.

Keywords: Machine learning, reinforcement learning, adaptive multi-agent systems, agent economies

1. INTRODUCTION

Reinforcement Learning (RL) procedures have been established as powerful and practical methods for solving Markov Decision Problems. One of the most significant and actively investigated RL algorithms is Q-learning (Watkins, 1989). Q-learning is an algorithm for learning to estimate the long-term expected reward for a given state-action pair. It has the nice property that it does not need a model of the environment, and it can be used for on-line learning. Strong convergence of Q-learning to the exact optimal value functions and policies has been proven when lookup table representations of the Q-function are used (Watkins and Dayan, 1992); this is feasible in small state spaces. In large state spaces where lookup tables are infeasible, RL methods can be combined with function approximators to give good practical performance despite the lack of theoretical guarantees of convergence to optimal policies.

Most real-world problems are not fully Markov in nature – they are often non-stationary, history-dependent and/or not fully observable. In order for RL methods to be more generally useful in solving such problems, they need to be extended to handle these non-Markovian properties. One important application domain where the non-Markovian aspects are paramount is the area of multi-agent systems. This area is expected to be increasingly important in the future, due to the potential rapid emergence of "agent economies" consisting of large populations of interacting software agents engaged in various forms of economic activity. The problem of multiple agents simultaneously adapting is in general non-Markov, because each agent provides an effectively non-stationary environment for the other agents. Hence the existing convergence guarantees do not hold, and in general, it is not known whether any global convergence will be obtained, and if so, whether such solutions are optimal.

Some progress has been made in analyzing certain special case multi-agent problems. For example, the problem of "teams," where all agents share a common objective function, has been studied, for example, in (Crites and Barto, 1996). Likewise, the purely competitive case of zero-sum objective functions has been studied in (Littman, 1994), where an algorithm called "minimax-Q" was proposed for two-player zero-sum games, and shown to converge to the optimal value function and policies for both players. Sandholm and Crites studied simultaneous Q-learning by two players in the Iterated Prisoner's Dilemma game (Sandholm and Crites, 1995), and found that the learning procedure generally converged to stationary solutions. However, the extent to which those solutions were "optimal" was unclear. Recently, Hu and Wellman proposed an algorithm for calculating optimal Q-functions in two-player arbitrary-sum games (Hu and Wellman, 1998). This algorithm is an important first step. However, it does not yet appear to be useable for practical problems, because it assumes that policies followed by both players will be Nash equilibrium policies, and it does not address the "equilibrium coordination" problem, i.e. if there are multiple Nash equilibria, how do the agents decide which equilibrium to choose? We suspect that this may be a serious problem, since according to the "folk theorem of iterated games" (Kreps, 1990), there can be a proliferation of Nash equilibria when there is sufficiently high emphasis on future rewards, i.e., a large value of the discount parameter γ. Furthermore, there may be inconsistencies between the assumed Nash policies, and the policies implied by the Q-functions calculated by the algorithm.

In this paper, we study simultaneous Q-learning in an economically motivated two-player game. The players are assumed to be two sellers of similar or identical products, who compete against each other on the basis of price. At each time step, the sellers alternately take turns setting prices, taking into account the other seller's current price. After the price has been set, the consumers then respond instantaneously and deterministically, choosing either seller 1's product or seller 2's product (or no product), based on the current price pair (p_1, p_2), leading to an instantaneous reward or profit (R_1, R_2) given to sellers 1 and 2 respectively. We assume initially that the sellers have full knowledge of the expected consumer demand for any given price pair, and in fact have full knowledge of both profit functions.

Our work builds on prior research reported in (Tesauro and Kephart, 1998; Tesauro and Kephart, 1999). Those papers examined the effect of including foresight, i.e. an ability to anticipate longer-term consequences of an agent's current action. Two different algorithms for agent foresight were presented: (i) a generalization of the minimax search procedure in two-player zero-sum games; (ii) a generalization of the Policy Iteration method from dynamic programming, in which both players' policies are simultaneously improved, until self-consistent policy pairs are obtained that optimize expected reward over two time steps. It was found that including foresight in the agents' pricing algorithms generally improved overall agent profitability, and usually damped out or eliminated the pathological behavior of unending cyclic "price wars," in which long episodes of repeated undercutting amongst the sellers alternate with large jumps in price. Such price wars were found to be rampant in prior studies of agent economy models (Kephart, Hanson and Sairamesh, 1998; Sairamesh and Kephart, 1998) when the agents use "myopically optimal" or "myoptimal" pricing algorithms that optimize immediate reward, but do not anticipate the longer-term consequences of an agent's current price setting.

Our motivation for studying simultaneous Q-learning in this paper is threefold. First, if Q-functions can be learned simultaneously and self-consistently for both players, the policies implied by those Q-functions should be self-consistently optimal. In other words, an agent will be able to correctly anticipate the longer-term consequences of its own actions, the other agents' actions, and will correctly model the other agents as having an equivalent capability. Hence the classic problem of infinite recursion of opponent models will be avoided. In contrast, in other approaches to adaptive multi-agent system, these issues are more problematic. For example, (Vidal and Durfee, 1998) propose a recursive opponent modeling scheme, in which level-0 agents do no opponent modeling, level-1 agents model the opponents as being level-0, level-2 agents model the opponents as being level-1, etc.. In both of these approaches, there is no effective way for an agent to model other agents as being at an equivalent level of depth or complexity.

The second advantage of Q-learning is that the solutions should correspond to deep lookahead: in principle, the Q-function represents the expected reward looking infintely far ahead in time, exponentially weighted by a discount parameter $0 < \gamma < 1$. In contrast, the prior work of (Tesauro and Kephart, 1999) was based on shallow finite lookahead. Finally, in comparison to directly modeling agent policies, the Q-function approach seems more extensible to the situation of very large economies with many competing sellers. Our intuition is that approximating Q-functions with nonlinear function approximators such as neural networks is more feasible than approximating the corresponding policies. Furthermore, in the Q-function approach, each agent only needs to maintain a single Q-function for itself, whereas in the policy modeling approach, each agent needs to maintain a policy model for every other agent; the latter seems infeasible when the number of sellers is large.

The remainder of this paper is organized as follows. Section 2 describes the structure and dynamics of our model two-seller economy, and presents three economically-based models of seller profit (Price-Quality, Information-Filtering, and Shopbot) which are known to be prone to price wars when agents myopically optimize their short-term payoffs.

We deliberately choose parameters to place each of these systems in a price-war regime. In section 3, we describe details of how we implement Q-learning in these model economies. As a first step, we examine the simple case of ordinary Q-learning, where one of the two sellers uses Q-learning and the other seller uses a fixed pricing policy (the myopically optimal, or "myoptimal" policy). We then examine the more interesting and novel situation of simultaneous Q-learning by both sellers. Finally, section 5 summarizes the main conclusions and discusses promising directions and challenges for future work.

2. MODEL AGENT ECONOMIES

Real agent economies are likely to contain large numbers of agents, with complex details of how the agents behave and interact with each other on multiple time scales. Our approach toward modeling and understanding such complexity is to begin by making a number of simplifying assumptions. We first consider the simplest possible case of two competing seller agents offering similar or identical products to a large population of consumer agents. The sellers compete on the basis of price, and we assume that prices are discretized and can lie between a minimum and maximum price, such that the number of possible prices is at most a few hundred. This renders the state space small enough that it is feasible to use lookup tables to represent the agents' pricing policies and expected profits. Time in the simulation is also discretized; at each time step, we assume that the consumers compare the current prices of the two sellers, and instantaneously and deterministically choose to purchase from at most one seller. Hence at each time step, for each possible pair of seller prices, there is a deterministic reward or profit given to each seller. The simulation can iterate forever, and there may or may not be a discounting factor for the present value of future rewards.

It is worth noting that the consumers are not regarded as "players" in the model. The consumers have no strategic role: they behave according to an extremely simple, fixed, short-term greedy rule (buy the lowest priced product at each time step), and are regarded as merely providing a stationary environment in which the two sellers can compete in a two-player game. This is clearly a simplifying first step in the study of multi-agent phenomena, and in future work, the models will be extended to include strategic and adaptive behavior on the part of the consumers as well. This will change the notion of "desirable" system behavior. In the present model, desirable behavior would resemble "collusion" between the two sellers in charging very high prices, so that both could obtain high profits. Obviously this is not desirable from the consumers' viewpoint.

Regarding the dynamics of seller price adjustments, we assume that the sellers alternately take turns adjusting their prices, rather than simultaneously setting prices (i.e. the game is extensive-form rather than normal-form). Our choice of alternating-turn dynamics is motivated by two considerations: (a) As the number of sellers becomes large and the model becomes more realistic, it seems more reasonable to assume that the sellers will adjust their prices at different times rather than at the same time, although they probably will not take turns in a well-defined order. (b) With alternating-turn dynamics, we can stay within the normal Q-learning framework where the Q-function implies a deterministic optimal policy: it is known that in two-player alternating turn games, there always exists a deterministic policy that is as good as any non-deterministic policy (Littman, 1994). In contrast, in games with simultaneous moves (for example, rock-paper-scissors), it is possible that no deterministic policy is optimal, and that the existing Q-learning formalism for MDPs would have to be modified and extended so that it could yield non-deterministic optimal policies.

We study Q-learning in three different economic models that have been described in detail elsewhere (Sairamesh and Kephart, 1998; Kephart, Hanson and Sairamesh, 1998; Greenwald and Kephart, 1999). The first model, called the "Price-Quality" model (Sairamesh and Kephart, 1998), models the sellers' products as being distinguished by different values of a scalar "quality" parameter, with higher-quality products being perceived as more valuable by the consumers. The consumers are modeled as trying to obtain the lowest-priced product at each time step, subject to threshold-type constraints on both quality and price, i.e., each consumer has a maximum allowable price and a minimum allowable quality. The similarity and substitutability of seller products leads to a potential for direct price competition; however, the "vertical" differentiation due to differening quality values leads to an asymmetry in the sellers' profit functions. It is believed that this asymmetry is responsible for the unending cyclic price wars that emerge when the sellers employ myoptimal pricing strategies.

The second model is an "Information-Filtering" model described in detail in (Kephart, Hanson and Sairamesh, 1998). In this model there are two competing sellers of news articles in somewhat overlapping categories. In contrast to the vertical differentiation of the Price-Quality model, this model contains a horizontal differentiation in the differing article categories. To the extent that the categories overlap, there can be direct price competition, and to the extent that they differ, there are asymmetries introduced that again lead to the potential for cyclic price wars.

The third model is the so-called "Shopbot" model described in (Greenwald and Kephart, 1999), which is intended to model the situation on the Internet in which some consumers may use a Shopbot to compare prices of all sellers offering a given product, and select the seller with the lowest price. In this model, the sellers' products are exactly identical and the profit functions are symmetric. Myoptimal pricing leads the sellers to undercut each other until the minimum price point is reached. At that point, a new price war cycle can be launched, due to buyer asymmetries rather than seller asymmetries. The fact that not all buyers use the Shopbot, and some buyers instead choose a seller at random, means that it can be profitable for a seller to abandon the low-price competition for the bargain hunters, and instead maximally exploit the random buyers by charging the maximum possible price.

An example profit function that we study, taken from the Price-Quality model, is as follows: Let p_1 and p_2 represent the prices charged by seller 1 and seller 2 respectively. Let q_1 and q_2 represent their respective quality parameters, with $q_1 > q_2$. Let $c(q)$ represent the cost to a seller of producing an item of quality q. Then assuming the particular model of consumer behavior described in (Sairamesh and Kephart, 1998) one can show analytically that in the limit of infinitely many consumers, the instantaneous profits per consumer R_1 and R_2 obtained by seller 1 and seller 2 respectively are given by:

$$R_1 = \begin{cases} (q_1 - p_1)(p_1 - c(q_1)) & \text{if } 0 \leq p_1 \leq p_2 \text{ or } p_1 > q_2 \\ (q_1 - q_2)(p_1 - c(q_1)) & \text{if } p_2 < p_1 < q_2 \end{cases} \tag{1}$$

$$R_2 = \begin{cases} (q_2 - p_2)(p_2 - c(q_2)) & \text{if } 0 \leq p_2 < p_1 \\ 0 & \text{if } p_2 \geq p_1 \end{cases} \tag{2}$$

A plot of the profit landscape for seller 1 as a function of prices p_1 and p_2 is given in figure 14.1, for the following parameter settings: $q_1 = 1.0$, $q_2 = 0.9$, and $c(q) = 0.1(1 + q)$. (These specific parameter settings were chosen because they are known to generate harmful price wars when the agents use myopic optimal pricing.) We can see in this figure that the myopic optimal price for seller 1 as a function of seller 2's price, $p_1^*(p_2)$, is obtained for each value of p_2 by sweeping across all values of p_1 and choosing the value that gives the highest profit. We can see that for small values of p_2, the peak profit is obtained at $p_1 = 0.9$, whereas for larger values of p_2, there is eventually a discontinuous shift to the other peak, which follows along the parabolic-shaped ridge in the landscape.

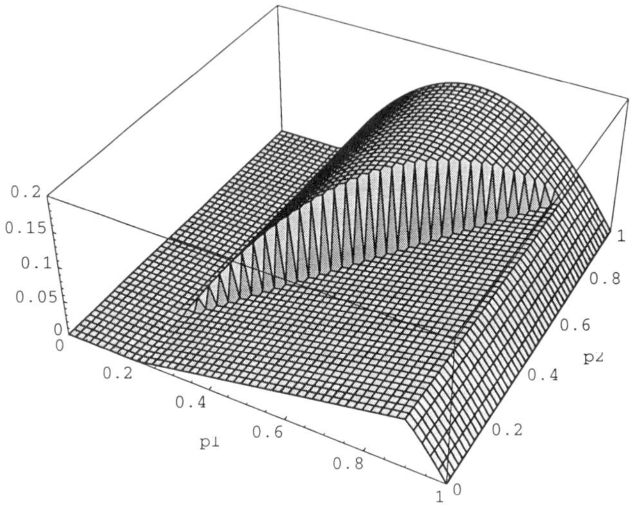

Figure 14.1. Sample profit landscape for seller 1 in Price-Quality model, as a function of seller 1 price p_1 and seller 2 price p_2.

An analytic expression for the myopic optimal price for seller 1 as a function of p_2 is as follows (defining $x_1 = q_1 + c(q_1)$ and $x_2 = q_2 + c(q_2)$):

$$p_1^*(p_2) = \begin{cases} q_2 & \text{if } 0 \leq p_2 < x_1 - q_2 \\ p_2 & \text{if } x_1 - q_2 \leq p_2 \leq \frac{1}{2}x_1 \\ \frac{1}{2}x_1 & \text{if } p_2 > \frac{1}{2}x_1 \end{cases} \tag{3}$$

Similarly, the myopic optimal price for seller 2 as a function of the price set by seller 1, $p_2^*(p_1)$, is given by the following formula (assuming that prices are discrete and that ϵ is the price discretization interval):

$$p_2^*(p_1) = \begin{cases} c(q_2) & 0 \leq p_1 \leq c(q_2) \\ p_1 - \epsilon & \text{if } c(q_2) \leq p_1 \leq \frac{1}{2}x_2 \\ \frac{1}{2}x_2 & \text{if } p_1 > \frac{1}{2}x_2 \end{cases} \tag{4}$$

We also note in passing that there are similar profit landscapes for each of the sellers in the Information-Filtering model and in the Shopbot model. In all three models, it is the existence of multiple, disconnected peaks in the landscapes, with relative heights that can change depending on the other seller's price, that leads to price wars when the sellers behave myopically.

Regarding the information set that is made available to the sellers, we have made a simplifying assumption as a first step that the players have essentially perfect information. They can model the consumer behavior perfectly, and they also have perfect knowledge of each other's costs and profit functions. Hence our model is thus a two-player perfect-information deterministic game that is very similar to games like chess. The main differences are that the profits in our model are not strictly zero-sum, and that there are no terminating or absorbing nodes in our model's state space. Also in our model, payoffs are given to the players at every time step, whereas in games such as chess, payoffs are only given at the terminating nodes.

As mentioned previously, we constrain the prices set by the two sellers to lie in a range from some minimum to maximum allowable price. The prices are discretized, so that one can create lookup tables for the seller profit functions $U(p_1, p_2)$. Furthermore, the optimal pricing policies for each seller as a function of the other seller's price, $p_1^*(p_2)$ and $p_2^*(p_1)$, can also be represented in the form of table lookups.

3. SINGLE-AGENT Q-LEARNING

We first consider ordinary single-agent Q-learning in the above two-seller economic models. The procedure for Q learning is as follows. Let $Q(s,a)$ represent the discounted long-term expected reward to an agent for taking action a in state s. The discounting of future rewards is accomplished by a discount parameter γ such that the value of a reward expected at n time steps in the future is discounted by γ^n. Assume that the $Q(s,a)$ function is represented by a lookup table containing a value for every possible state-action pair, and assume that the table entries are initialized to arbitrary values. Then the procedure for solving for $Q(s,a)$ is to infinitely repeat the following two-step loop:

1. Select a particular state s and a particular action a, observe the immediate reward r for this state-action pair, and observe the resulting state s'.

2. Adjust $Q(s,a)$ according to the following equation:

$$\Delta Q(s,a) = \alpha[r + \gamma \max_b Q(s',b) - Q(s,a)] \tag{5}$$

where α is the learning rate parameter, and the max operation represents choosing the optimal action b among all possible actions that can be taken in the successor state s' leading to the greatest Q-value. A wide variety of methods may be used to select state-action pairs in step 1, provided that every state-action pair is visited infinitely often. For any stationary Markov Decision Problem, the Q-learning procedure is guaranteed to converge to the correct values, provided that α is decreased over time with an appropriate schedule.

We first consider using Q-learning for one of the two sellers in our economic models, while the other seller maintains a fixed pricing policy. In the simulations described below the fixed policy is in fact the myoptimal policy $p*$ represented for example in the Price-Quality model by equations 3 and 4.

In our pricing application, the distinction between states and actions is somewhat blurred. We will assume that the "state" for each seller is sufficiently described by the other seller's last price, and that the "action" is the current price decision. This should be a sufficient state description because no other history is needed either for the determination of immediate reward, or for the calculation of the myoptimal price by the fixed-strategy player. We have also modified the concepts of immediate reward r and next-state s' for the two-agent case. We define s' as the state that is obtained, starting from s, of one action by the Q-learner and a response action by the fixed-strategy opponent. Likewise, the immediate reward is defined as the sum of the two rewards obtained after those two actions. These modifications were introduced so that the state s' would have the same player to move as state s. (A possible alternative to this, which we have not investigated, is to include the side-to-move as additional information in the state-space description.)

In the simulations reported below, the sequence of state-action pairs selected for the Q-table updates were generated by uniform random selection from amongst all possible table entries. The initial values of the Q-tables were generally set to the immediate reward values. (Consequently the initial Q-derived policies corresponded to myoptimal policies.) The learning rate was varied with time according to:

$$\alpha(t) = \alpha(0)/(1 + \beta t) \tag{6}$$

where the initial learning rate $\alpha(0)$ was usually set to 0.1, and the constant $\beta \sim 0.01$ when the simulation time t was measured in units of N^2, the size of the Q-table. (N is the number of possible prices that could be selected by either player.) A number of different values of the discount parameter γ were studied, ranging from $\gamma = 0$ to $\gamma = 0.9$.

Results for single-agent Q-learning in all three models indicated that Q-learning worked well (as expected) in each case. In each model, for each value of the discount parameter, exact convergence of the Q-table to a stationary optimal solution was found. The convergence times ranged from a few hundred sweeps through each table element, for smaller values of γ, to at most a few thousand updates for the largest values of γ. In addition, once Q-learning converged, we then measured the expected cumulative profit of the policy derived from the Q-function. We ran the Q-policy against the other player's myopic policy from 100 random starting states, each for 200 time steps, and averaged the resulting cumulative profit for each player. We found that, in each case, the seller achieved greater profit against a myopic opponent by using a Q-derived policy than by using a myopic policy. (This was true even for $\gamma = 0$, because, due to the redefinition of Q updates summing over two time steps, the case $\gamma = 0$ effectively corresponds to a two-step optimization, rather than the one-step optimization of the myopic policies.) Furthermore, the cumulative profit obtained with the Q-derived policy monotonically increased with the increasing γ (as expected).

It was also interesting to note that in many cases, the expected profit of the myopic opponent also increased when playing against the Q-learner, and also improved monotonically with increasing γ. The explanation is that, rather than better exploiting the myopic opponent, as would be expected in a zero-sum game, the Q-learner instead reduced the region over which it would participate in a mutually undercutting price war. Typically we find in these models that with myopic vs. myopic play, large-amplitude price wars are generated that start at very high prices and persist all the way down to very low prices. When a Q-learner competes against a myopic opponent, there are still price wars starting at high prices, however, the Q-learner abandons the price war more quickly as the prices decrease. The effect is that the price-war regime is smaller and confined to higher average prices, leading to a closer approximation to cooperative or collusive behavior, with greater expected utilites for both players.

An illustrative example of the results of single-agent Q-learning is shown in figure 14.2. Figure 14.2(a) plots the average profit for both sellers in the Shopbot model, when one of the sellers is myopic and the other is a Q-learner. (As the model is symmetric, it doesn't matter which seller is the Q-learner.) Figure 14.2(b) plots the myopic price curve of seller 2 against the Q-derived price curve (at $\gamma = 0.5$) of seller 1. We can see that both curves have a maximum price of 1 and a minimum price of approximately 0.58. The portion of both curves lying along the diagonal indicates undercutting behavior, in which case the seller will respond to the opponent's price by undercutting by ϵ, the price discretization interval.

The system dynamics for the state (p_1, p_2) in figure 14.2(b) can be obtained by alternately applying the two pricing policies. This can be done by a simple iterative graphical construction, in which for any given starting point, one first holds p_2 constant and moves horizontally to the $p_1(p_2)$ curve, and then one holds p_1 constant and moves vertically

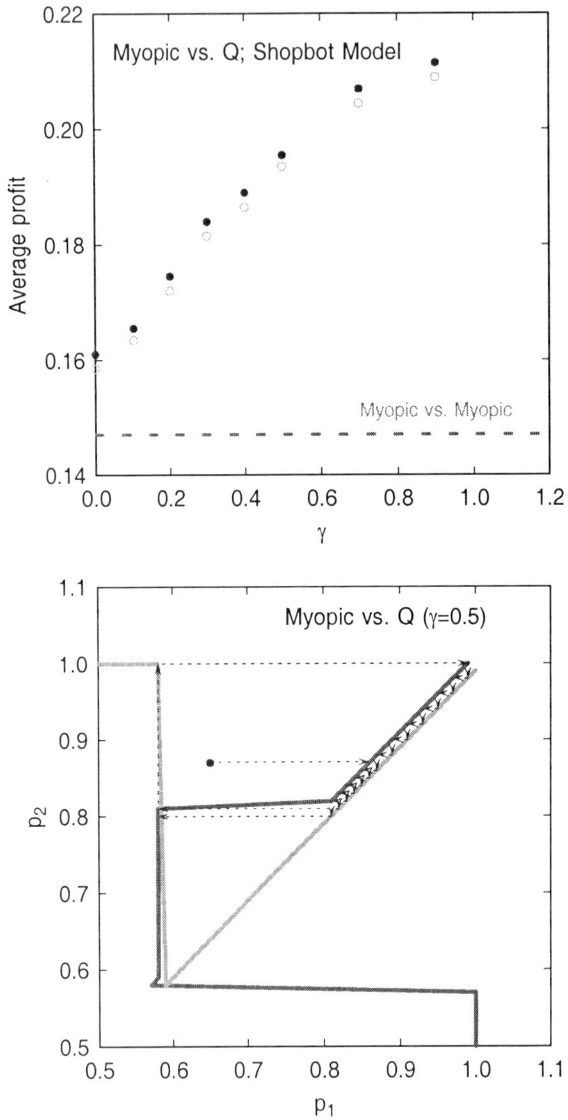

Figure 14.2. Results of single-agent Q-learning in the Shopbot model. (a) Average profit per time step for Q-learner (seller 1, filled circles) and myopic seller (seller 2, open circles) vs. discount parameter γ. Dashed line indicates baseline expected profit when both sellers are myopic. (b) Cross-plot of Q-derived price curve (seller 1) vs. myopic price curve (seller 2) at $\gamma = 0.5$. Dashed line and arrows indicate a temporal price-pair trajectory using these policies, starting from filled circle.

to the $p_2(p_1)$ curve. We see in this figure that the iterative graphical construction leads to an unending cyclic price war, whose trajectory is indicated by the dashed line. Note that the price-war behavior begins at the price pair (1, 1), and persists until a price of approximately 0.83. At this point, seller 1 abandons the price war, and resets its price to 1, leading once again to another round of undercutting.

The amplitude of this price war is diminished compared to the situation in which both players use a myopic policy. In that case, seller 1's curve would be a mirror image of seller 2's curve, and the price war would persist all the way to the minimum price point, leading to a lower expected profit for both sellers.

4. MULTI-AGENT Q-LEARNING

We now examine the more interesting and challenging case of simultaneous training of Q-functions and policies for both sellers. Our approach is to use the same formalism presented in the previous section, and to alternately adjust a random entry in seller 1's Q-function, followed by a random entry in seller 2's Q-function. As each seller's Q-function evolves, the seller's pricing policy is correspondingly updated so that it optimizes the agent's current Q-function. In modeling the two-step payoff r to a seller in equation 5, we use the opponent's current policy as implied by its current Q-function. The parameters in the experiments below were generally set to the same values as in the previous section. In most of the experiments, the Q-functions were initialized to the instantaneous payoff values (so that the policies corresponded to myopic policies), although other initial conditions were explored in a few experiments.

For simultaneous Q-learning in the Price-Quality model, we find robust convergence to a unique pair of pricing policies, independent of the value of γ, as illustrated in figure 14.3(b). This solution also corresponds to the solution found by generalized minimax and by generalized DP in (Tesauro and Kephart, 1999). We note that repeated application of this pair of price curves leads to a dynamical trajectory that eventually converges to a fixed-point located at $(p_1 = 0.9, p_2 = 0.4)$. A detailed analysis of these pricing policies and the fixed-point solution is presented in (Tesauro and Kephart, 1999). In brief, for sufficiently low prices of seller 2, it pays seller 1 to abandon the price war and to charge a very high price, $p_1 = 0.9$. The value of $p_2 = 0.4$ then corresponds to the highest price that seller 2 can charge without provoking an undercut by seller 1, based on a two-step lookahead calculation (seller 1 undercuts, and then seller 2 replies with a further undercut). We note that this fixed point does not correspond to a Nash equilibrium, since both players have

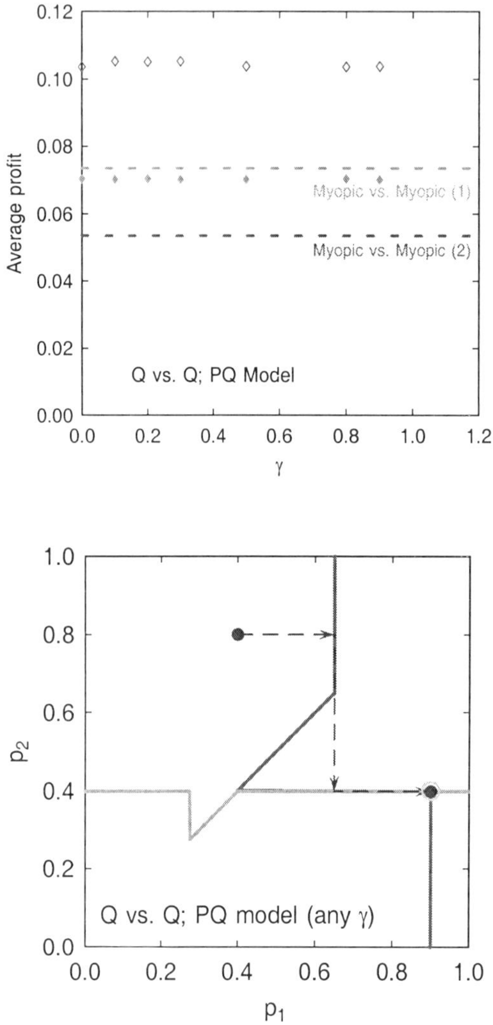

Figure 14.3. Results of simultaneous Q-learning in the Price-Quality model. (a) Average profit per time step for seller 1 (solid diamonds) and seller 2 (open diamonds) vs. discount parameter γ. Dashed line indicates baseline myopic vs. myopic expected profit. Note that seller 2's profit is higher than seller 1's, even though seller 2 has a lower quality parameter. (b) Cross-plot of Q-derived price curves (at any γ). Dashed line and arrows indicate a sample price dynamics trajectory, starting from the filled circle. The price war is eliminated and the dynamics evolves to a fixed point indicated by an open circle.

an incentive to deviate, based on a one-step lookahead calculation. It was conjectured in (Tesauro and Kephart, 1999) that the solution observed in figure 14.3(b) corresponds to a subgame-perfect equilibrium (Fudenberg and Tirole, 1991) rather than a Nash equilibrium.

The cumulative profits obtained by the pair of pricing policies are plotted in figure 14.3(a). It is interesting that seller 2, the lower-quality seller, actually obtains a significantly higher profit than seller 1, the higher-quality seller. In contrast, with myopic vs. myopic pricing, seller 2 does worse than seller 1.

In the Shopbot model, we did not find exact convergence of the Q-functions for each value of γ. However, in those cases where exact convergence was not found, we did find very good approximate convergence, in which the Q-functions and policies converged to stationary solutions to within small random fluctuations. Different solutions were obtained at each value of γ. We generally find that a symmetric solution, in which the shapes of $p_1(p_2)$ and $p_2(p_1)$ are identical, is obtained at small γ, whereas a broken symmetry solution, similar to the Price-Quality solution, is obtained at large γ. We also found a range of γ values, between 0.1 and 0.2, where either a symmetric or asymmetric solution could be obtained, depending on initial conditions. The asymmetric solution was counter-intuitive to us, because we expected that the symmetry of the two sellers' profit functions would lead to a symmetric solution. In hindsight, we can apply the same type of reasoning as in the Price-Quality model to explain the asymmetric solution. A plot of the expected profit for both sellers as a function of γ is shown in figure 14.4(a). Plots of the symmetric and asymmetric solution, obtained at $\gamma = 0$ and $\gamma = 0.9$ respectively, are shown in figures 14.4(b) and 14.4(c).

Finally, in the Information-Filtering model, we found that simultaneous Q-learning produced exact or good approximate convergence for small values of γ ($0 \leq \gamma \leq 0.5$). For large values of γ, no convergence was obtained. The simultaneous Q-learning solutions yielded reduced-amplitude price wars, and montonically increasing profitability for both sellers as a function of γ, at least up to $\gamma = 0.5$. A few data points were examined at $\gamma > 0.5$, and even though there was no convergence, the Q-policies still yielded greater profit for both sellers than in the myopic vs. myopic case. A plot of the Q-derived policies and system dynamics for $\gamma = 0.5$ is shown in figure 14.5(b). The expected profits for both players as a function of γ is plotted in figure 14.5(a).

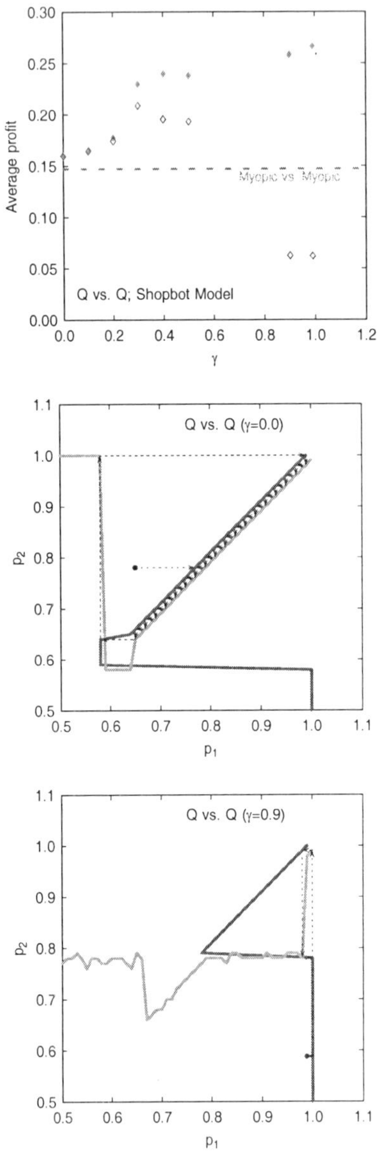

Figure 14.4. Results of simultaneous Q-learning in the Shopbot model. (a) Average profit per time step for seller 1 (solid diamonds) and seller 2 (open diamonds) vs. discount parameter γ. Dashed line indicates baseline myopic vs. myopic expected profit. (b) Cross-plot of Q-derived price curves at $\gamma = 0$; the solution is symmetric. Dashed line and arrows indicate a sample price dynamics trajectory. (c) Cross-plot of Q-derived price curves at $\gamma = 0.9$; the solution is asymmetric.

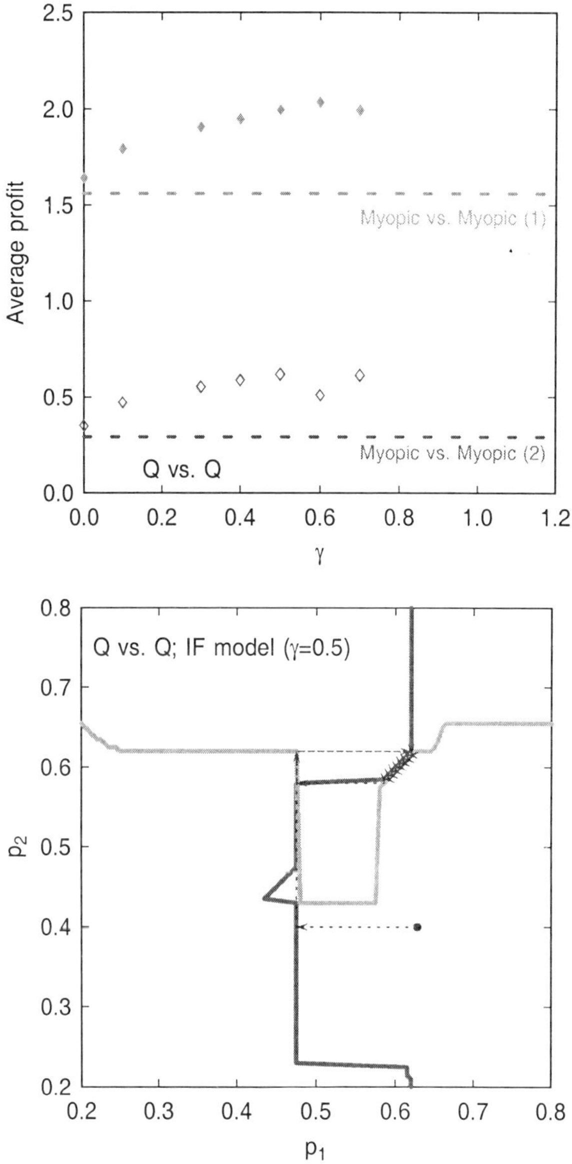

Figure 14.5. Results of multi-agent Q-learning in the Information-Filtering model. (a) Average profit per time step for seller 1 (solid diamonds) and seller 2 (open diamonds) vs. discount parameter γ. (The data points at $\gamma = 0.6, 0.7$ represent unconverged Q-functions and policies.) Dashed lines indicates baseline expected profit when both sellers are myopic. (b) Cross-plot of Q-derived price curves at $\gamma = 0.5$.

5. CONCLUSIONS

We have examined single-agent and multi-agent Q-learning in three models of a two-seller economy in which the sellers alternately take turns setting prices, and then instantaneous profits are given to both sellers based on the current price pair. Such models fall into the category of two-player, alternating-turn, arbitrary-sum Markov games, in which both the rewards and the state-space transitions are deterministic. The game is Markov because the state space is fully observable and the rewards are not history dependent.

In all three models (Price-Quality, Information-Filtering, and Shopbot), large-amplitude cyclic price wars are obtained when the sellers myopically optimize their instantaneous profits without regard to longer-term impact of their pricing policies. We find that, in all three models, the use of Q-learning by one of the sellers against a myopic opponent invariably results in exact convergence to the optimal Q-function and optimal policy against that opponent, for all allowed values of the discount parameter γ. The use of the Q-derived policy yields greater expected profit for the Q-learner, with monotonically increasing profit as γ increases. In many cases, it has a side benefit of also enhancing the welfare of the myopic opponent. This comes about by reducing the amplitude of the undercutting price-war regime, or in some cases, eliminating it completely.

We have also studied the more interesting and challenging situation of simultaneously training Q-functions for both sellers. This is more difficult because as each seller's Q-function and policy change, it provides a non-stationary environment for adaptation of the other seller. No convergence proofs exist for such simultaneous Q-learning by multiple agents. Nevertheless, despite the absence of theoretical guarantees, we do find generally good behavior of the algorithm in our model economies. In two of the models (Shopbot and Price-Quality), we find exact or very good approximate convergence to simultaneously self-consistent Q-functions and optimal policies for any value of γ, whereas in the Information-Filtering model, simultaneous convergence was found for $\gamma \leq 0.5$. In the Information-Filtering and Shopbot models, monotonically increasing expected profits for both sellers were also found for small values of γ. In the Price-Quality model, simultaneous Q-learning yields an asymmetric solution, corresponding to the solution found in (Tesauro and Kephart, 1999), that is highly advantageous to the lesser-quality seller, but slightly disadvantageous to the higher-quality seller, when compared to myopic vs. myopic pricing. A similar asymmetric solution is also found in the Shopbot model for large γ, even though the profit functions for both players are symmetric.

For each model, there exists a range of discount parameter values where the solutions obtained by simultaneous Q-learning are self-consistently optimal, and outperform the solutions obtained in (Tesauro and Kephart, 1999). This is presumably because the previously published methods were based on limited lookahead, whereas the Q-functions in principle look ahead infinitely far, with appropriate discounting.

It is intruiging that simultaneous Q-learning works well in our models, despite the lack of theoretical convergence proofs. Sandholm and Crites also found that simultaneous Q-learning generally converged in the Iterated Prisoner's Dilemma game. These empirical findings suggest that a deeper theoretical analysis of simultaneous Q-learning may be worth investigating. There may be some underlying theoretical principles that can explain why simultaneous Q-learning works, for at least certain classes of arbitrary-sum profit functions.

Several important challenges will also be faced in extending our approach to larger-scale, more realistic simulations. While there are some economic situations in the real world where there are only two dominant sellers, in general the number of sellers can be much greater. The situation that we foresee in agent economies is that the number of competing sellers will be very large. In this case, the seller profits and pricing functions will have such high input dimensionality that it will be infeasible to use lookup table state-space representations, and most likely some sort of compact representation combined with a function approximation scheme will be necessary. Furthermore, with many sellers, the concept of sellers taking turns adjusting their prices in a well-defined order becomes problematic. This could lead to an additional combinatorial explosion, if the mechanism for calculating expected reward has to anticipate all possible orderings of opponent responses.

Furthermore, while our economic models have a moderate degree of realism in their profit functions, they are unrealistic in the assumptions of knowledge and dynamics. In the work reported here, the state space was fully observable infinitely frequently at zero cost and with zero propagation delays. The expected consumer demand for a given price pair was instantaneous, deterministic and fully known to both players. Indeed, the players' exact profit functions were fully known to both players. It was also assumed that the players would alternately take turns equally often in a well-defined order in adjusting their prices. Under such assumptions of knowledge and dynamics, one could hope to develop an algorithm that could calculate in advance something like a game-theoretic optimal pricing algorithm for each agent.

However, in realistic agent economies, it is likely that agents will have much less than full knowledge of the state of the economy. Agents may not know the details of other agents' profit functions, and indeed an agent may not know its own profit function, to the extent that buyer behavior is unpredictable. The dynamics of buyers and sellers may also be more complex, random and unpredictable than what we have assumed here. There may also be information delays for both buyers and sellers, and part of the economic game may involve paying a cost in order to obtain information about the state of the economy faster and more frequently, and in greater detail. Finally, we expect that buyer behavior will be non-stationary, so that there will be a more complex co-evolution of buyer and seller strategies.

While such real-world complexities are daunting, there are reasons to believe that learning approaches such as Q-learning may play a role in practical solutions. The advantage of Q-learning is that one does not need a model of either the instantaneous payoffs or of the state-space transitions in the environment. One can simply observe actual rewards and transitions and base learning on that. While the theory of Q-learning requires exhaustive exploration of the state space to guarantee convergence, this may not be necessary when function approximators are used. In that case, after training a function approximator on a relatively small number of observed states, it may then generalize well enough on the unobserved states to give decent practical performance. Several recent empirical studies have provided evidence of this (Tesauro, 1995; Crites and Barto, 1996; Zhang and Dietterich, 1996).

Acknowledgements

The authors thank Amy Greenwald for helpful discussions regarding the Shopbot model.

References

R. H. Crites and A. G. Barto, "Improving elevator performance using reinforcement learning." In: D. Touretzky et al., eds., Advances in Neural Information Processing Systems 8, 1017-1023, MIT Press, 1996.

D. Fudenberg and J. Tirole, *Game Theory.* Cambridge, MA: MIT Press, 1991.

A. Greenwald and J. O. Kephart, "Shopbots and pricebots." To appear in: Proceedings of IJCAI-99, 1999.

J. Hu and M. P. Wellman, "Multiagent reinforcement learning: theoretical framework and an algorithm." Proceedings of ICML-98, 1998.

J. O. Kephart, J. E. Hanson and J. Sairamesh, "Price-war dynamics in a free-market economy of software agents." In: Proceedings of ALIFE-VI, Los Angeles, 1998.

D. Kreps, *A Course in Microeconomic Theory.* Princeton Univ. Press, Princeton, NJ, 1990.

M. L. Littman, "Markov games as a framework for multi-agent reinforcement learning," Proceedings of the Eleventh International Conference on Machine Learning, 157–163, Morgan Kaufmann, 1994.

J. Sairamesh and J. O. Kephart, "Dynamics of price and quality differentiation in information and computational markets." Proceedings of the First International Conference on Information and Computation Economics (ICE-98), 28–36, ACM Press, 1998.

T. W. Sandholm and R. H. Crites, "On multiagent Q-Learning in a semi-competitive domain." 14th International Joint Conference on Artificial Intelligence (IJCAI-95), Workshop on Adaptation and Learning in Multiagent Systems, Montreal, Canada, 71–77, 1995.

G. Tesauro, "Temporal difference learning and TD-Gammon." *Comm. of the ACM*, **38:3**, 58-67, 1995.

G. J. Tesauro and J. O. Kephart, "Foresight-based pricing algorithms in an economy of software agents." Proceedings of the First International Conference on Information and Computation Economics (ICE-98), 37–44, ACM Press, 1998.

G. J. Tesauro and J. O. Kephart, "Foresight-based pricing algorithms in agent economies." *Decision Support Sciences*, to appear, 1999.

J. M. Vidal and E. H. Durfee, "Learning nested agent models in an information economy," J. of Experimental and Theoretical AI, to appear, 1998.

C. J. C. H. Watkins, "Learning from delayed rewards." Ph. D. thesis, Cambridge University, 1989.

C. J. C. H. Watkins and P. Dayan, "Q-learning." *Machine Learning* **8**, 279–292, 1992.

W. Zhang and T. G. Dietterich, "High-performance job-shop scheduling with a time-delay TD(λ) network." In: D. Touretzky et al., eds., Advances in Neural Information Processing Systems 8, 1024-1030, MIT Press, 1996.

Chapter 15

USING HYPERGAMES TO INCREASE PLANNED PAYOFF AND REDUCE RISK

Russell Vane
Veridian Systems,
Arlington,
Virginia 22209, USA.
russ.vane@veridian.com

Paul Lehner
The MITRE Corporation,
McLean,
Virginia 22102, USA
plehner@mitre.org

Abstract Hypergames are an emerging tool for combining decision theoretic and game theoretic information. We explain a new approach to the standard two-player zero-sum single-stage normal game that maximizes expected gain while quantifying possible loss. A DTGT agent can use this formulation to select a plan based on its assessment of an opponent's intent, its assessment of an opponent's unpredictability and its utility model of the situation.

Keywords: Decision theory, game theory, hypergame theory

1. INTRODUCTION

Plan selection research has proceeded primarily in two directions: decision theory and game theory. Both theories share a common definition of the usefulness of a plan in a given situation, called its utility. Both theories use a common definition of plan desirability, called expected utility, that is based on the probability of the occurrence of a situation and the utility of the plan.

Game theory and decision theory disagree sharply about how to assess the probability that a given situation will arise. They also disagree about how to choose the most desirable plan. Game theory recommends that the planner determine these probabilities *implicitly* from an expert's view of the situation, called the full game. In contrast, decision theory recommends that the planner use knowledge of the opponent to explicitly determine the situational probabilities. Even when the assessed probabilities are the same, the theories recommend selecting different plans: Game theory recommends choosing the plan (or mix of plans) with the least vulnerability; decision theory recommends choosing the plan with the highest expected utility.

Decision theory often recommends a plan with a higher expected utility than game theory, but a plan that a clever adversary may be able to exploit, commonly called outguessing. Our research allows the planner to quantify the risk of exploitation in terms of probability and utility. This technique often allows the planner to employ decision theory's higher expected utility solution because the risk is understood. When the risk is too great then game theory's solution to the full game becomes preferred. This solution is also known as the Nash Equilibrium Mixed Strategy (NEMS) (Nash, 1951).

Hypergame theory was invented by Peter G. Bennett (Bennett and Dando, 1979; Bennett and Huxham, 1982). Hypergame theory evaluates two different game theoretic formulations about the same planning scenario. We use the term "planning scenario" to represent the aggregation of all of the plan and situational information affecting the plan selection. First, it records a hypothetical game that is deemed by the planner to represent the opponent's thinking. This *subgame* may exclude plans of either player that are in the planner's full game. The second formulation excludes none of the planner's plans, but is otherwise similar to the first formulation. The planner solves the second formulation to select a plan. This two-phase approach frequently results in a beneficial change in expected utility for the planner.

Even though presented as a game theoretic approach, hypergame theory is fundamentally heretical to the concepts of game theory. Game theory requires a shared knowledge of the planning scenario (the full game) and a shared knowledge of rationality which is used to derive a consistent alignment of beliefs. Hypergame theory discards consistent alignment of beliefs. Hypergame theory asserts (as does our work) that a consistent alignment of beliefs is a fiction in many cases, requiring high levels of expertise among the competitors and time for them to think.

The primary contribution of hypergame theory is that it considers more than one view of the planning scenario. This innovation matches our intuition that it is only under very unusual circumstances that the planner and the opponent would actually be considering the same game. While it is likely that the planner and opponent would share the same concept of utility, the differing expertise and information of the adversaries is very likely to change their perception of the planning scenario. Hypergame theory can account for this difference, unlike game theory.

As mentioned before, game theory and decision theory make different plan selection recommendations. In this article we extend hypergame theory to incorporate decision theory. We also define an opponent behavior that can be substituted into the planner's reasoning process when there is uncertainty. Additionally, we can account for the planner's fear of being outguessed in a different expected utility formulation called Hypergame Expected Utility (HEU). These three contributions are captured in a generalized hypergame representation. These improvements enable the planner to apply both knowledge about an opponent as in decision theory and knowledge about the planning scenario as in game theory to create a mathematical bridge that provides an integrated plan selection technique.

2. DISCUSSION

Starting with game theory, a representation of the plan selection problem is developed using an extended hypergame formulation.

2.1. PLAN SELECTION IN TWO-PLAYER ZERO-SUM SINGLE-STAGE GAMES

Selecting a plan from a list of candidate plans requires situational awareness and a model of how the plans will interact with the situation to produce outcomes. When the model produces a single value for each plan and situation pair that represents the value of that outcome to the planner, then that value is called a utility. When the plan being used by an opponent is one of the primary factors which influences the planner's choice, then that planning scenario is called a two-player game. When the utility value of each outcome for the planner is the additive inverse of the opponent's utility, then that is called a zero-sum game. When the decision is a one-time event that will not be repeated, then that is called a single-stage game.

The normal form of the two-player zero-sum single-stage game includes a set of plans that the planner is considering, a set of plans that the opponent may be considering, and the utility to the planner of the

	col 1	col 2	col 3	...	col n
row 1	u_{11}	u_{12}	u_{13}	...	u_{1n}
row 2	u_{21}	u_{22}	u_{23}	...	u_{2n}
...	...	...	...	...	...
row m	u_{m1}	u_{m2}	u_{m3}	...	u_{mn}

Figure 15.1. An $m \times n$ normal game.

outcomes of all of the combinations of such plans. The normal form is usually recorded in a two dimensional matrix with the planner's plans listed in the rows and the opponent's plans listed in the columns. The value recorded at the intersection of each row and column represents the utility to the planner. An example of a game with m rows and n columns is shown in figure 15.1.

In this representation there is no place to record any beliefs that the planner has about the situation. Such beliefs are based on the planner's experience against this or other opponents in similar situations. Furthermore, the planner is aware neither opponent wants to be predictable. There is also no place to record the planner's concern about being outguessed in the normal game. Some of a plan's expected utility may be lost when an opponent surprises the planner. Outguessing can occur when the planner selects a plan that has an unexpected low-valued outcome.

2.2. PROBLEM REPRESENTATION

The new representation has four quadrants as seen in figure 15.2. The lower right quadrant is the normal game.

Another three quadrants were added to allow more situational information to be recorded and represent a model of the opponent's behavior. The opponent is the Column player and will be called Column throughout the article. The planner or DTGT agent will be called Row. A final synopsis of Column's expected play is summarized in the top row and labeled C_Σ. Once C_Σ has been constructed all of the other information may be ignored, except the full game NEMS comprised of RMS $m \times n$ (or R_0) and CMS $m \times n$ (or C_0). This section provides the instructions for Row to construct a consistent prediction of Column's expected play.

The belief-context in the upper left quadrant is Row's belief (as a probability) about the likelihood that Column is reasoning using the row directly to the right and the column directly underneath each entry. Belief-contexts are recorded in a diagonalized matrix with components that sum to one. Let P_k, at the intersection of C_k and R_k, be the value

				C_Σ	S_1	S_2	S_3	...	S_n
P_{K-1}				C_{K-1}	c_{k1}	c_{k2}	c_{k3}	...	c_{kn}
	...			...	...	...	...	...	...
		P_1		C_1	c_{11}	c_{12}	c_{13}	...	c_{1n}
			$P_0 = 1 - \sum_{k=1}^{K-1} P_k$	C_0	c_{01}	c_{02}	c_{03}	...	c_{0n}
R_{K-1}	...	R_1	R_0 = full game		col 1	col 2	col 3	...	col n
r_{k1}	...	r_{11}	r_{01}	row 1	u_{11}	u_{12}	u_{13}	...	u_{1n}
r_{k2}	...	r_{12}	r_{02}	row 2	u_{21}	u_{22}	u_{23}	...	u_{2n}
...	...	...	...	...	...	...	...	...	...
r_{km}	...	r_{1m}	r_{0m}	row m	u_{m1}	u_{m2}	u_{m3}	...	u_{mn}

Figure 15.2. Generalized hypergame without selection strategies.

of Row's belief about Column reasoning by using reasoning context k. From Row's perspective K accounts for all of the possible reasoning situations that Column might use. Therefore, all of Row's belief-contexts sum to one as in equation (1).

$$\Sigma_{k=0}^{K-1} P_k = 1 \tag{1}$$

In the upper right quadrant are the Column Mixed Strategies (CMS). There are only two kinds of CMSs expressed as probability vectors used in the hypergame. In the first, an explicit CMS, Row's belief about Column's expected behavior is recorded directly into the CMS. In the second, a subgame CMS, Row believes that Column is thinking about a μ(mu) $\times \nu$ (nu) subgame, where $\mu \leq m$ and $\nu \leq n$. Row believes that the NEMS solution for Column for the μ subgame fairly represents Column's reasoning context k. As a result, Row records each CMS as probability entries in C_k. C_0 is reserved for the CMS associated with the $m \times n$ full game NEMS. In either CMS, let c_{kj} be the components of a probability vector, C_k, which represents Row's belief about Column's expected play in situational context k. Each subgame CMS, C_k, has n components that sum to one as shown in equation (2):

$$\Sigma_{j=1}^{n} c_{kj} = 1 \tag{2}$$

The CMS Summary, C_Σ is the final assessment of Row's belief about Column's upcoming choice. CMS Summary is a probability vector as shown in equation (3). Row's belief that Column is using column j is S_j , which is component j of the CMS Summary. S_j is determined by using the belief-context of each reasoning context to weight CMSs as shown in equation (4). The CMSs, and CMS Summary have n components.

CMS Summary $= C_\Sigma = [S_1, \ldots, S_n]$

$$\Sigma_{j=1}^{n} S_j = 1 \tag{3}$$

$$S_j = \Sigma_{k=0}^{K-1} P_k c_{kj} \tag{4}$$

In the lower left quadrant of the generalized hypergame Row has a place to record information associated with each CMS. In the situation where Column is deemed to be using a $\mu \times \nu$ subgame CMS then Row has an associated $\mu \times \nu$ subgame Row Mixed Strategy (RMS), also called R_k. This is only recorded for completeness. The RMS $\mu \times nu$ are almost never used even for reference purposes. In the situation where Row has recorded an explicit CMS which only refers to Column's play, Row does not have an associated RMS. Thus this quadrant is the least important part of the hypergame. R_0 is reserved for the RMS $m \times n$, which is Row's part of a NEMS $m \times n$ and is always recorded adjacent to the left side of the full game. R_k has m components, r_{ki}, which sum to one as in equation (5).

$$\Sigma_{i=1}^{m} r_{ki} = 1 \tag{5}$$

The hypergame normal form generalizes game theory by considering subgames that may be used by opponents similar to the original hypergame theory as a subgame CMS. The explicit CMS meets the representational needs of decision theory as well.

An important property for DTGT agents is the ease with which this representation can accommodate new plans (by adding new rows). This means that an incremental plan generation approach could be supported by this technique. Each row addition does change CMS $m \times n$, but would only change the other CMSs if they were deemed to include the new row in their respective $\mu \times \nu$ subgames. There are numerous reasons why new rows may be available to Row but are unlikely to be known by Column. Some reasons might be plans which incorporate new inventions, surprise reinforcements, or extra resources (such as money).

Finding 1 *The generalized hypergame representation allows a decision-maker to record a consistent situation context about an opponent's choices (see figure 15.2).*

Row's precise method for assessing his beliefs about Column's expected play has purposefully been ignored. For DTGT agents any domain dependent pattern recognition approach is possible, but it should observe the constraints of probability to retain the desired consistency. Although this article does not address how Row would collect information on Column before selecting a plan, it does however provide two mechanisms to account for Row's lack of certainty about Column.

The first mechanism is to assign all uncommitted situational belief to the Column player's NEMS solution for the full game, CMS $m \times n$. For an $m \times n$ game the NEMS solution may also be called NEMS $m \times n$. Its expected utility (EU) is the expected utility of the Row Mixed Strategy for the full game and the Column Mixed Strategy for the full game, or $EU(RMS\ m \times n, CMS\ m \times n)$. The rationale is provided by Vane (Vane, 2000) and summarized as follows.

1. For any explicit CMS $\neq$ CMS $m \times n$, there exists an RMS such that:

   $$EU(RMS, \text{explicit } CMS) > EU(RMS\ m \times n, CMS\ m \times n)$$

   This means that assuming any CMS other than CMS $m \times n$, allows Row to choose an RMS that increases his EU over the $EU(NEMSm \times n)$.

2. Assuming an unknown CMS that is exploitable is contrary to the idea of having uncommitted belief. Therefore the unknown CMS *should* be set to CMS $m \times n$, so that Row will not increase his expected utility above the $EU(NEMS\ m \times n)$.

The second mechanism will be explained during the evaluations of candidate plans in the next section. It provides a way to incorporate Column's most effective counterstrategy weighted by a term that reflects Row's fear of being outguessed.

The hypergame formulation is a richer problem representation than game theory's normal form. It should be used in most cases to record the planner's knowledge about the Column's expected behavior or even to some extent what Column is thinking. A default behavior is proposed to account for Row's lack of certainty about Column's reasoning process. For example Row may be willing to assert that Column is 0.3 likely to choose column 2, but be unwilling to allocate the remaining $(1 - 0.3) = 0.7$ belief to any other subgame. This should be assigned to the NEMS probability distribution for Column (CMS $m \times n$).

Finding 2 *If a decision-maker's beliefs in the belief-context (upper left quadrant) sum to less than one, then the remaining belief is associated with an unknown CMS that should be assigned to CMS $m \times n$, the full game NEMS for the opponent.*

2.3. PLAN EVALUATION USING HYPERGAME EXPECTED UTILITY (HEU)

Up to this point in the discussion, we have not provided any way to evaluate the information in the new hypergame normal form. Row has recorded situational beliefs that have led to an expected Column behavior encapsulated in C_Σ. All of the information except the C_Σ and the full game NEMS comprised of RMS $m \times n$ (or R_0) and CMS $m \times n$ (or C_0) is unneeded except for background purposes. The additions that are made to the lower left quadrant provide the candidate plans and the evaluations of their desirability. While Vane has also explained how to evaluate specific mixes of plans (Vane, 2000), only single row candidate plans are addressed in this article. Each candidate plan is called a hyperstrategy as in figure 15.3. For this work, one of the h_{a_i} terms in each hyperstrategy, H_a, will have a value of one corresponding to a single plan. The other terms will be zero. Thus, each new hyperstrategy is a probability vector and is entered in a column to the left of the hypergame.

					C_Σ	S_1	S_2	S_3	...	S_n
	P_{K-1}				C_{K-1}	c_{k1}	c_{k2}	c_{k3}	...	c_{kn}
		...			...	...	...	...	...	...
			F_1		C_1	c_{11}	c_{12}	c_{13}	...	c_{1n}
				$P_0 = 1 - \sum_{k=1}^{K-1} P_k$	C_0	c_{01}	c_{02}	c_{03}	...	c_{0n}
Hyperstrategy H_1	R_{K-1}	...	R_1	R_0 = full game		col 1	col 2	col 3	...	col n
h_{11}	r_{k1}	...	r_{11}	r_{01}	row 1	u_{11}	u_{12}	u_{13}	...	u_{1n}
h_{12}	r_{k2}	...	r_{12}	r_{02}	row 2	u_{21}	u_{22}	u_{23}	...	u_{2n}
...	...	...	...	...	...	...	...	...	...	...
h_{1m}	r_{km}	...	r_{1m}	r_{0m}	row m	u_{m1}	u_{m2}	u_{m3}	...	u_{mn}
$EU(H_1, C_\Sigma)$	NA	NA	NA	$EU(R_0), C_\Sigma)$	Expected					
$EU(H_1, G)$	NA	NA	NA	$EU(R_0, G)$	Worst case					

Figure 15.3. A hypergame with hyperstrategies.

The previously determined C_Σ and the utility entries in the full game are needed to complete the EU evaluation. As shown in equation (7), this allows us to determine the EU of any hyperstrategy using standard EU equations. The hyperstrategy is matrix multiplied with the C_Σ to determine a probability matrix of the same cardinality as the $m \times n$ full game. This probability matrix represents the expected probability (a weight) of each entry in the full game. The EU is then determined by performing a dot product of this matrix with the utility values in the full game. Both decision theory and game theory use the same method to compute EU.

$$
\begin{aligned}
EU(H, C_\Sigma) &= (H, C_\Sigma).U && (6)\\
&= \Sigma_{i=1}^{m} h_i.\left(\Sigma_{j=1}^{n} S_j.u_{ij}\right) && (7)
\end{aligned}
$$

This is not enough to fairly evaluate the situation, because the EU of a hyperstrategy does not account for Row's willingness to consider the possibility of Column outguessing Row's choice. With a little reflection, it can be seen that another concept is needed to complete the evaluation. For a given hyperstrategy, the worst case condition for Row is the entry in that row which yields the least utility for Row. Let that CMS Worst Case be called G, a probability vector with a 'one' in the column that causes the least utility in the row. Row always uses the $m \times n$ full game to find G, because we are quantifying the possible.

The Hypergame Expected Utility (HEU) represents the continuum over the interval $[0, 1]$ based on Row's belief about being outguessed. A belief, g, (stated as a probability) represents Row's concern that Column is going to choose the most deleterious column for a given hyperstrategy. This yields a graph which starts on the left side with the EU of the hyperstrategy and C_Σ, and the line descends as g increases until it reaches EU of the hyperstrategy and G on the right side as in figure (15.4).

Also plotted in figure (15.4) is the $HEU(NEMS\ m \times n)$. Sometimes the HEU of a hyperstrategy is more desirable than the NEMS and sometimes it is not. The NEMS by definition provides the maximum of the minimums and as expected $EU(RMS\ m \times n, G) = EU(NEMS\ m \times n)$.

$$
HEU(H) = (1 - g).EU(H, C_\Sigma) + g.EU(H, G) \qquad (8)
$$

where H is the hyperstrategy being evaluated, CS is the CMS Summary (Column's expected behavior), G is the worst case strategy for H, and g is the value for Row's fear of being outguessed in the interval [0,1].

Thus, if $g = 1$, the NEMS is always a good choice. In fact it is the best choice. Our formulation of HEU is provided in equation (8). The $EU(RMS\ m \times n, C_\Sigma)$ is always greater than or equal to $EU(NEMSm \times n)$

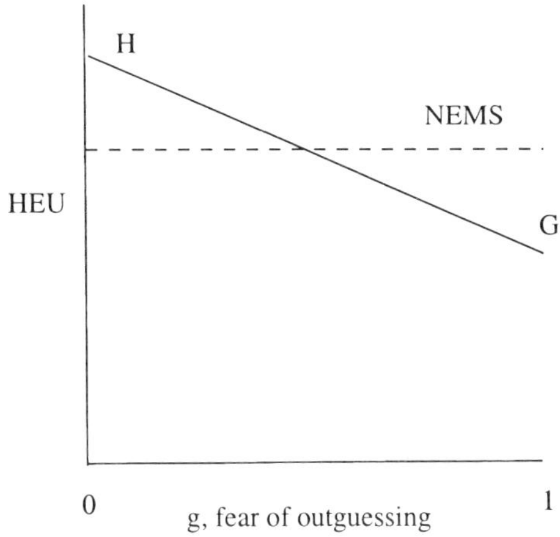

Figure 15.4. Outguessing and strategy selection.

because of reasoning similar to finding 2. These facts lead naturally to the concept of hyperstrategy effectiveness.

2.4. PLAN EFFECTIVENESS USING HEU

When we started this research (Vane and Lehner, 1990), it was because game theory seemed too conservative and different from what world-class competitors considered when selecting a plan. Since game theory provided a conservative "guaranteed" expected utility, it seemed important to use the NEMS in an effectiveness criterion. An *effective hyperstrategy* should provide at least the expected utility of the NEMS for the full game as shown in equation (9).

$$Eff(H) \text{ iff } HEU(H) \geq HEU(RMS\ m \times n) \tag{9}$$

The NEMS $m \times n$ is constructed by Row using RMS $m \times n$ while Column uses CMS $m \times n$. An effective hyperstrategy should provide an HEU greater than or equal to whatever the HEU is for RMS $m \times n$. Because the HEU is constructed over the interval $[0, 1]$ for g, EU criteria need to be met for a hyperstrategy to be fully effective. These correspond to the graph in figure 15.4. The two points on each line corresponds to $EU(*, C_\Sigma)$ and $EU(*, G)$ as in equations (10) and (11).

$$Eff(H) \text{ iff } EU(H, C_\Sigma) \geq EU(R_0, C?) \tag{10}$$

$$Eff(H) \text{ iff } EU(H, G) \geq EU(R_0, G) \tag{11}$$

In equation (10), both H and R_0 are being compared with C_Σ. Since the value of $EU(R_0, C_\Sigma)$ is always greater than or equal to $EU(R_0, C_0)$, this

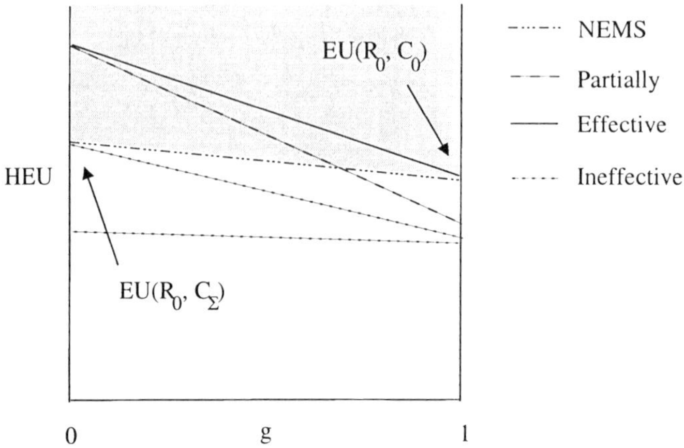

Figure 15.5. Plots of various $EU(H, C_\Sigma)$ to $EU(H, G)$.

Effective	$EU(H, C_\Sigma) > EU(R_0, C_\Sigma)$	$EU(H, G) = EU(R_0, G)$
Partially	$EU(H, C_\Sigma) > EU(R_0, C_\Sigma)$	$EU(H, G) < EU(R_0, G)$
NEMS	$EU(H, C_\Sigma) = EU(R_0, C_\Sigma)$	$EU(H, G) = EU(R_0, G)$
Ineffective	$EU(H, C_\Sigma) \leq EU(R_0, C_\Sigma)$	$EU(H, G) \leq EU(R_0, G)$

Figure 15.6. Relationships of $EU(H, *)$ to $EU(R_0, *)$.

formulation is more stringent and harder to meet. However, $EU(R_0 0, G)$ is always greater than or equal to $EU(H, G)$. From equation (11) the $EU(H, G)$ is at best equal with the $EU(NEMS\ m \times n)$. Figure 15.5 illustrates three kinds of effectiveness: effective, partially effective, and ineffective. Effective hyperstrategies lie entirely in the shaded region. Ineffective hyperstrategies lie entirely in the unshaded region. The NEMS is the boundary between effective and ineffective. Partially effective hyperstrategies cross the NEMS similar to the dashed line labeled 'Partially.'

No hyperstrategy is better than or equal to the NEMS for an $m \times n$ game and all values of g. Thus the NEMS $m \times n$ is the only effective hyperstrategy because no other hyperstrategy has as high an EU as R_0 against G. This result is not unexpected because G is exactly the criterion for which the NEMS was invented. However numerous hyperstrategies may have a better value against C_Σ. So the question becomes how valuable are the partially effective hyperstrategies or is there a way to use this analysis to come up with an effective combination of hyperstrategies?

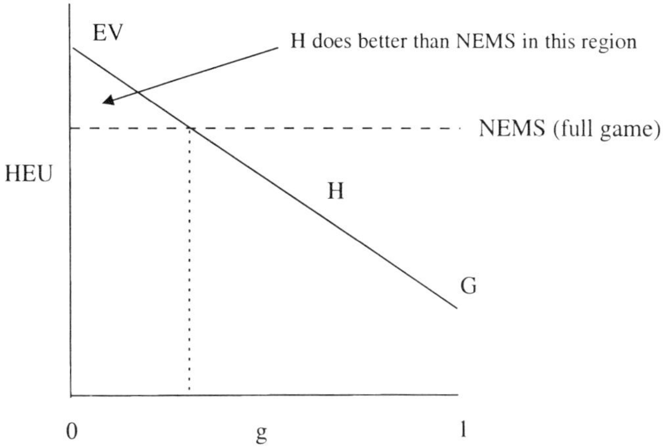

Figure 15.7. This figure represents uncertainty in the range of 0–1 as g. Both a partially effective hyperstrategy, H, and the NEMS are plotted. As Row becomes more fearful of Column's capability to outguess (as g increases), the full game NEMS solution becomes more attractive. The down arrow shows where HEU(NEMS) = HEU(H)

As seen in figure 15.7, in any partially effective hyperstrategy there is a point where the $HEU(H) = HEU(R_0)$. By solving for g, the planner has a criterion for when to choose the proposed hyperstrategy over the NEMS.

3. RESULTS

Two examples are provided as a demonstration of the applicability of hypergames to solving the two-player, zero-sum games. Both show a $\mu \times \nu$ subgame as Row's belief. In the first game the $\mu \times \nu$ game is 3×3 and the full game is 4×4. In the second game the $\mu \times \nu$ game is 3×3 and the full game is 6×6. When constructing a hypergame about the planning scenario, having more belief in a non-NEMS CMS increases the eventual $EU(H, C_\Sigma)$.

3.1. A 3×3 IN A 4×4

In the game in figure 15.8, Row believes that Column is thinking that the contest is a 3×3 game. Row also believes that this is 0.7 likely as shown in the belief-context associated with the CMS 3×3. Row solves the NEMS 3×3 to determine CMS 3×3 and computes the dual for Row to solve RMS 3×3. Row's not sure of what else to believe, so the remaining 0.3 is assigned to the full game NEMS, which is also solved.

			C_Σ	0.416	0.398	0.108	0.078
	0.7		CMS 3×3	0.451	0.431	0.118	
		0.3	CMS 4×4	0.334	0.320	0.087	0.259
H	NEMS 3×3	NEMS 4×4		col 1	col 2	col 3	col 4
0 (0.112)	0.275	0.305	row 1	0	−1	4	1
0 (0.016)	0.353	0.311	row 2	−2	3	−3	1
0 (0.114)	0.372	0.241	row3	2	−2	0	1
1 (1.34)		0.143	row 4	3	1	0	−4
1.34	0.039	0.288	Expected				
−4	NA	0.288	Worst case				

Figure 15.8. 3×3 subgame in a 4×4 full game.

Row's belief about Column's expected behavior is $0.7(CMS\ 3 \times 3) + 0.3(CMS\ 4 \times 4)$. In the hypergame shown in figure 15.8,

$$C_{\Sigma} = [0.4160.3980.1080.078]$$

This can be used to evaluate every row and the EU of each row is place in parentheses in the hyperstrategy column on the left of the diagram. These are:

$$\begin{aligned}
EU(row\ 1, C_{\Sigma}) &= 0.416(0) + 0.398(-1) + 0.108(4) + 0.078(1) \\
&= .112 \\
EU(row\ 2, C_{\Sigma}) &= 0.416(-2) + 0.398(3) + 0.108(-3) + 0.078(1) \\
&= .016 \\
EU(row\ 3, C_{\Sigma}) &= 0.416(2) + 0.398(-2) + 0.108(0) + 0.078(1) \\
&= .114 \\
EU(row\ 4, C_{\Sigma}) &= 0.416(3) + 0.398(1) + 0.108(0) + 0.078(-4) \\
&= 1.34
\end{aligned}$$

Since the HEU of rows 1–3 are less than the HEU(NEMS) they are ineffective strategies. Row 4 however is partially effective and in order to use Row 4 instead of the NEMS, Row must have less than the uncertainty expressed by g, by solving when $HEU(row4)$ equals $HEU(NEMS)$.

For certain values of g, the Row 4 is better than the NEMS. The equations for the two lines are:

$$\begin{aligned}
HEU(NEMS) &= .228 - (0)g \\
HEU(Row\ 4) &= 1.34 - (5.34)g
\end{aligned}$$

$HEU(Row\ 4)$ intersects $HEU(NEMS)$ at 0.209. Therefore, Row 4 is dominant in the region $0 \leq g \leq 0.209$ and NEMS where $0.209 \geq g \geq 1$.

3.2. A 3 × 3 IN A 6 × 6

This case will further demonstrate the work's generality. Row believes that Column is reasoning on a 3×3 subgame, but personally views this planning scenario as a 6×6 full game. Thus, Column is expected to be reasoning on 9 cells in a game of 36 cells.

The following scenario is a possible model of Allied decision-making during the Desert Storm War. The Iraqi High Command is the Column player. The Allied Command is the Row player. The utility values in the game vary from -5 (decisive defeat) to 5 (decisive victory).

1. (a) The Row Player sees the possible attacks and defenses for each row and column pair. This is a very knowledge intensive

	Defend	Defend Left	Defend Right	Defend w/Res.	Screen	Counterattack
Attack	−1	−3	0	1	1	1

Figure 15.9. The attack right row.

task. It requires knowing the domain: having an estimation function based on possible plans and probable results. Every number in this example was determined from first military principles. Using one row as an illustration may help make this process more understandable.

This subsection may be skipped without any loss of knowledge of the technique.

All of these discussions are against an *Attack Right* plan shown in figure 15.9. *Attack Right* is a weighted attack of about 2/3 of the attacking forces in the right half of a battle field.

Defend is a frontal defense, evenly distributed.
The result of these two plans is that the Allies will take a significant number of casualties, hence the -1 result. They will accomplish the mission of ejecting the Iraqis from Kuwait.

Defend Left is a frontal defense, weighted to stop an *Attack Right* .
The result of these two plans is that the Allies will take a large number of casualties and may fail the mission (-3), but may succeed in ejecting the Iraqis.

Defend Right is a frontal defense weighted against an *Attack Left.*
The result of these two plans is that the Allies will take a small number of casualties and will succeed in ejecting the Iraqis (0). This column (plan) is weakest where the the 'Attack Right' row is aimed, so it yields the best expected result of the three frontal attacks.

Defend with Reserve is an area defense, designed to control Kuwait and stop Paratroopers and Helicopter penetrations.
The result of these two plans is that the Allies will take very few casualties because even less forces will be fighting at the front lines (1). They will eject Iraq by defeating the Iraqi forces in numerous small battles.

Screen is a thin forward defense designed to give up ground rather than receive casualties.
The result of these two plans is that the Allies will take very few casualties, but so will Iraq (1). They will eject Iraq.

Counterattack is a frontal defense, weighted to stop an *Attack Right.*
The result of these two plans is that the Allies will take some casualties, but will destroy all forces committed to the counterattack (4). They will eject Iraq and may topple the regime.

The rest of the of the rows are determined in the same way. The rest of the steps refer to figure 10.

(b) Row believes that the Column Player is reasoning (0.8 belief-context) that entrenched defenses are the only strategies that will work against Row given Allied air superiority. These plans are Defend, Defend Weighted Left, Defend Weighted Right. Row believes that Column also reasons that the Allies' stress on Kuwaiti sovereignty will keep them within the bounds of Kuwait, therefore Column is deemed to believe that only three frontal attacks are possible. These are Attack, Attack Right, Attack Left.

2. Row calculates both subgame CMS's, the CMS 3×3 and the CMS $m \times n$, and combines them to form C_Σ which is placed in the top row.

$$\begin{aligned} C_\Sigma &= [0.8 \times 1 + 0.2 \times 0 \; 0.2 \times 0.125 \; 0.2 * 0.25 \; 0.2 \times 0 \\ &\qquad 0.2 \times 0.625 \; 0.2 \times 0] \\ &= [0.8 \; 0.025 \; 0.05 \; 0 \; 0.125 \; 0] \end{aligned}$$

3. Row computes the EUs for all of the rows and determines that Envelop Left is a partially effective hyperstrategy. Envelop Left's G is $[0\;0\;0\;0\;0\;1]$ since column 6 results in -2. Figure 15.11 is the graph of HEU.

4. The Row player is willing to believe that Column will outguess him 1 out of 5 but no more than that and assigns $g = 0.2$. Row checks the graph in figure 15.11 and determines that row 4 dominates until g is nearly 0.5.

5. The Row player chooses the "Envelop Left" row with an expected value of 3. In this scenario the hyperstrategy dominates, until g is very close to 0.5.

			C_Σ	0.8	0.025	0.05	0	0.125	0
	0.8		3×3	1	0	0	-	-	-
		0.2	6×6	0	0.125	0.25	0	0.625	0
$H1^*$ (row 4)	NEMS 3×3	NEMS 6×6		Defend	Defend Left	Defend Right	Defend w/Res	Screen	Counterattack
0	1	0.125	Attack	−1	−1	−1	1	1	4
0	0	0.5	Attack Right	−1	−3	0	1	1	1
0	0	0	Attack Left	−1	0	−3	1	1	1
1	-	0.375	Envelop Left	3	5	1	2	−1	−2
0	-	0	Invade Beach	1	−3	2	−1	−1	−2
0	-	0	Envelope Vertical	2	0	0	−3	−2	−1
2.45	−1	0.25	Expected						
−2	−1	0.25	Worst case						

Figure 15.10. An excursion: 3×3 in a 6×6.

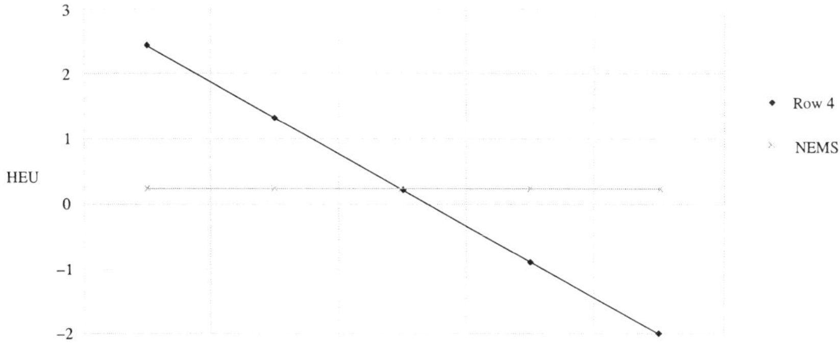

Figure 15.11. Utility with outguessing: 3×3 in a 6×6.

4. CONCLUSIONS

Overall this paper has shown that the plan selection problem can be represented as a hypergame that quantifies the possibilities for gain with the risk of loss in competitive situations. Hypergame Expected Utility is a robust, useful evaluation of the desirability of any hyperstrategy. There exists a solution using a combination of hyperstrategies (one of which is always the NEMS for the full game) for Row's beliefs about a situation as synopsized in C_Σ and his risk of being outguessed expressed as g.

There are many areas of future research, the foremost being this technique's extension into non-zero sum games. The rest of this paper contains some observations on these areas.

Non-Zero Sum games must modify the representation to use both Row's and Column's utilities. The Hypergame approach already uncouples the decisions of the Row player (the selection strategy) from the understandings of the Column player. This separation promotes non-zero sum interactions, because Column still has a decipherable strategy-vector to replace the standard C_Σ and Row can still evaluate his pure and even mixed hyperstrategies. Additionally, the worst-case condition also still exists and is computable.

Yet some behavior remains unknown. For example, g may be better modeled by maximizing Column's EU, rather than minimizing Row's EU. Additionally, in cooperative games, the notion of hidden reasoning by Row may not make as much sense. Also the idea of fearing an opponent's trickery might change to accounting for a belief in the inherent amount of cooperation that one could expect from the other player (or players) as an ally, unaligned participant, or competitor. There is no evidence that this approach has some fundamental shortfall with respect to non-zero sum games.

The second area would be gathering empirical evidence about the effectiveness of the approach in human decision problems. Many human decision-making domains would benefit from a more versatile theory of plan selection. While this approach has a lot of theoretical applicability, it is still a model. It may be a practical model for human decision-makers, if they are inclined to remain objective while using it. Additionally, it is not easy to generalize from non-repeating situations with a large number of factors, so careful experimental design will be needed to provide any generalizable results.

The third would be applying the technique to problems already being addressed by AI approaches and DTGT agents. There are numerous applications from automated speech understanding to robotic planning that would benefit from this research. Developing and testing effective utility functions to measure usefulness in each application domain will be an important part of the new work. This approach has a lot of dependence on effective evaluation of the utility (or 'goodness') of predicted planning outcomes to determine the desirability of a plan.

This work provides a mechanism for tuning the quality of the artificial intelligence in computer games on the fly. The idea is that regardless of how one initializes the AI engine, columns can be added or subtracted from the AI reasoning process according to rules by subject matter experts. Automated learning is not implied in the classical sense, instead observations of opponent play are compared with the observable characteristics of the strategies represented by columns in a full game. Such reasoning would include more columns if the opponent exhibits better play (more diverse) and exclude columns, if the opponent appears to be locked into an approach. Additionally, this allows the AI to "detune", that is, discard rows to reduce the EU of the game. This will make the battles in computer games more exciting or as challenging as one could desire. It may become a new criterion of "good" AI in electronic games—to significantly improve game play by providing a model of the opponent that causes the games to be more fun, not necessarily to have the AI win more often (increase the EU).

We believe that the greatest potential comes from applying this technique to DTGT agents on large complex tasks. Exploration of this approach to establish a hierarchical reasoning system is a large and domain dependent task. The precepts are these:

- Each row is a scheme of maneuver regardless of which column has been chosen.

- Often these schemes break up into lower level tasks (or missions) which are also subject to adversarial reasoning or risks which need to be bounded.

- During implementation of the selected plan, subschemes may begin to fail (or to succeed better), which may cause higher level plans to fail (or to succeed better); leading to an overall change in strategy (Hsu and Lehner, 1997).

This approach could become a very large system with the downside of some pathological behaviors, but the result could be an excellent general planning system.

Additionally, employing the techniques to lessen the computational complexity of artificial intelligence approaches in very complex and large games could yield benefits to military AI research. The structure of the systems described above would have to be explored with these goals in mind.

The fourth is automated instructional delivery. In this case the expert's full game would encompass all of the lessons and the state of the learner. The next lesson to be taught to address learner remediation or to pace learner progress would be selected by the hypergame approach. The instructional software would be the planner and the learner would be the other player. The software would be constantly attempting to recognize which information that the student had assimilated and try to pick the next piece of information it would attempt to teach. When the developer accounts for boredom as a negative utility, I believe that the game would be interesting and valuable to solve.

Hypergame use in any of these four areas: non-zero-sum games, support for complex human decision-making, AI applications including hierarchical knowledge-based planning, and adaptive instructional delivery; would validate this technique's importance.

References

Bennett, P. G. and Dando, M. R. (1979). Complex strategic analysis: A hypergame study of the fall of France. *Journal of the Operations Research Society.*

Bennett, P. G. and Huxham, C. S. (1982). Hypergames and what they do: a 'Soft OR' approach. *Journal of the Operations Research Society.*

Hsu, W. and Lehner, P. E. (1997). Strategic patterns of play in Go. *Transactions on Pattern and Machine Intelligence.*

Nash, J. (1951). Non-cooperative games. *Annals of Mathematics.*

Vane, R. R. (2000). *Using Hypergames to Select Plans in Competitive Environments.* PhD thesis, George Mason University, Fairfax, Virginia.

Vane, R. R. and Lehner, P. E. (1990). Hypergames in AI planning. In *Innovative Approaches to Planning, Scheduling and Control: Proceedings of a Workshop Held at San Diego, California, November 5–8, 1990.*

Chapter 16

BILATERAL NEGOTIATION WITH INCOMPLETE AND UNCERTAIN INFORMATION

Trading Help in a Distributed Peer Help Environment

Julita Vassileva

Department of Computer Science, University of Saskatchewan, 57 Campus Drive, Saskatoon S7N 5A9 Canada.

jiv@cs.usask.ca

Chhaya Mudgal

Institute of Marine and Coastal Science, Rutgers University, 71 Dudley Road, New Brunswick, NJ 08901, USA.

cmudgal@hotmail.com

Abstract The paper describes a bilateral negotiation approach with incomplete and uncertain information that has been implemented in the context of a distributed multiagent peer help system, I-Help, supporting students in a university course. Personal agents keep models of student preferences and negotiate on their behalf to acquire resources (help) from other agents. The agents negotiate iteratively using an influence diagram, a decision theoretic tool. To cope with the uncertainty inherent in a dynamic market with self-interested participants, the agents create models of their opponents during negotiation, which help them predict better their opponents' actions. The evaluation shows the advantages and disadvantages of the proposed negotiation mechanisms.

Keywords: Negotiation, bargaining, decision-theory, personal agents.

1. INTRODUCTION

Despite the convenience that e-commerce has brought, humans are still involved in most of the important processes of business, for example, in making decisions in all phases of buying and selling. Negotiation is one of the key factors in commerce systems, which involves decision-making and considering tradeoffs between multiple factors. The benefit of dynamically negotiating a price is that the resource is allocated to a consumer who values it the most. Negotiation varies in duration and complexity and can extend over a long period of time, which is a disadvantage for time-bounded consumers. Agent technology has helped human users by supporting or even replacing them in negotiation (Maes et al., 1997). Therefore, it is very important that users understand and trust the mechanism by which agents negotiate on their behalf.

A basic ingredient of the negotiation process is the correct anticipation of the opponent's actions. In open multi agent systems (i.e. the systems in which the agents can come and go at any time) there is always an element of uncertainty about the participants.

We propose a negotiation mechanism that imitates human behaviour in the process of negotiation, is time-sensitive, anticipates the opposing party's actions and considers the agent's risk attitude towards money. This mechanism has been implemented in personal agents trading peer help within a multi-agent based web-based system called I-Help (Vassileva et al., 1999). The paper presents the negotiation mechanism, an evaluation showing the advantage of modelling the opponent and finally talks about the real-world test of negotiation within I-Help.

2. THE APPLICATION AREA

I-Help is a web-based multiagent system that provides a student of a university course with human and electronic help resources. Human help and time can be considered as non-tangible differentiated goods, which are traded for money. As a part of the system, there is a matchmaking service to find a competent peer-student online to help (Vassileva et al., 1999). When the students in the class request peer help their personal agents contact a centralized matchmaker who knows who is online and has the required knowledge resources; it provides a ranked list of potential helpers. It also provides a standard market price for the help request, which is calculated based on the difficulty of the topic and the number of users who are knowledgeable on this topic and who are on line at the moment. The agent of the person requesting help starts a bilateral negotiation (bargaining) with the agent of the first potential helper from the list about the price (in our case this is the payment rate per unit of help time) and when a deal is made both agents inform their users. If the knowledgeable user agrees to help, a chat window opens for both sides and the help session is started. If the agents fail to achieve a deal, the agent of the person seeking help starts a negotiation with the second agent on the list, etc. In this way, similarly to Kasbah (Maes et al., 1997) a one-to-many or many-to-many negotiation problem is modelled as a series of disconnected one-to-one negotiation problems.

The reason for choosing a bilateral negotiation model over an auction is that it seems to be more understandable and natural from a human user point of view. First, it allows for considering multiple factors in the negotiation, not only the price (even though there are multi-value auctions that allow this too). Second, it allows a more personal way of making deals, since the agent is not obliged to give the help-request to the lowest-price bidder (helper in our case). Users always have the final say and they may decide not to accept/give help to a given person or at a given moment. Finally, bilateral negotiation avoids the problem with building cliques and corrupting the bidding process.

Agent negotiation in I-Help reduces the burden on the user (in our case the students) by helping them concentrate on their work rather than thinking about how to get a better deal. Viewed on a macro-level, I-Help's personal agents participate in an economy, designed to motivate the knowledgeable students to help their peers. For this they receive payment in virtual currency, called ICU (I-Help Currency Unit), which can later be exchanged for some real world equivalent (gift certificates, marks, etc.). In I-Help the personal agents make decisions on behalf of their users about the price to offer and how to increase or decrease the price to strike a better deal depending on user specified constraints, comprising the users' preference model, such as the urgency of their current task, importance of money and the risk behaviour. This model is initialized by the user and can be updated by the user, after receiving feedback from the agent about the success rate in negotiation. The I-Help environment is dynamic (users log in and out all the time, and when they are on line, they have different tasks, priorities, etc.) and since the personal agents represent real users, it is hard to predict the priorities of the opponent agent on the basis of its past behaviour (since the user's preferences can change from session to session). However, it is useful to try to model the opponent's behaviour during one session, since this can help predict better the opponent's reaction.

3. THE NEGOTIATION MECHANISM

We define bilateral negotiation as an iterative process in which the agents make offers and counteroffers based on the preferences of their users. The negotiation protocol is a straightforward iterative process of agents making offers and counteroffers. During negotiation the agent can be in *Offer* or *Counter-offer* state repeatedly. The final state will be *Accept* or *Reject.* Once the agent is in a final state, it cannot retreat back. In order to do so the whole negotiation process has to start again.

Thus, from a microscopic (i.e. from an individual agent's) point of view, we can consider negotiation as a decision problem that requires a decision-maker to weigh preferences and to choose an action (from the set of actions allowed by the negotiation protocol) that gives the maximum utility.

3.1. MODELING DECISION

Decision making in the negotiation process is modelled using an influence diagram (ID) (Shachter, 1988). ID is a Bayesian network, extended with utility functions and with variables representing decisions and especially suited for modelling decision-making problems. IDs can represent multiple objectives and allow tradeoffs in one area against costs in another. IDs allow accounting for uncertainty and are able to represent it in a quantitative way. Unlike decision trees, IDs do not grow exponentially; they suppress minute details and hence are suitable for getting an overview of a complex problem. IDs have been widely used in modelling decision-making processes (for example, in (Suryadi and Gmytrasiewicz, 1999)), however, not in the context of agent negotiation.

The agent decision making (the ID) takes into account the preferences of the user, which depend on the domain of the negotiation. The preferences in the domain of I-Help include:

- the maximum price of the buyer (i.e. the helpee),
- the urgency of demand of the resource for the buyer, or the urgency of the seller's current work (which has to be interrupted in order to help),
- the importance that either agent attaches to money (either to get more money or to save money), and
- the user's risk behaviour (risk-averse or a risk-seeking person).

We have incorporated utility in order to represent how the decision-maker values different outcomes and objectives. The utility of a decision depends on the role in which the agent is at the moment of decision making (each agent in I-Help can be in the role of a buyer or a seller of help). The utilities for the buyer (helpee) and the seller (helper) for the actions *accept*, *reject* and *counter-propose* vary according to their risk behaviour.

It is important to note that the agent's risk behaviour does not overlap with the money importance. In literature these factors have often been considered as tightly connected, but in our case this is not necessarily true. Money importance and risk-behaviour are two different variables in the user preference model and they are set by the user. Risk behaviour affects the increment and the decrement of the proposed/asking price.

Similarly to (Zheng and Sycara, 1997), we use "negotiation strategies" to denote the actions, which the agents take in every iteration depending on the preference model. The functions that the agents use to increase or decrease their offers and counteroffers as a buyer and as a seller are defined as follows:

For Buyers

If ($max\_price > std\_price$) then

    Offered price := $std\_price - \Delta$

Else

    Offered price := $max\_price - \Delta$

For Sellers

If ($min\_price > std\_price$) then

    Offered price := $min\_price + \Delta$

Else

    Offered price := $std\_price + \Delta$

where std_price is the market price provided by the matchmaker. It is calculated based on the current situation of the market of help on this topic and on the difficulty of the topic, which provides some measure for the actual worth of the resource. For both the buyer and the seller the values of Δ should not exceed their preferred prices, R. Δ is determined as follows (x is the offered price).

For Buyers

If ($urgency = very\ urgent$) then

    If ($risk\_behavior = risk\ seeking$) then

        $\Delta := 1 - e^{\frac{-x}{R}}, x > R$

    If ($risk\_behavior = risk\ averse$) then

        $\Delta := 1 - e^{\frac{-x}{R}}, x < R$

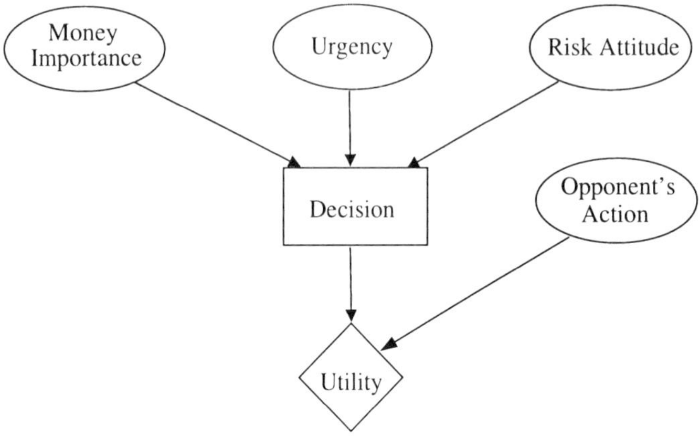

Figure 16.1. An influence diagram for the decision model.

For Sellers

If ($urgency$ = $very\ urgent$) then
    If ($risk\_behavior$ = $risk\ seeking$) then
        $\Delta := \sqrt{min\_price}$
    If ($risk\_behavior$ = $risk\ averse$) then
        $\Delta := \log(min\_price)$

We use an influence diagram that has a conditional node representing the uncertainty about the opponent's action (see Figure 16.1). The outcomes of this node are the probabilities that an opponent can decide to accept, reject and counteroffer. At every step the agents choose between these protocol actions by calculating the maximum expected utility for the actions, which are represented as the possible choices for the decision node in the influence diagram. Before the decision is made the factors that are already known and affect the decision (deterministic nodes) are taken into account as they affect the actions to be made. If the agent does not know anything about the opponent, the node corresponding to the opponent's action can be considered conditional and the states in which the opponent are equally likely. However, if the agent has some knowledge about the opponent (either from pervious encounters or from learning from the opponent's behaviour during the negotiation), the probability in the node representing the opponent's action will be calculated using the model of the opponent.

3.2. MODELLING THE OPPONENT

Ideally (as often is assumed in cooperative environments (Zlotkin and Rosenschein, 1991)) negotiating parties have full knowledge about the opponent. This, however, is not the case when agents are self-interested. In I-Help, it is unlikely that the user will be willing to share her private preferences with other users (or their agents). In order to negotiate more effectively, the personal agents model the preferences of the opponent using a probabilistic influence diagram. Unfortunately, in a dynamic market where new buyers and sellers can enter and leave the system at any time, it is very costly for agents to create and maintain models of all agents they have ever encountered. For this reason, the agents do not keep opponent models across sessions (when the user logs out). After the first round of offers in the negotiation each agent creates a model of its opponent and uses it to predict its reaction to the counteroffer that it is going to make. However, unlike (Gmytrasiewicz and Durfee, 1995), for computational efficiency reasons, there is no recursive agent modelling.

Figure 16.2 shows a probabilistic influence diagram (Shachter, 1988); the oval nodes are conditional and the double-circled node is deterministic. Conditional probability distribution of the conditional nodes over the outcomes is assessed on the basis of the first offer. Probability distribution of the Opponent's action node can be calculated by performing reductions over the nodes (Shachter, 1986). For instance, arc reversal from the Money Importance node to the Opponent's Action node makes Money Importance a barren node. Hence, it can be removed from the diagram and a new conditional probability distribution is calculated. Conditional predecessors of the nodes (if any) are inherited. The diagram can be simplified by using arc reversal operation and barren node removal, which finally gives the probability distribution for the Opponent's Action node. If the next move of the opponent does not match with the predicted action, Bayes' update rule is used to recalculate the probabilities. Evaluating both influence diagrams and probabilistic influence diagrams is NP-hard. In our case the number of nodes is relatively small and there are no complexity problems. However, if the number of nodes increases, negotiation might become computationally inefficient.

4. EVALUATION

The goals of the evaluation are to test the performance of the negotiation mechanism in terms of the number of successful negotiations and the quality of deals that agents make for their users. Another goal is to see if modelling the opponent brings benefits in negotiation. Finally,

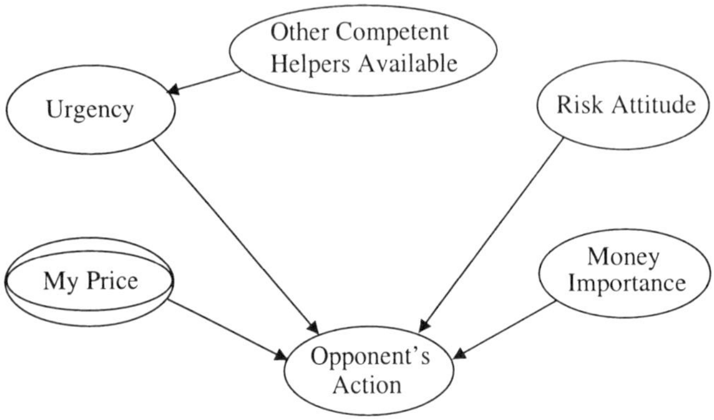

Figure 16.2. An influence diagram for the opponent model.

we want to see if the agent negotiation approach is acceptable for their human users. In order to answer the first two questions, we compare the deals obtained by agents using the proposed approach (decision theoretic with modelling the opponent) with deals obtained by the same agents under identical circumstances using other negotiation approaches.

4.1. EXPERIMENTAL SETUP

Since negotiation is done on behalf of the user, it is necessary to take into account her preferences. There are an immense number of possible preference combinations. For this evaluation, 5 different combinations of user preferences for helper and helpee are tested with 3 different combinations of preferred prices: one, in which the preferred price intervals of the negotiators overlap widely (Session 1/S1), another one in which the preferred prices overlap partially (Session 2/S2) and one where there is no overlap in the preferred prices (Session 3/S3). The experiments were carried out between agents implemented as Java servlets using simple messaging to communicate. The experimental setup follows several assumptions related to the peer-help application. First, the issue of negotiation is the price per unit of help-session time. The matchmaker provides to both parties the standard price for help on each topic, which is calculated by a central component based on the difficulty of the topic and the availability of competent helpers/helpees requesting help on the topic. The agent whose user needs help (helpee/buyer) begins the negotiation with the agent of the user who is on the top of the list of possible helpers. The agents do not know anything about the other agents in the system at the start of the negotiation process. Each agent makes an offer depending on the user's risk attitude. The price-offer decay and raise functions are dependent on the money importance and the urgency of the current task/help-request. The agents are allowed to offer the same price more than once — in this way the agents stay rigid on a price, if they do not want to increment or decrement it. However, a seller is not allowed to ask for price greater than his previous offered value and a buyer is not allowed to ask for price less than his previous value within one negotiation session. Each agent ensures that it does not exceed the preferred price limit set by the user (the maximum price for the helpee/buyer and minimum price for helper/seller).

We compared three negotiation approaches. In Approach 1 the agents do not use decision theoretic approach for negotiation. Both parties make offers and counteroffers by making an increment or a decrement by a fixed amount based on the difference between their preferred prices. This corresponds to an "ideal" case of fair trade between agents that are not self-interested, reveal their preferred prices and use the same price offer increment/decrement strategy. In this way no agent wins and no one looses. In Approach 2 the agents use an influence diagram in their decision-making processes, with a chance node corresponding to the opponent's action (thus, they are uninformed about the opponent; the probabilities of the opponent's actions are taken as equal). In Approach 3 the agents create and take into account the model of the opponent in their own decision-making, as explained in section 3.2.

4.2. PERCENTAGE OF REJECTIONS (FAILED NEGOTIATION)

Our experiments showed that increasing the "intelligence" of the negotiating parties leads to increase in the percentage of rejections in negotiation (as can be seen in Table 16.1). This could be interpreted as a negative result since the percentage of failed negotiations increases. However, it can be viewed also positively, since it means that instead of accepting, the agents are rejecting deals that are not profitable. As it was expected, most of the rejections happened in session 3, where there is no overlap between the preferred price intervals of the negotiators. That means that the agents are able to correctly infer that there is no big benefit to be gained in continuing negotiation. From the results one could generalize that the agent using the more advanced approach is usually the one that rejects, which is explainable, since the more "intelligent" negotiator decides to quit when he sees no sense to continue. However, there is an exception — in case (S:2, B:3), where the seller, though using less advanced approach, is responsible for 66.6% of the

Configuration	Total Rejections (in %)	% Rejected by Helpee (Buyer)			% Rejected by Helper (Seller)		
		S1	S2	S3	S1	S2	S3
S:1, B:1	0	-	-	-	-	-	-
S:1, B:2	0	-	-	-	-	-	-
S:1, B:3	20	-	-	100	-	-	0
S:2, B:1	0	-	-	-	-	-	-
S:2, B:2	6.67	-	-	100	-	-	0
S:2, B:3	20	-	-	33.3	-	-	66.6
S:3, B:1	26.6	-	-	0	-	-	100
S:3, B:2	26.6	-	0	0	-	100	100
S:3, B:3	53.3	50	0	66.6	50	100	33.3

Table 16.1 . Rejections in the various configurations. The first column describes the negotiation approaches used by the agents in the configuration, e.g. S:1, B:1 means that both the seller and the buyer use approach 1.

rejections (in session 3 when there is no overlap between the prices). It is hard to explain this result delving into speculations. The fact that in the case when both agents use the approach 3, the one that starts the negotiation (the buyer) is responsible for most of the rejections in session 3 is also hard to explain, since the buyer is one step behind the seller in modelling the opponent, i.e. the seller should be able to predict better that there is no point to continue and to reject.

4.3. QUALITY OF DEAL

A series of experiments were made to investigate the influence of the negotiation mechanism proposed (Approach 3) on the quality of deals achieved by using it. The preferences of the helpee and the helper were kept constant for all configurations and across the three sessions only the negotiation approach of one of the agents was changed. Again, 15 experiments were carried out for each negotiation approach, for each of the 5 different settings of user preferences and in 3 sessions depending on the preferred prices. The deals were analysed from the point of view of the helpee/buyer and of the helper/seller (i.e. depending on which agent changes its negotiation approach). The results (see Table 16.2) show that using the decision theoretic approach and modelling the opponent (Approach 3) brings better quality deals in most of the cases when there is a large preferred price overlap (session 1 and session 2).

Approach 1 is only better when there is little or no scope for negotiation (session 3), which is easy to explain: the agents are in fact

Case comparisons from Buyer's viewpoint	Better deals in S1 (in %)	Better deals in S2 (in %)	Better deals in S3 (in %)
B: 1→2, S: 1	60	25	20
B: 1→2, S: 2	100	80	0
B: 1→2, S: 3	60	0	0
B: 2→3, S: 1	100	80	100
B: 2→3, S: 2	60	40	50
B: 2→3, S: 3	100	50	0
B: 1→3, S: 1	80	60	50
B: 1→3, S: 2	80	80	50
B: 1→3, S: 3	66.6	0	0

Case comparisons from Seller's viewpoint	Better deals in S1 (in %)	Better deals in S2 (in %)	Better deals in S3 (in %)
S: 1→2, B: 1	60	40	25
S: 1→2, B: 2	0	33.3	0
S: 1→2, B: 3	80	60	0
S: 2→3, B: 1	80	100	100
S: 2→3, B: 2	100	100	100
S: 2→3, B: 3	33.3	100	100
S: 1→3, B: 1	100	100	100
S: 1→3, B: 2	100	100	100
S: 1→3, B: 3	66.6	100	100

Table 16.2 . Comparison of the percentage of better deals achieved by using a different strategy. The top part of the table shows the better deals achieved by the helpee/buyer when switching strategies, and the bottom part shows the percentage of better deals achieved by the helper/seller when switching strategies.

co-operative since they reveal their preferred prices. When both agents use Approach 3, the helper/seller gets always a better deal (the last three lines in the right side of the table), because of its advantageous position (one step ahead) in modelling its opponent. The agent that starts the negotiation is in a disadvantaged position, since it reveals its starting price first.

There are some anomalies in the cases when the seller switches to a more advanced approach while the buyer uses the same more advanced approach. For example, when the seller switches from Approach 1 to Approach 2 while the buyer uses Approach 2 in both cases, we see (in the 2nd line on the right side of Table 2) that the more advanced approach does not bring any benefits for the seller in sessions 1 and 3 and brings only a minor benefit (better deals in 1/3 of the cases) in session 2. In the case that the seller switches from Approach 2 to Approach 3 while the buyer uses Approach 3 all the time, we also see that there is not a big benefit in session 1 (only in 33.3% of the cases the seller achieves a better deals). However, in sessions 2 and 3, it achieves always better deals than with Approach 2.

4.4. USER ACCEPTANCE

This negotiation mechanism has been applied in two different deployments of the I-Help system at the University of Saskatchewan (available at `http://www.cs.usask.ca/i-help/`). Deployment 1 in fall 1999 was in a third year computer science class (operating systems) with about 60 students, and deployment 2 took place in fall 2000, in an introductory computer science class (Programming in C++) with 350 second year engineering students. The goal of both experiments was to see if the economy introduced by making personal agents trade help resources for currency helps to motivate human users to offer help. We also wanted to know if human users find negotiation on their behalf by their agents useful and if they trust their agents that they arrange the best deals for them. In both deployments trace data has been collected by I-Help itself as learners interact with the system. In addition we passed out questionnaires at the end to collect data about the students subjective feelings about the system. Unfortunately, the students rarely used the system and most of the usage concerned the public discussion forum provided by I-Help and the electronic help resources. Very few peer-help requests were made in both of the experiments, which did not allow us to test at all the user acceptance of the negotiation mechanism.

A number of social factors seem to affect I-Help usage. The choice of the group had a strong influence on the amount of use. Smaller or more cohesive groups do not need the system. Deployment 1 was with 3rd year students who knew each other very well, had established multiple ways of interacting with one another in the course and in the labs and hence they did not find any need to login to the system to get help. The reasons for this choice were purely pragmatic: time until the beginning of the term was short and implementation for this course required the least adaptation effort, for the domain representation and student modelling were already developed. A similar effect, however, appeared in deployment 2, event though we had a large group (3 parallel sections) of second year Engineering students. Due to the culture of the College of Engineering, involving a lot of group work and extra-curricular activities, the students knew each other well and had established knowledge networks, and they shared the same laboratory space with one another so there was ready access to face-to-face help. In addition, there were technical issues that prevented the students to use the peer-help component of the system. They usually accessed the system from a lab where the software required for their course assignments (Visual C++) is installed, rather than from their own computers. Unfortunately, the lab is using old and slow computers (Pentium I). Running the programming environment simultaneously with a browser consumes the operative memory entirely, which unacceptably slows down the performance in both I-Help and the programming environment.

Has the economy worked? There were a number of help-requests in deployment 1, but it isn't clear that the economy was the motivating factor. Respondents to the questionnaire administered after deployment 1 were evenly split as to whether they found the virtual currency motivating or not. Some students mentioned that it would be good to be able to exchange the accumulated help-currency in marks towards their final grade in the course. Two people were particularly negative. One problem may be that the currency exchange rate of I-Help credit units into things of value in the real world isn't very favourable (minimal prizes have been given for top helpers). Another problem may be that rewarding students solely on the level of their bank account does not take into account the quality of the help given. For example, in deployment 1, one student who was involved in many help sessions in the role of helper (17), abandoned 5 of his helpees during an ongoing discussion. Finally, perhaps the I-Help currency has to be converted into other things than material goods. Several students revealed that their main motivation for posting answers on the public discussion forum was "glory", that in this way they became recognised as "authorities" among their peers. Some other students mentioned that they hoped to attract the attention of the instructor, another form of recognition. Perhaps the I-Help currency has to be mapped onto fame and social status, not prizes. In fact, it seems to be generally recognised that social recognition is an efficient reward system also in many newsgroups and on the Internet for the developers of free software (Raymond, 1999). Though our I-Help data is inconclusive, we believe that some form of reward is necessary to stimulate student participation. The crucial question is the choice of the real world equivalent. The reward should be based on the social values of the group.

5. RELATED WORK ON NEGOTIATION

Earlier work on negotiation in DAI (Durfee and Lesser, 1987; Sasthi and Fox, 1987) is concerned with bringing cooperation and coordination among distributed nodes to improve the global efficiency of the system where the goals and the information of the system are not centralized. The application areas for this type of negotiation include manufacturing, planning, scheduling, meeting scheduling, task and resource allocation in subcontracting networks. DAI approaches typically assume cooperative agents. However, in commerce applications, agents are usually self-interested.

Negotiation protocols and strategy optimality have been studied extensively (Zlotkin and Rosenschein, 1991), from a game theoretic perspective. Game-theoretic approaches are based on a sound mathematical model, but they have some important limitations:

1. the payoff matrix size grows exponentially with the number of agents;
2. it is very hard to find equilibrium points for strategies to be favourable,
3. they tend to focus on the "macro-level", thus modelling the general behaviour of the system, but not on the "micro-level", i.e. the reasoning and decision making process of an individual agent;
4. not much work in this area can be applied to build agents working in practical market based systems,
5. game theory assumes a win-lose situation, while in reality negotiation usually aims at achieving a win-win situation.

Micro-level negotiation mechanisms based on case-based reasoning, argumentation and persuasion techniques have been proposed by Sycara (Sycara, 1988). Recently, the focus of automated negotiation has become saving the user's time and lifting the burden of information and decision overload, as well as the studying the impact of different negotiation mechanisms on the outcome of negotiation. For ensuring success in this type of negotiation it is essential that the parties are aware of each other's moves and motives. In the past researchers have used the history of past negotiations to learn about the opponent. However, using the history of negotiation to learn about the opponent is not efficient when the environment is dynamic, i.e. when new agents appear and leave, and agents change their strategies frequently. Various approaches for modelling the opponent, learning the opponent's strategies and modelling the environment of the system have been studied to see their effect on negotiation. Modelling the opponent in negotiation has also been proposed in game theory and DAI (Carmel and Markovitch, 1998; Vidal and Durfee, 1996). The agents of Carmel and Markovich (Carmel and Markovitch, 1998) use model-based learning and explicit models (based on deterministic finite automata) of their opponents' strategies to generate expectations about their behaviour. Durfee and his students (Gmytrasiewicz and Durfee, 1995; Vidal and Durfee, 1996) have studied the impact of agent modelling each other (recursive modelling) using reinforcement learning in an information economy. Combined approaches to model the opponent and the environment have been proposed, for example, using a Bayesian belief update mechanism (Zheng and Sycara, 1997) and stochastic modelling using Markov chains (Park et al., 1996).

Dealing with risk attitudes is an important feature of decision-making. A few researchers (Zlotkin and Rosenschein, 1991) have taken into account risk attitudes for negotiation, but their work has mainly utilized the Zeuthen's Principle (Harsanyi, 1977) in a game theoretic situation to determine who is the person more willing to make concessions. We utilize the risk attitudes in a different way, as discussed in section 3.

Finally, we are not aware of much work aimed at evaluating agents negotiating on behalf of human users in a social context and in a real application with respect to user acceptance, trust and appropriateness of the economic interaction. Even though, we currently have no strong results in this area, we believe our work is an important first step in the right direction.

6. CONCLUSIONS

In open environments with self-interested agents, a decision-theoretic approach to negotiation with modelling the opponent has proven to be quite advantageous. In this paper we have presented a negotiation mechanism that utilizes an influence diagram taking into account the preferences and the risk behaviour of the user. Such a decision model allows to take into account and to handle tradeoffs among the factors that affect decision-making in negotiation and can be applied to any domain.

We have extended this negotiation mechanism to create and use a model of the opponent, represented with a probabilistic influence diagram. Our experimental results show that this mechanism finds usually a better deal for the agent who uses it when there is space for negotiation. We have implemented this negotiation mechanism in personal agents representing human users in an Internet-based virtual market environment for peer help (I-Help). The negotiation mechanism proposed in this paper is not restricted to intangible goods. It can be generalized to other market domains.

Acknowledgements

The authors would like to thank Silvia Breban, a graduate student at the University of Saskatchewan, for re-implementing the negotiation mechanism described in this paper so that it could be incorporated in the I-Help agents. Thanks to Lori Kettel and Susan Bull for analysing the data from the two deployments of I-Help. This work has been partially supported by Telelearning Network of Centres of Excellence under Project 6.28.

References

Carmel, D. and Markovitch, S. (1998). Learning models of intelligent agents. In *Proceedings of the 3rd International Conference on Multi-agent Systems*, pages 64–71.

Durfee, E. and Lesser, V. (1987). Negotiating task decomposition and allocation using partial global planning. In Huhns, M. and Gasser, L., editors, *Distributed Artificial Intelligence, Volume 2*, pages 229–243. Morgan Kaufmann, San Mateo, CA.

Gmytrasiewicz, P. and Durfee, E. (1995). A rigorous operational formalization of recursive modelling. In *Proceedings of the 1st International Conference on Multiagent Systems*, pages 125–132.

Harsanyi, J. (1977). *Rational Behaviour and Bargaining Equilibrium in Games and Soocial Situations.* Cambridge University Press, Cambridge, UK.

Maes, P., Guttman, R., and Moukas, G. (1997). Agents that buy and sell. *Communications of the ACM*, 42(3):81–83.

Park, S., Durfee, E., and Birmingham, W. (1996). Advantages of strategic thinking in multi-agent contracts. In *Proceedings of the 2nd International Conference on Multiagent Systems*, pages 259–266.

Raymond, E. (1999). Homesteading the Noosphere, the Cathedral and the Bazaar: Musings on Linux and Open Source by an Accidental Revolutionary. O'Reilly and Associates. available at `http://www.tuxedo.org/~esr/writings`.

Sasthi, A. and Fox, S. (1987). Constraint directed negotiation of resource allocation. In Huhns, M. and Gasser, L., editors, *Distributed Artificial Intelligence, Volume 2*, pages 163–194. Morgan Kaufmann, San Mateo, CA.

Shachter, R. (1986). Evaluating influence diagrams. *Operations Research*, 34(36):871–882.

Shachter, R. (1988). Probabilistic inference and influence diagrams. *Operations Research*, 36(4):589–604.

Suryadi, D. and Gmytrasiewicz, P. (1999). Learning models of other agents using influence diagrams. In *Proceedings of the 7th International Conference on User Modelling*, pages 223–232.

Sycara, K. (1988). Resolving goal conflicts via negotiation. In *Proceedings of the 7th National Conference on Artificial Intelligence.*

Vassileva, J., Greer, J., McCalla, G., Deters, R., Zapata, D., Mudgal, C., and Grant, S. (1999). A multi-agent approach to the design of peer-help environments. In *Proceedings of the International Conference on AI in Eductaion*, pages 38–45.

Vidal, J. and Durfee, E. (1996). The impact of nested agent models in an information economy. In *Proceedings of the 2nd International Conference on Multiagent Systems*, pages 377–384.

Zheng, D. and Sycara, K. (1997). Benefits of learning in negotiation. In *Proceedings of the 14th National Conference on Artificial Intelligence*, pages 36–41.

Zlotkin, G. and Rosenschein, J. (1991). Cooperation and conflict resolution via negotiation among autonomous agents in non cooperative domains. *IEEE Transactions on Systems, Man and Cybernetics*, 21(6):1317–1332.

Chapter 17

ROBUST COMBINATORIAL AUCTION PROTOCOL AGAINST FALSE-NAME BIDS

Makoto Yokoo
NTT Communication Science Laboratories, NTT Corporation.
yokoo@cslab.kecl.ntt.co.jp

Yuko Sakurai
NTT Communication Science Laboratories, NTT Corporation.
yuko@cslab.kecl.ntt.co.jp

Shigeo Matsubara
NTT Communication Science Laboratories, NTT Corporation.
matsubara@cslab.kecl.ntt.co.jp

Abstract In this paper, we develop a new combinatorial auction protocol to thwart the fraudulent practice called *false-name bids.* Auctions conducted over the Internet have become immensely popular and form a pillar of Electronic Commerce (EC) as we know it today. In particular, the Internet offers promising opportunities for executing complicated types of auctions such as combinatorial auctions, where multiple goods are sold at the same time. However, this potential also opens the door to a new method of cheating called false-name bids, where an agent attempts to profit from submitting several bids under fictitious names.

When there is no possibility of false-name bids, the Generalized Vickrey Auction protocol (GVA) is suitable because it satisfies three key conditions, i.e., individual rationality, Pareto efficiency, and incentive compatibility. However, if agents can submit false-name bids, it is theoretically impossible for a combinatorial auction protocol to simultaneously satisfy all of the three conditions.

Considering the above drawback, we have modified the GVA to a new protocol called the Leveled Division Set protocol (LDS). The LDS adaptively decides whether to sell goods in a bundle or separately based on reservation prices of the auctioned goods. The LDS can guarantee individual rationality and incentive compatibility but cannot always guarantee to achieve a Pareto efficient social surplus. The results of simulations show that the LDS can achieve a superior social surplus compared with a a protocol that sells all goods in one bundle.

Keywords: Mechanism design, auction, game theory

1. INTRODUCTION

Internet auctions have become a popular part of Electronic Commerce (EC), and there has already been much theoretical and practical research on them (Monderer and Tennenholtz, 1998; Sandholm, 1996; Wurman et al., 1998). In particular, studies on combinatorial auctions have recently attracted much attention (Fujishima et al., 1999; Klemperer, 1999; Rothkopf et al., 1998; Sandholm, 1999). In combinatorial auctions, multiple items with interdependent values are simultaneously sold. As opposed to conventional single-item auctions, bidders in a combinatorial auction may bid on any combination of items.

The bidder in a combinatorial auction can declare either complemental or substitutional preferences in making multiple bids. For example, in a spectrum auction held by the Federal Communications Commission (FCC) (McMillan, 1994), a bidder may try to buy licenses for adjoining regions simultaneously (regions are complemental) but be indifferent to which particular channel is bought (channels are substitutional). By designing a system that takes into account complemental and substitutional preferences, we can increase both the participants' utilities (social surplus) and the revenue of the seller.

The Internet provides an excellent infrastructure for combinatorial auctions, but it also raises the problem of new types of cheating. For example, an agent may try to gain profit by submitting several bids under fictitious names. Since it is virtually impossible to identify each auction participant on the Internet, such fraud is very difficult to detect. We call a bid made under a fictitious name a *false-name bid.* The issue of collusion has been studied by many researchers (Klemperer, 1999; Rasmusen, 1994; Sandholm, 1996; Varian, 1995). However, a false-name bid is even more insidious than collusion because it can be done alone, while a bidder has to seek out and persuade other bidders to join in collusion.

In previous work, the authors have analyzed the effects of false-name bids on auction protocols (Sakurai et al., 1999; Yokoo et al., 2000). The following summarizes our findings.

- The Generalized Vickrey Auction protocol (GVA) (Varian, 1995) has been proven to satisfy individual rationality, Pareto efficiency, and incentive compatibility when no false-name bid can be made. However, it fails to satisfy incentive compatibility and Pareto efficiency when there is the possibility of making false-name bids.

- No combinatorial auction protocol can ensure that incentive compatibility, Pareto efficiency, and individual rationality are simultaneously satisfied when agents can submit false-name bids.

- The revelation principle (Mas-Colell et al., 1995) still holds even when false-name bids are possible.

This paper concentrates on private value auctions (Mas-Colell et al., 1995; Rasmusen, 1994), in which each agent knows its own preference and has an evaluation value of the goods that is independent of the values of the other agents. Also, we assume an agent's utility is represented as the difference between its true evaluation value of the allocated goods and the payment made for the allocated goods. Such a utility is called a *quasi-linear* utility (Mas-Colell et al., 1995; Rasmusen, 1994). These two assumptions are commonly used to make theoretical analyses tractable.

An auction protocol is incentive compatible if bidding the true private values of goods is the dominant strategy for each agent, which means the optimal strategy regardless of other agents' actions. According to the revelation principle, we can restrict our attention to incentive compatible protocols when designing an auction protocol without the loss of generality (Mas-Colell et al., 1995). This implies that if a certain property (e.g., Pareto efficiency, individual rationality) can be achieved in a dominant strategy equilibrium by using some auction protocol, that property can also be satisfied in a dominant strategy equilibrium by using an incentive compatible auction protocol. Auction protocols are considered robust against false-name bids if no agent can profit from submitting false-name bids. If an auction protocol is not robust in this way, it does not provide incentive compatibility.

A Pareto efficient protocol allocates goods to the bidders with the highest evaluation values and maximizes the sum of all participants' utilities (including that of the seller), i.e., the social surplus. In a more general setting, Pareto efficiency does not necessarily ensure a maximized social surplus. In an auction, however, agents can transfer money among themselves, and each agent's utility is quasi-linear. Therefore, a Pareto efficient allocation always maximizes the sum of the utilities.

An auction protocol is individually rational if no participant suffers any loss, that is, if the payment never exceeds the participant's evaluation value of the auctioned goods. In a private value auction, a protocol must satify individual rationality because no agent would participate in an auction where it might be charged more money than it is willing to pay.

Since it is impossible to simultaneously satisfy these three properties, we must abandon Pareto efficiency and devise an auction protocol that satisfies incentive compatibility and individual rationality while still achieving a good social surplus. This is because individual rationality is indispensable and we can restrict our attention to incentive compatible protocols without sacrificing generality. Moreover, an incentive compatible protocol has practical advantages in Internet auctions. As discussed in (Varian, 1995), a participant must explicitly declare his/her preferences to a computerized agent to make it function effectively. However, if we use a standard auction protocol, such as the first-price sealed-bid auction protocol (Rasmusen, 1994), the user must ensure that his/her preferences are not revealed to the agents of the other participants. On the other hand, an incentive compatible protocol renders any knowledge of other participants' preferences useless, and privacy is no longer an issue.

In the next section, we describe the GVA and show an example where it is not robust against false-name bids (Section 2). Then, we describe our newly developed protocol, the *Leveled Division Set (LDS) protocol* (Section 3), and prove that it can satisfy incentive compatibility (Section 4). Next, we present simulation results to demonstrate that the LDS can achieve a better social surplus than a protocol that always sells goods in one bundle (Section 5). Finally, we weigh the advantages and disadvantages of our new protocol (Section 6).

2. GENERALIZED VICKREY AUCTION PROTOCOL (GVA)

The GVA protocol is a type of the Clarke mechanism (Clarke, 1971; Mas-Colell et al., 1995) and a generalized version of the Vickrey, or second-price sealed-bid, auction protocol (Mas-Colell et al., 1995; Varian, 1995). In the second-price sealed-bid auction protocol, each agent bids for a single item, and the agent with the highest bid wins and pays the amount of the second highest bid.

Let us consider a set of agents $N = \{1, 2, \ldots, n\}$ and a set of all auctioned goods $M = (1, 2, \ldots, m)$. If we allow the possibility of not selling a good to any agent, there are $(n+1)^m$ possible allocations. Let us represent one possible allocation as m-element vector $G =< g_1, g_2, \ldots, g_m >$, where each element g_i is an integer $0 \leq g_i \leq n$, which means that good i is allocated to agent g_i. If $g_i = 0$, we assume that good i is not sold to any agent. The set of all possible allocations can be represented as SG.

The outline of the GVA can be described as follows.

1. Each agent declares evaluation values for each possible allocation $G \in SG$. Let $v_x(G)$ denote agent x's declared evaluation value for allocation G. The reported evaluation values may or may not be true.

2. The GVA chooses the allocation G^* that maximizes the social surplus, i.e., the sum of all agents' declared evaluation values defined as
$$G^* = \arg \max_{G \in SG} (\sum_{y \in N} v_y(G)). \tag{1}$$

3. The payment of agent x (represented as p_x) is calculated as
$$p_x = \sum_{y \neq x} v_y(G^*_{\sim x}) - \sum_{y \neq x} v_y(G^*). \tag{2}$$

Here, $G^*_{\sim x}$ is the allocation that maximizes the sum of all agents' evaluation values except agent x and is defined as
$$G^*_{\sim x} = \arg \max_{G \in SG} (\sum_{y \neq x} v_y(G)). \tag{3}$$

In the GVA, we can assume that agent x pays the decreased amount of the other agents' evaluation values resulting from its participation.

The GVA has been proven to be incentive compatible if there is no false-name bid (Mas-Colell et al., 1995; Varian, 1995). The proof is as follows. Since we assume that the utility of each agent is quasi-linear, agent x's utility after payment is given by the following formula. $u_x(G^*)$ denotes agent x's true evaluation value for G^*.
$$u_x(G^*) - p_x = u_x(G^*) + \sum_{y \neq x} v_y(G^*) - \sum_{y \neq x} v_y(G^*_{\sim x}) \tag{4}$$

The third term in formula (4), i.e., $\sum_{y \neq x} v_y(G^*_{\sim x})$, is independent of agent x's declaration (assuming that there is no false-name bid); thus agent x can maximize its utility when the sum of the first and second items in formula (4), i.e., $u_x(G^*) + \sum_{y \neq x} v_y(G^*)$, is maximized.

On the other hand, the optimal allocation G^* is chosen so that the sum of the agents' declared evaluation values are maximized, i.e.,

$$G^* = \arg\max_G (v_x(G) + \sum_{y \neq x} v_y(G)). \tag{5}$$

Therefore, agent x can maximize its utility by submitting its true evaluation value, i.e., by setting $v_x(G) = u_x(G)$.

In the following example, we show how the GVA works.

Example 1 *Let us assume three agents are participating in an auction of two different goods, A and B, and declare the following evaluation values. The evaluation values of an agent are expressed as a triplet: (the value for A alone, the value for B alone, the value for A and B together).*

- *agent 1: (6, 0, 6)*
- *agent 2: (0, 0, 8)*
- *agent 3: (0, 6, 6)*

The evaluation values of agent 2 are all-or-nothing, i.e., having only one good is useless. In this case, A is allocated to agent 1 and B is allocated to agent 3. The payment of agent 1 is calculated as follows. If agent 1 does not participate, both goods are allocated to agent 2, and the social surplus is 8. *When agent* 1 *does participate, agent 3 obtains B and the social surplus except for agent 1 is* 6. *Therefore, agent 1 pays the difference* $8 - 6 = 2$. *The obtained utility of agent 1 is* $6 - 2 = 4$. *The payment and utility of agent 3 are identical to those of agent 1.*

In the next example, the GVA is not robust against false-name bids.

Example 2 *Let us assume there are only two agents.*

- *agent 1: (6, 6, 12)*
- *agent 2: (0, 0, 8)*

In this case, both goods are allocated to agent 1. Its payment is calculated as 8, since if agent 1 does not participate, agent 2 obtains both goods and the social surplus is 8, and when agent 1 does participate, agent 1 obtains all goods and the social surplus except for agent 1 is 0.

However, agent 1 can use a false-name 3 and create a situation identical to Example 1. In this case, agent 1 can obtain both goods by paying only $2 + 2 = 4$. *Therefore, its utility becomes* $12 - 4 = 8$, *which means that agent 1 can make a profit by submitting a false-name bid.*

3. ROBUST PROTOCOL AGAINST FALSE-NAME BIDS

3.1. BASIC IDEAS

A trivial protocol can satisfy incentive compatibility even if agents can submit false-name bids. Here, all goods are sold in a bundle and the second-price sealed-bid auction protocol is used to determine the winner and its payment. We call this simple protocol the *set protocol.* Selling goods in a bundle makes sense if goods are complementary[1] for all agents, i.e., the utility of a set of goods is larger than the sum of the utilities of having each good separately. However, if goods are substitutional for some agents, i.e., the utility of a set of goods is smaller than the sum of the utilities of having each good separately, the set protocol is wasteful. In fact, the social surplus and the revenue of the seller can be significantly worse than those achieved by the GVA. Consider a simple case where there are two goods A and B. To increase the social surplus, we must design a protocol where the goods can be sold separately in some cases, and to guarantee that the protocol is robust against false-name bids, we must satisfy the following condition.

Proposition 1 *If A and B are sold separately to different agents, the sum of the payments must be larger than the highest declared evaluation value for the set of A and B.*

If this condition is not satisfied, a single agent may be using two false-names to obtain the goods. However, designing an incentive compatible protocol satisfying this condition is rather difficult. For example, the following simple protocol does not satisfy incentive compatibility.

Vulnerable Protocol: Determine tentative winners and payments using the GVA. If A and B are sold separately to different agents, and the sum of the payments does not satisfy the condition of Proposition 1, then sell the goods in a bundle, otherwise, use the results of the GVA.

Assume the situation in Example 1. Since the result of the GVA does not satisfy the condition in Proposition 1, both goods are sold to agent 2. However, agent 1 can create the following situation by using false-names 4 and 5.

- agent 1: (6, 0, 6)
- agent 2: (0, 0, 8)
- agent 3: (0, 6, 6)

- agent 4: (3, 0, 3)
- agent 5: (0, 5.9, 5.9)

In this case, the GVA allocates A to agent 1 and B to agent 3, and agent 1 pays 3 and agent 3 pays 5.9. Accordingly, the condition in Proposition 1 is satisfied. If an agent can submit false-name bids, manipulating the payments of other agents is rather easy. In this case, agent 1 manipulates the result by forcing agent 3 to pay more by submitting a false-name bid. In essence, we must solve the difficult dilemma of satisfying the above condition on payments without using the second highest evaluation values, which is essential to calculating the payments.

Our new protocol solves this dilemma by utilizing the *reservation prices* of goods (Rasmusen, 1994). The seller does not sell a good if the payment for the good is smaller than the reservation price. Let us assume that the reservation prices of A and B are r_A and r_B, respectively. If we sell goods separately only when the highest declared evaluation value for the set is smaller than $r_A + r_B$, we can satisfy the condition of Proposition 1. In the following, we show how this idea can be incorporated in the GVA.

3.2. LEVELED DIVISION SET PROTOCOL

In the following, we define several terms and notations. To help readability, we use three different types of parentheses, i.e., (),{}, and [], to represent sets.

- a set of agents $N = \{1, 2, \ldots, n\}$
- a set of all auctioned goods $M = (1, 2, \ldots, m)$
- a division of goods $D = \{S \mid S \subseteq M\}$, where $\forall S, S' \in D$ and $S \neq S'$, $S \cap S' = \emptyset$ holds[2]
- For each good j, the reservation price r_j is defined.
- For a set of goods S, we define $R(S)$ as $\sum_{j \in S} r_j$.

A leveled division set is defined as follows.

- Levels are defined as $1, 2, \ldots, max\_level$.
- For each level i, a division set $SD_i = [D_{i1}, D_{i2}, \ldots]$ is defined.

A leveled division set must satisfy the following three conditions.

- $SD_1 = [\{M\}]$: the division set of level 1 contains only one division, which consists of a set of all goods.

	case 1	case 2	case 3
level 1	[{(A,B)}]	[{(A,B,C)}]	[{(A,B,C,D)}]
level 2	[{(A),(B)}]	[{(A,B)}, {(B,C)}, {(A,C)}]	[{(A,B,C)}, {(B,C,D)}, {(A,D)}]
level 3		[{(A),(B),(C)}]	[{(A),(D),(B,C)}]

Figure 17.1. Examples of leveled division sets.

- For each level and its division set, a union of multiple sets of goods in a division is always included in a division of a smaller level, i.e., $\forall i > 1, \forall D_{ik} \in SD_i, \forall D' \subseteq D_{ik}$, where $\mid D' \mid\geq 2, S_u = \bigcup_{S \in D'} S$, then there exists a level $j < i$, with a division set SD_j, where $D_{jl} \in SD_j$ and $S_u \in D_{jl}$.

- For each level and its division set, each set of goods in a division is not included in a division of a different level[3], i.e., $\forall i, \forall D_{ik} \in SD_i, \forall S \in D_{ik}, \forall j \neq i, \forall D_{jl} \in SD_j, S \notin D_{jl}$.

Examples of leveled division sets are shown in Figure 17.1. In case 1, there are two goods (A and B), and in cases 2 and 3, there are three and four goods, respectively.

Let us assume agent 0 is a *dummy agent*, whose evaluation value of the good j is equal to the reservation price r_j. For a division $D = \{S_1, S_2, .., S_l\}$ and one possible allocation of goods $G =< g_1, g_2, .., g_m >$, we say G is allowed under D if the following conditions are satisfied.

1. Multiple goods that belongs to the same set in the division must be allocated to the same agent, i.e., $\forall S \in D$, $\forall i, j \in S$, $g_i = g_j$ holds.

2. If two goods belong to different sets in the division, they must be allocated to different agents, except when they are allocated to the dummy agent 0, i.e., $\forall S, S' \in D$, where $S \neq S'$, $\forall i \in S, \forall j \in S'$, $g_i \neq g_j$ or $g_i = g_j = 0$ hold.

3. If a good does not belong to any set in the division, it must be allocated to the dummy agent 0, i.e., $\forall i$, if $\forall S \in D, i \notin S$ holds, then $g_i = 0$.

Note that we allow some set of goods to be allocated to the dummy agent 0. For each level i and its division set $SD_i = [D_{i1}, D_{i2}, \ldots]$, we represent the union of all allowed allocations for each element of SD_i as SG_i.

Before executing the leveled division set protocol (LDS protocol), the auctioneer must pre-define the leveled division set and the reservation prices of goods. Each agent x declares its evaluation value $B(x, S)$ for each subset of goods S, which may or may not be true. The declared evaluation value of agent x for an allocation G (represented as $v_x(G)$) is defined as $B(x, S)$ if S is allocated to agent x in G, otherwise $v_x(G) = 0$. Also, we define the evaluation value of the dummy agent 0 for an allocation G as the sum of the reservation prices of goods allocated to agent 0.

The winners and payments are determined by calling the procedure LDS(1), which is defined as follows.

Procedure LDS(i)

Step 1: If there is only one agent $x \in N$ whose evaluation values satisfy the following condition: $\exists D_{ik} \in SD_i, \exists S_x \in D_{ik}$, where $B(x, S_x) \geq R(S_x)$, then compare the result obtained by the GVA(i) defined below and that by LDS$(i+1)$ and choose the procedure[4] that gives the larger utility for agent x. Here, we call agent x a *pivotal* agent. If we choose the result of LDS$(i + 1)$, we don't assign any good, nor transfer money, to agents other than x, although the assigned goods for agent x and its payment are calculated as if goods were allocated to the other agents.

Step 2: If there are at least two agents $x_1, x_2 \in N, x_1 \neq x_2$ whose evaluation values satisfy the following condition: $\exists D_{ik} \in SD_i, \exists D_{il} \in SD_i, \exists S_{x_1} \in D_{ik}, \exists S_{x_2} \in D_{il}$, where $B(x_1, S_{x_1}) \geq R(S_{x_1})$, $B(x_2, S_{x_2}) \geq R(S_{x_2})$, then apply the GVA$(i)$.

Step 3: Otherwise: call LDS$(i + 1)$, or terminate if $i = max\_level$.

Procedure of GVA(i)**:** Choose an allocation $G^* \in SG_i$ such that it maximizes $\sum_{y \in N \cup \{0\}} v_y(G)$. The payment of agent x (represented as p_x) is calculated as $\sum_{y \neq x} v_y(G^*_{\sim x}) - \sum_{y \neq x} v_y(G^*)$, where $G^*_{\sim x} \in SG_i$ is the allocation that maximizes the sum of all agents' (including the dummy agent 0) evaluation values except that of agent x.

The procedure of the GVA(i) is equivalent to that of the GVA, except that the possible allocations are restricted to SG_i. We say that the *applied level* of the LDS protocol is i if the result of GVA(i) is used.

3.3. EXAMPLES OF PROTOCOL APPLICATION

Example 3 *Let us assume there are two goods A and B, the reservation price of each good is 50, the leveled division set is defined as case 1 in Figure 17.1, and the evaluation values of agents are defined as follows.*

	A	*B*	*AB*
agent 1	*80*	*0*	*110*
agent 2	*0*	*80*	*105*
agent 3	*60*	*0*	*60*

Since there are two agents whose evaluation values for the set are larger than the sum of the reservation prices (i.e., 100), the condition in Step 2 of LDS(1) *is satisfied; agent 1 obtains both goods by paying 105. This allocation is not Pareto efficient; in a Pareto efficient allocation, agent 1 would obtain A and agent 2 would obtain B.*

Example 4 *The problem setting is basically the same as Example 3, but the evaluation values are defined as follows.*

	A	*B*	*AB*
agent 1	*80*	*0*	*80*
agent 2	*0*	*80*	*80*
agent 3	*60*	*0*	*60*

There is no agent whose evaluation value of the set is larger than 100. In this case, the condition in Step 3 of LDS(1) *is satisfied, and then the condition in Step 2 of LDS*(2) *is satisfied. As a result, agent 1 obtains A and agent 2 obtains B. Agent 1 pays 60, and agent 2 pays the reservation price 50.*

Example 5 *The problem setting is basically the same as Example 3, but the evaluation values are defined as follows.*

	A	*B*	*AB*
agent 1	*80*	*0*	*110*
agent 2	*0*	*80*	*80*
agent 3	*60*	*0*	*60*

There is only one agent whose evaluation value of the set is larger than 100. The condition in Step 1 of LDS(1) *is satisfied, and agent 1 is the pivotal agent. Agent 1 prefers obtaining only A (with the payment 60) to obtaining both A and B (with the payment 100). Therefore, agent 1 obtains A and pays 60.*

In Example 5, B is not allocated to any agent. This might seem wasteful, but it is necessary to guarantee incentive compatibility. In Example 3, if agent 2 declares its evaluation value for the set as 80, the situation becomes identical to Example 5. If we allocate the remaining good B to agent 2, under-bidding becomes profitable for agent 2.

Example 6 *There are three goods A, B, and C. The reservation price for each is 50, and the leveled division set is defined as case 2 in Figure 17.1. The evaluation values of agents are defined as follows.*

	A	*B*	*C*	*AB*	*BC*	*AC*	*ABC*
agent 1	*60*	*30*	*30*	*90*	*60*	*90*	*120*
agent 2	*30*	*60*	*30*	*90*	*90*	*60*	*120*
agent 3	*30*	*30*	*60*	*60*	*90*	*90*	*120*

The condition in Step 2 of LDS(3) is satisfied. Agents 1, 2, 3 obtain A, B, C, respectively, and each pays the reservation price 50.

4. PROOF OF INCENTIVE COMPATIBILITY

The LDS protocol clearly satisfies individual rationality. In this section, we prove that it also satisfies incentive compatibility.

Theorem 1 *The LDS protocol satisfies incentive compatibility even if agents can submit false-name bids.*

To prove Theorem 1, we use the following lemmas.

Lemma 2 *In the LDS protocol, the payment of agent x, who obtains a set of goods S, is larger than (or equal to) the sum of the reservation prices $R(S)$.*

The proof is as follows. Assume that the applied level is i. The payment of agent x (represented as p_x) is defined as $p_x = \sum_{y \neq x} v_y(G^*_{\sim x}) - \sum_{y \neq x} v_y(G^*)$. The set of allocations SG_i considered at level i contains an allocation G', which is basically the same as G^*, but all goods in S are allocated to the dummy agent 0 rather than x. The following formula holds.

$$\sum_{y \neq x} v_y(G') = \sum_{y \neq x} v_y(G^*) + R(S) \tag{6}$$

Since $G^*_{\sim x}$ is the allocation that maximizes the sum of all agents' evaluation values (including the dummy agent) except x in SG_i, $\sum_{y \neq x} v_y(G') \leq$

$\sum_{y \neq x} v_y(G^*_{\sim x})$ holds. Thus, the following formula also holds.

$$\begin{aligned} p_x &= \sum_{y \neq x} v_y(G^*_{\sim x}) - \sum_{y \neq x} v_y(G^*) \\ &\geq \sum_{y \neq x} v_y(G') - \sum_{y \neq x} v_y(G^*) = R(S) \end{aligned} \tag{7}$$

Lemma 3 *In the LDS protocol, an agent cannot increase its utility by submitting false-name bids.*

The proof is as follows. Assume that agent x uses two false names x' and x'' to obtain two sets of goods $S_{x'}$ and $S_{x''}$, respectively. Also, assume that the applied level is i. From Lemma 2, the payments $p_{x'}$ and $p_{x''}$ satisfy $p_{x'} \geq R(S_{x'})$ and $p_{x''} \geq R(S_{x''})$. Now, assume that agent x declares the evaluation value $R(S)$ for the set $S = S_{x'} \cup S_{x''}$ by using a single identifier. From the condition of a leveled division set, there exists a level $j < i$, where $S \in D_{jl}$, $D_{jl} \in SD_j$ holds. In this case, the condition in Step 1 of LDS(j) is satisfied, i.e., only agent x declares evaluation values that are larger than or equal to the sum of reservation prices. Thus, $\sum_{y \neq x} v_y(G^*_{\sim x}) = R(M)$, and $\sum_{y \neq x} v_y(G^*) = R(M) - R(S)$ hold. As a result, the payment becomes $R(S) \leq p_{x'} + p_{x''}$, i.e., the payment of agent x becomes smaller than (or equal to) the case where agent x uses two false names. Similarly, we can show that even when an agent uses more than two identifiers, the payment of this agent would become smaller (or remain the same) if it used only one identifier.

Lemma 3 states that false-name bids are not effective in the LDS protocol. The following shows that truth-telling is the dominant strategy for each agent assuming that each agent uses a single identifier.

The following lemma holds.

Lemma 4 *When there is no false-name bid, and the applied level of the LDS protocol remains the same, an agent can maximize its utility by declaring its true evaluation values.*

The proof is as follows. As long as the applied level does not change, the possible allocation set SG_i does not change. The payment of agent x is defined as $\sum_{y \neq x} v_y(G^*_{\sim x}) - \sum_{y \neq x} v_y(G^*)$. The true evaluation value of agent x of an allocation G is represented as $u_x(G)$. The utility of agent x is represented as $u_x(G^*) + \sum_{y \neq x} v_y(G^*) - \sum_{y \neq x} v_y(G^*_{\sim x})$, i.e., the difference between the evaluation value and the payment. The third item of this formula is determined independently of agent x's declaration if there is no false-name bid. Therefore, agent x can maximize its utility by maximizing the sum of the first two items. On the other hand, the allocation G^* is chosen to maximize $\sum_{y \in N \cup \{0\}} v_y(G) = v_x(G) + \sum_{y \neq x} v_y(G)$. Therefore, agent x can maximize its utility by declaring $v_x(G) = u_x(G)$, i.e., declaring its true utility.

Next, we show that an agent cannot increase its utility by changing the applied level.

Lemma 5 *An agent cannot increase its utility by over-bidding so that the applied level decreases.*

The proof is as follows. Assume that when agent x truthfully declares its utility, the applied level is i, and by over-bidding, the applied level is changed to $j < i$. In this case, for every set of goods S included in the divisions of level j, agent x's evaluation value of S must be smaller than the sum of the reservation prices $R(S)$; otherwise, level j is applied when agent x tells the truth. On the other hand, by Lemma 2, the payment for a set S is always larger than the sum of the reservation prices $R(S)$, which means that agent x cannot obtain a positive utility by over-bidding.

Lemma 6 *An agent cannot increase its utility by under-bidding so that the applied level increases.*

The proof is as follows. Agent x can increase the applied level only in the following two cases.

1. Agent x is the pivotal agent when agent x truthfully declares its evaluation values.

2. By under-bidding, another agent y becomes the pivotal agent.

In the first case, if agent x tells the truth, agent x is the pivotal agent and the larger level is applied if agent x prefers the result of that level. Consequently, under-bidding is useless. In the second case, agents other than y cannot obtain any good, and the utility of agent x becomes 0. In both cases, agent x cannot increase its utility by under-bidding.

From these lemmas, we derive Theorem 1. □

5. EVALUATION

In the LDS protocol, the social surplus and the revenue of the seller can vary significantly according to the leveled division set and reservation prices. In this section, we show how the social surplus changes according to the reservation prices through a simulation in a simple setting where there are only two goods A and B, and the leveled division set is defined as case 1 in Figure 17.1.

We use the following method to determine the evaluation values of agent x.

- Determine whether the goods are substitutional or complementary for agent x. With probability p, the goods are substitutional, and with probability $1 - p$, the goods are complementary.
  - When the goods are substitutional: for each good, randomly choose its evaluation value from within the range of $[0, 1]$. The evaluation value of the set is the maximum of the evaluation value of A and that of B (having only one good is enough).
  - When the goods are complementary: the evaluation value of A or B is 0. Randomly choose the evaluation value of the set from within the range of $[0, 2]$ (all-or-nothing).

Figure 17.2 shows the result where $p = 0.5$ and the number of agents N is 10. We created 100 different problem instances and show the average of the social surplus while varying the reservation price. Both A and B have the same reservation price. For comparison, we show the social surplus of the GVA on the assumption that there is no false-name bid, i.e., the Pareto efficient social surplus[5], and the social surplus of the set protocol.

When the reservation price is small, the results of the LDS protocol are identical to the set protocol. We can see that by setting an appropriate reservation price the obtained social surplus becomes larger than that for the set protocol. When the probability that the goods are substitutional becomes high, the difference between the set protocol and the GVA and that between the set protocol and the LDS protocol, also become large.

6. DISCUSSION

As far as the authors know, the LDS protocol is the first non-trivial protocol that is robust against false-name bids. One shortcoming of this protocol is that when the leveled division set and reservation prices are not determined appropriately, some goods may not get sold. In that case, the social surplus and the revenue of the seller might be smaller than that for the set protocol.

One advantage of the LDS protocol over the GVA is that it has lower communication/computation costs. To execute the GVA, the bidder must declare its evaluation values for all possible subsets of the goods. Also, the seller must solve a complicated optimization problem to determine the winners and their payments (Fujishima et al., 1999; Rothkopf et al., 1998; Sandholm, 1999). In the LDS protocol, the allowed divisions

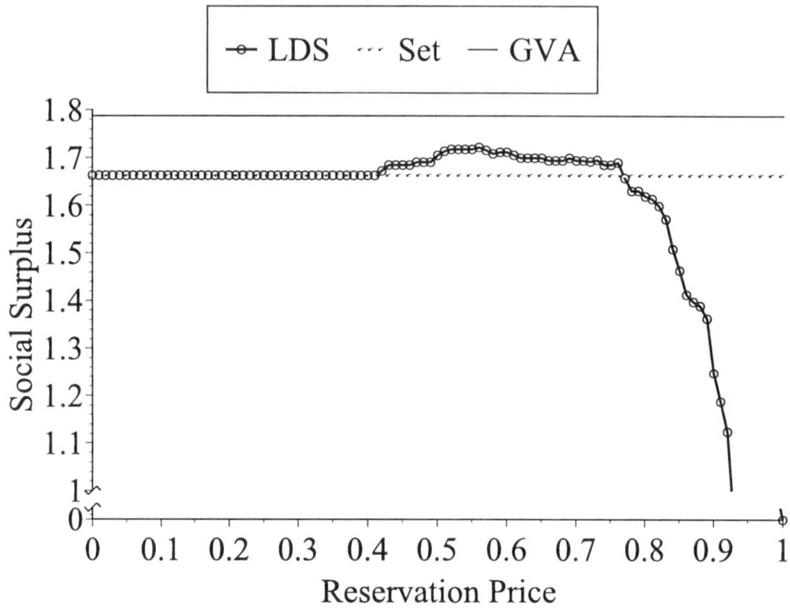

Figure 17.2. Comparison of social surplus ($p = 0.5$).

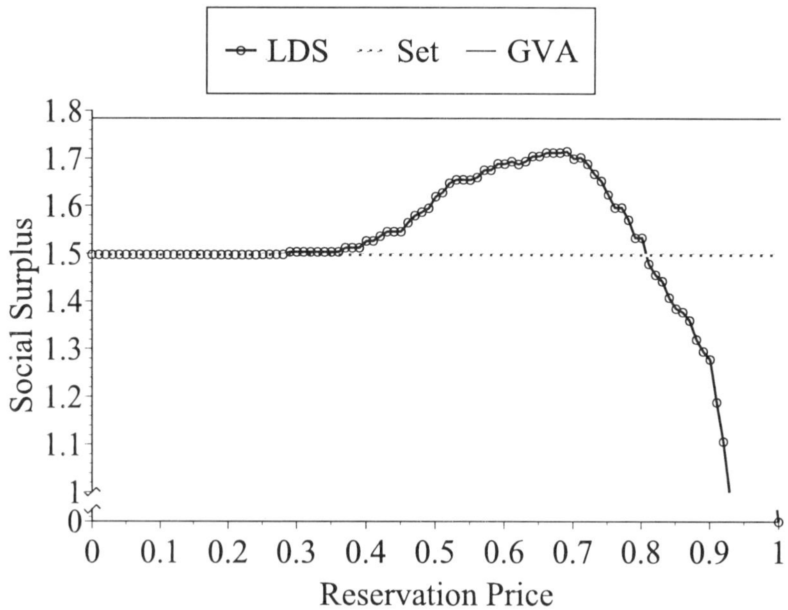

Figure 17.3. Comparison of social surplus ($p = 0.7$).

are pre-determined, and bidders only have to submit bids for these subsets. Furthermore, the search space of the possible allocations is much smaller than that in the GVA[6]. In (Rothkopf et al., 1998), possible sets where agents could submit bids were restricted to efficiently determine winners in combinatorial auctions. The leveled division set introduced in this paper is intended to guarantee robustness against false-name bids, and there is no direct relationship between this leveled division set and the methods of dividing goods described in (Rothkopf et al., 1998).

7. CONCLUSIONS

This paper presented the Leveled Division Set protocol (LDS), a new combinatorial auction protocol that provides robustness against false-name bids. This protocol achieves individual rationality, incentive compatibility, and a comparatively good, though not always Pareto efficient, social surplus. The LDS introduces the ability to adaptively decide whether to sell goods in a bundle or separately based on reservation prices of the auctioned goods. Simulation results showed that the social surplus achieved by the LDS is superior to that of a trivial protocol that always sells goods in one bundle.

Future work includes developing a method to determine the leveled division set and reservations prices that maximize the social surplus or revenue of the seller. We are now developing such a method based on certain assumptions of the bidders' evaluation values.

Acknowledgments

The authors wish to thank Atsushi Kajii, Yoshikatsu Tatamitani, Fredrik Ygge, and Katsumi Tanaka for their helpful discussions.

Notes

1. In some microeconomic studies, the definition that item A and item B are complementary states that if the price of item B increases, the demand for item A decreases, and vice versa. This definition is stricter than the definition used in this paper.
2. We don't require that $\bigcup_{S \in D} S = M$ hold, i.e., satisfying $\bigcup_{S \in D} S \subseteq M$ is sufficient.
3. This condition is not contradictory to the second condition, since the second condition involves the union of multiple sets.
4. If the condition of Step 1 is also satisfied for LDS($i+1$), then also compare with the results of GVA($i+1$) and LDS($i+2$), and so on.
5. We cannot tell the result of the GVA when agents can submit false-name bids, since there exists no dominant strategy equilibrium.
6. Naturally, we cannot achieve a Pareto efficient allocation since we restrict the possible allocations.

References

Clarke, E. H. (1971). Multipart pricing of public goods. *Public Choice*, 2:19–33.

Fujishima, Y., Leyton-Brown, K., and Shoham, Y. (1999). Taming the computation complexity of combinatorial auctions: Optimal and approximate approaches. In *Proceedings of the Sixteenth International Joint Conference on Artificial Intelligence (IJCAI-99)*, pages 548–553.

Klemperer, P. (1999). Auction theory: A guide to the literature. *Journal of Economics Surveys*, 13(3):227–286.

Mas-Colell, A., Whinston, M. D., and Green, J. R. (1995). *Microeconomic Theory.* Oxford University Press.

McMillan, J. (1994). Selling spectrum rights. *Journal of Economics Perspectives*, 8(3):145–162.

Monderer, D. and Tennenholtz, M. (1998). Optimal auctions revisited. In *Proceedings of the Fifteenth National Conference on Artificial Intelligence (AAAI-98)*, pages 32–37.

Rasmusen, E. (1994). *Games and Information.* Blackwell.

Rothkopf, M. H., Pekeč, A., and Harstad, R. M. (1998). Computationally manageable combinatorial auctions. *Management Science*, 44(8):1131–1147.

Sakurai, Y., Yokoo, M., and Matsubara, S. (1999). A limitation of the Generalized Vickrey Auction in Electronic Commerce: Robustness against false-name bids. In *Proceedings of the Sixteenth National Conference on Artificial Intelligence (AAAI-99)*, pages 86–92.

Sandholm, T. (1996). Limitations of the Vickrey auction in computational multiagent systems. In *Proceedings of the Second International Conference on Multiagent Systems (ICMAS-96)*, pages 299–306.

Sandholm, T. (1999). An algorithm for optimal winner determination in combinatorial auction. In *Proceedings of the Sixteenth International Joint Conference on Artificial Intelligence (IJCAI-99)*, pages 542–547.

Varian, H. R. (1995). Economic mechanism design for computerized agents. In *Proceedings of the First Usenix Workshop on Electronic Commerce.*

Wurman, P. R., Wellman, M. P., and Walsh, W. E. (1998). The Michigan Internet AuctionBot: A configurable auction server for human and software agents. In *Proceedings of the Second International Conference on Autonomous Agents (Agents-98)*, pages 301–308.

Yokoo, M., Sakurai, Y., and Matsubara, S. (2000). The effect of false-name declarations in mechanism design: Towards collective decision making on the Internet. In *Proceedings of the Twentieth International*

Conference on Distributed Computing Systems (ICDCS-2000), pages 146–153.

Index

Accession game, 160, 163, 174
Agent economy, 219, 232, 294, 296, 349
Agent-based computing, xv, xxv, xxxi
Agent-based systems, xxv, xxxi
Agents
  Altruistic, 46, 48, 50
  Benevolent, 57
  Egoistic, 46, 48
  Interactive, 83
  Panicked, 83
  Personality, 88
  Rational, 82, 84
  Self-interested, 10, 82, 93, 254, 351
  Selfish, 57
  Socially competent, 84
Air defense, 103
Altruistic agents, 46, 48, 50
Amazon, 268
Argumentation, 350
Auction
  Combinatorial, 356
  Generalized Vickrey, 358–362, 364, 369
  Internet, 356
  Private value, 357
  Protocol, xxxii, xxxiii, 357
  Sealed bid, 358
  Second-price, 358
  Vickrey, 357
Autonomy as selfishness, 57

BargainFinder, 120
Basilisk, xxxiv
Basin of attraction, 135, 137
Bayesian networks, xv, 4, 101, 102, 340
BDI, 66, 67, 71, 72
Belief context, 318
Belief, Desire, Intention, 66, 67, 71, 72
Bellman equation, 233
Bellman error, 225
Benevolence
  Evolution of, 57
Benevolent agents, 57
Blocks world, xxv, 201
Boldness, 247
Bounded rationality, 198, 236
Brownie points, 269–272, 280, 281, 284–287, 289

Catch 22, 24
Chance node, 8
CMS, 320, 321, 323–325, 331
Co-evolution, 134
Coalition formation, xxx, 30, 31, 160
Cockatrice, xxxiv
Cognitive modelling, 60
Collusion, 187, 191
Column mixed strategy, 320, 321, 323–325, 331
Column player, 13
Combinatorial auctions, 356
Common knowledge, 100
Comparative paradigm, 240
Complementary subintermixture, 251
Complete information, 202
Computer games, 334
Condition-action rules, 82
Conditional independence, 251
Conditional interdependence mass function, 251
Conditional probability, 3
Conflicting desires, 71
Constrained optimization, 236
Conventions, 59
Coordination
  With communication, 112
  Without communication, 110
Cults, 59

Dealtime, 133, 153
Decision making, xxv, 128
  Qualitative, xvi
  Unilateral versus multilateral, 252
  Univariate versus multivariate, 259
Decision node, 8
Decision theory
  As game against nature, xxvii, 2
  Classical, 2

Introduced, 1–10
Rationality, 2, 82
Satisficing, 261
Social competence, 98
Decommitment penalties, xxxii, 269, 270, 272, 274–277, 281, 284–287, 289
Desert Storm, 329
Desires, 66, 68
Dichotomy-based techniques, 261
Disagreement strategy, 163
Dissociation, 255
Dominance
Strong, 14
Weak, 15
Dominant coalition, 160
Dominant strategy, 202
Dummy agent, 363
Dynamic logic, 74–76
Dynamic programming, 106, 107, 295

E-commerce, 120, 191, 268, 289, 338, 356
Eastern Bloc, 161
eBay, 268
Economic integration, 159
Economic welfare, 155
Effective hyperstrategy, 325, 326
EFTA, 164
Egoistic agents, 46, 48
Electronic
Buyers, 121
Commerce, 120, 191, 268, 289, 338, 356
Markets, xvi, xxix, xxxii, 120, 177, 180, 218, 219, 232, 294, 296
Sellers, 121
Emotional stance, 57
Emotions, 58, 82
Taxonomy, 87
Enlightened liberals, 255
Epistemic logic, 241
Epistemic utility, 242, 244, 245
Equilibrium
Coordination, 295
Satisficing, 238, 247, 253
EU, 168
European Free Trade Association, 164
European integration, 159
European Union, 161, 164, 168
Evolutionary computation, xxx, 51, 52
Expected praxeic utility, 246
Expected realized utility, 178
Expected utility, xxvi, xxvii, xxxii, 8, 36, 82, 86, 98, 258, 316, 317, 324–326, 329
Hypergame, 317, 324, 327, 329, 333
Extrinsic rationality, 240

False-name bid, xxxii, 356, 361, 366, 367
FCC, 356
Federal Communications Commission, 356
Folk theorem, 295
Future payoff, 262

Game theory
Introduced, 2, 10–26
Gatekeeper, 151, 155
GDP, 167
Generalized Vickrey auction, 358–362, 364
Gibbard-Satterthwaite theorem, 206
Go, 335
Goals, 66
Greedy algorithm, 201
Greedy consumers, 297
Gross domestic product, 167
GVA, 358

HCI, 83, 84
HEU, 317
Human-Computer Interaction, 82, 84
Hypergame, 319–323
Expected utility, 317, 324, 327, 329, 333
Theory, xxxii, 316, 317
Hyperstrategy, 322–324
Effective, 325, 326

I-Help, 338, 339, 342, 348, 349, 352
ICU, 339
Incentive compatibility, 359
Indicator function, 245
Individual rationality
Principle of, 247
Infinite interactions, 36
Influence diagrams, 8, 102, 340, 343, 345
Information-Filtering model, 220, 221, 227, 296, 298, 300, 307, 309, 310
Intermixture, 250
Complemetary sub-, 251
Joint, 250
Sub-, 251
Internet auctions, 356
Intrinsic rationality, 239, 241, 254, 259–261
Iterated Prisoner's Dilemma, xxx, 24, 25, 44, 295, 311

Joint intermixture, 250

Kasbah, 268
Knowledge representation, xv
KQML, 109

Kripke frame, 73
Kripke models, 70, 76

Laissez-faire markets, 269
LDS, 358
Levelled division set, 358, 362, 364, 366–369, 371
Lifting problem, 69
LISP, 110
Liverpool, xxvi

Machine learning, xvi, 218, 232, 294
Markov chains, 183, 351
Markov decision process, 9, 86, 218, 222, 294, 298, 301
  Partially observable, 9, 86
Markov games, 310
Markov structure, 252
Maskin's theorem, 209
Mass function
  Conditional interdependence, 251
  Rejectability, 245, 250
  Selectability, 246, 250
Matchmaker, 181
MAXSAT, 199–201, 205–207, 210–214
MDP, 9, 86, 218, 222, 294, 298, 301
Mechanism
  Design, 2, 99, 115, 198–200, 212
  Polynomial time, 204
MICE, 110
Minimal error complexity, 226
Minimax-Q, 294
Mixed equilibrium, xxix, 17–20, 151
Mixed strategy, xxix
Mixture, 249
Monopoly, 125
Moral sentiments, 45, 47, 52
Multi agent Q-learning, 218, 226, 309
Multi-attribute utility theory, 86
Multilateral decision making, 259
Multiple equilibria, 145
Multivariate decision making, 259
Myopic optimality, 219, 296
Myopic strategy, 191, 219, 296, 310
Myoptimality, 219, 221, 296, 302, 303

Nash equilibrium, xxviii, 13–16, 100, 123, 124, 182, 183, 190, 199, 202, 208, 212, 238, 239, 295, 305, 307, 316
  Asymmetric, 151
  Mixed strategy, 316, 318, 322–326, 329
  Multiple, 145
Negotiation, xxxii, 72, 338–340, 342–345, 348, 350, 351
  Protocol, 340
  Strategy, 341
NEMS, 316, 318, 322–326, 329
Neural networks, 225
Non-determinism, xxv, xxvi, 7, 85
Non-Markov properties, 294
Non-transferable utility, 164
Norms, 59
NP-hardness, 199, 210, 213, 343

Optimality
  Strict, 237
Optimization
  Constrained, 236
Outguessing, 316, 333

Panic, 83
Pareto efficiency, 357, 358, 365–367, 371
Pareto equilibrium, 238, 239, 257
Pareto optimality, xxviii, xxix, 198
Partial commitment, 268
Partially observable Markov decision process, 9, 86
Partnership, 29
Payoff, 104
  Future, 262
Payoff matrix, 13
Payoff structures, 32–34, 36
Performance criterion, 262
Perpetrators, 161
Personality, 88
Persuasion, 350
Policy, 9
Policy iteration, 295
Political integration, 159
Polynomial time mechanism, 204
POMDP, 9, 86
Positive paradigm, 240
Pragmatics, 109
Praxeic likelihood, 260
Praxeic utility, 243–245, 259
  Expected, 246
Preferences, xxix, xxx, 5, 6, 12, 260
  Strict, 6
Prevailing price, 184
Price wars, xxxii, 191, 192, 221, 296, 305, 310
Price-Quality model, 220, 221, 223, 227, 228, 231, 296, 298, 299, 302, 305–307, 310
Pricebots, xvi, xxxi, xxxii, 178, 182, 183, 188
Principle of individual rationality, 247
Prisoner's Dilemma, xxix, xxx, 20, 22–25, 44, 233, 237–239, 248, 254–258
  Iterated, xxx, 24, 25, 44, 295, 311
Probability distribution, xxvi
Probability theory
  Bayesian interpretation, 3

Conditional probability, 3
Frequentist interpretation, 3
Independence, 3
Introduced, 2–5
Semantics, 3
Temporal projection, 85
Probability vector, 18
Procedural rationality, 240, 259
Profit landscape, 220, 299, 300
Property Q, 212
Protocol
Auction, xxxii, xxxiii, 357
Design, 2
Negotiation, 340
Vulnerable, 361

Q-learning, 220, 223, 231, 294, 295, 298, 302, 303, 305, 307, 310, 312
Explained, 222
Multi agent, 218, 226, 309
Simultaneous, 219, 224, 296, 306, 308, 311
Single agent, 218, 227, 301, 304
Qualitative decision making, xvi

Rank-ordering, 261
Rational agents, 82
Rational behaviour, 162
Rational communication, 109
Rational fools, 45, 57
Rationality, 2, 82, 98
Bounded, 236
Extrinsic, 240
Individual, 247, 357
Intrinsic, 239, 241, 254, 259–261
Procedural, 240, 259
Substantive, 240, 258, 259
Recursive modelling method, 101, 104, 106, 110–115
Regression trees, 225, 226, 231, 232
Reinforcement learning, 232, 294
Rejectability mass function, 245, 250
Rejectability mixture, 249
Reliability, 271, 272, 280, 282, 283, 287, 289
Reputation, xxxii
Reservation price, 362
Residuals, 161
Revelation mechanism, 207
Revelation principle, 205, 357
RMM, 101, 104, 106, 110–115
RMS, 320, 322
Row mixed strategy, 320, 322
Row player, 13

Saijo's mechanism, 210
Satisficing, 236, 237, 252–254, 260
Prisoner's Dilemma, 254
Satisficing equilibrium, 247, 252, 258
Satisficing set
Multilateral, 252
Univariate, 253
Selectability mass function, 246, 250
Selectability mixture, 249
Self-interested agents, 10, 82, 93, 254, 351
Selfish agents, 57
Service model, 177
Shopbot model, 220, 221, 225, 227–231, 296, 299, 300, 304, 307, 308, 310
Shopbots, xvi, xxxi, 120, 131, 178, 182, 184, 187, 188
Economic motivation, 138
Linear prices, 138–147
Non-linear prices, 147–150
Profit, 143
Single agent Q-learning, 218, 227, 301, 304
Social
Action, 44
Choice, 44, 198, 200, 201
Competence, 84
Coordination, 59
Modelling, 60
Welfare, 188
Sociological theory, 248
Spyproof strategy, 17
Stable teams, 287
Strategies, 2, 14
Strong dominance, 14
Subgame, 316
Subintermixture, 251
Substantive rationality, 240, 258, 259
Superlative paradigm, 240

Team formation, 270, 288, 289
Temporal projection, 85
The fall of France, 335
Time horizon, 85
Tit-for-Tat, 47
Tit-for-Two-Tats, 89, 90
Trustworthiness, 268
Truth assignment, 201

Unilateral decision making, 259
Univariate decision making, 259
Utility, 5, 6, 67, 316
Epistemic, 242, 244, 245
Expected, xxvi, xxvii, xxxii, 8, 36, 82, 86, 98, 258, 316, 317, 324–326, 329
Expected realized, 178
Function, 5, 11, 91, 200
Multi-attribute, 86
Non-transferable, 164

Praxeic, 243–245, 259
Quasi-linear, 357
Versus money, 7

Value node, 8
Vickrey auction, 357
Generalized, 358–362, 364, 369
Vulnerability, 255
Vulnerable protocol, 361

Wald's criterion, 69
Washing machine, 77
Weak dominance, 15
Welfare
Economic, 155
Social, 188

Yellow pages, 181

Zero-sum game, 295, 317
Zero-sum interactions, 16, 17
Zeuthen's Principle, 351